User Interfaces for Electronic Appliances

Konrad Bauman and Bruce Thomas

Foreword by Brenda Laurel

User Interface Design for Electronic Appliances

Konrad Bauman and Bruce Thomas

Foreword by Brenda Laurel

CRC Press
Taylor & Francis Group
Boca Raton London New York

CRC Press is an imprint of the
Taylor & Francis Group, an **informa** business

CRC Press
Taylor & Francis Group
6000 Broken Sound Parkway NW, Suite 300
Boca Raton, FL 33487-2742

First issued in paperback 2019

© 2001 by Taylor & Francis Group, LLC
CRC Press is an imprint of Taylor & Francis Group, an Informa business

No claim to original U.S. Government works

ISBN-13: 978-0-415-24335-3 (hbk)
ISBN-13: 978-0-367-39741-8 (pbk)

Library of Congress Cataloging-in-Publication Data

User interface design of electronic appliances/Konrad Baumann and Bruce Thomas.
    p. cm.
Includes bibliographical references and index.
ISBN 0-415-24335-1
    1. Electronic apparatus and appliances—Design and construction. 2. Human
engineering. I. Baumann, Konrad, 1966-   II. Thomas, Bruce, 1954-
TK7870.U82 2000
621.385'4—dc21   00-053264

Library of Congress Card Number 00-053264

Visit the Taylor & Francis Web site at
http://www.taylorandfrancis.com

and the CRC Press Web site at
http://www.crcpress.com

Contents

List of figures   *viii*
List of tables   *xv*
List of contributors   *xvii*
Addresses of contributors   *xx*
Foreword by Brenda Laurel   *xxiii*
Acknowledgements   *xxv*

PART I
Introduction   **1**

1 Background   **3**
BRUCE THOMAS

2 Introduction   **6**
KONRAD BAUMANN

3 Interaction design process   **29**
GEORG RAKERS

PART II
User interface design   **49**

4 Creativity techniques   **51**
IRENE MAVROMMATI

5 Design principles   **77**
ADRIAN MARTEL AND IRENE MAVROMMATI

6 Design of on-screen user interfaces   **108**
IRENE MAVROMMATI

PART III
Input devices 129

7 **Controls** 131
KONRAD BAUMANN

8 **Keyboards** 162
KONRAD BAUMANN

9 **Alternative interaction techniques** 173
CHRISTOPHER BABER AND KONRAD BAUMANN

10 **Speech control** 190
CHRISTOPHER BABER AND JAN NOYES

11 **Wearable computers** 209
CHRISTOPHER BABER

PART IV
Output devices 217

12 **Visual displays** 219
KONRAD BAUMANN

13 **Auditory displays** 253
OTHMAR SCHIMMEL

14 **Tactile displays and speech output** 268
LEO POLL

PART V
Important issues 283

15 **Standards in user-interface design** 285
JENNIFER WESTON

16 **Usability evaluation** 295
BRUCE THOMAS

17 **Pleasure with products – New human factors** 303
PATRICK W. JORDAN

18 **National cultures and design** 329
PATRICK W. JORDAN

APPENDICES 343

Summary of guidelines 345
KONRAD BAUMANN

Guide to further reading 355
SUSAN COLES

References 366
Index of authors 385
Index of companies and products 389
Index of subjects 393

Figures

2.1   (a) Siemans mobile; (a) Philips mobile; (c) Sony mobile   9

2.2   List of features and functions of a corded telephone in the 1970s   10

2.3   List of features and functions of a mobile telephone in 2000   10

2.4   Types of customers of electronic products   10

2.5   Training length for bus drivers in Germany and in the USA   11

2.6   (a) Bus panel for expert bus drivers; (b) bus panel for less qualified bus drivers   12

2.7   Gauss' normal distribution of body size of American soldiers   13

2.8   C. G. Jung's basic types of human character   14

2.9   (Left) Four parts of the brain according to the HBDI model; (right) an example of a person's preferred way of thinking   15

2.10   User interface variants for a food-manufacturing machine   17

2.11   User interface of a microwave oven   18

2.12   Automation should increase over time as quickly as the overall number of features in a specific appliance does. In a product range, high-featured products ideally should not have a higher user interface complexity than low-end products   19

2.13   The technology life cycle   19

2.14   Visible simplicity of small hi-fi UI   20

2.15   UI of a car-navigation system   21

2.16   (Left) Combined controls on a car hi-fi and (right) on a home hi-fi   22

2.17   Example of state diagrams   23

2.18   Block diagram, state diagram, characteristic   24

2.19   Analogue display: for quick rough reading of a variable, for discerning changes in the value   25

2.20   Digital display: for precise reading of a variable   25

2.21   Analogue control: for quick adjustment of a value, gives tactile feedback about the status   25

2.22   Digital control + digital display + keypress repetition function: moderately good for precise adjustment of a value   25

2.23   Analogue control + digital display: better for quick and
       precise adjustment of a value   26
2.24   Analogue control + analogue variable: for control tasks (e.g.
       vehicle or plane steering)   26
2.25   Digital control + analogue variable: for regulation tasks
       with strongly integrating behaviour of the variable   26
2.26   Analogue representation of several variables in a diagram:
       for quick overview and comparison of values   26
2.27   Digital control + digital binary display: for possible
       automatic mode (i.e. setting of the variable by the appliance) 27
2.28   Analogue control + digital binary control: for on-screen
       cursor movement, drag-and-drop, pointing and clicking   27
2.29   Analogue display + digital display: for flexibility in different
       contexts of use   27
2.30   Analogue display + digital display: for maximal flexibility in
       reading a value   27
2.31   Symbol: for quick grasping of information   28
2.32   Analogue activity sign (progress bar): for processes of known
       (or estimatable) duration   28
2.33   Animation as an activity sign for processes of unknown
       duration   28
3.1   Tri-partition of the interaction design process   33
3.2   Mirroring perspectives for determining goals, roles and
       responsibilities   34
3.3   Transition from the current situation to the next situation   36
3.4   Transition with the domain situation   37
3.5   Full transition story   38
3.6   Key stages in the interaction design process   42
3.7   The user interface in the task-situation model   44
3.8   Structure of models   45
4.1   Workshop   55
4.2   Keywords   57
4.3   Trend board   59
4.4   (a) Axes; (b) visual mapping   60
4.5   Brainstorming meeting   62
4.6   Role playing   64
4.7   Sample storyboard   66
4.8   Screen shots taken from an animated storyboard   67
4.9   Flowchart   71
4.10   Detailed flowcharts   72
4.11   Other types of flowchart   73
4.12   Protype of a car-navigation system   75
5.1   Examples of frequently used controls on portable appliances:
       (a) CD player buttons; (b) portable cassette player   80
5.2   Example of a power button on a portable appliance   81

5.3   Analogue sliders   82
5.4   Conceptual models: (a) wheels attached by an axle; (b) gear-
       wheel arrangement   87
5.5   Virtual folders   89
5.6   Desktop   90
5.7   Elements providing help in the user interface   92
5.8   Inserting a floppy disk   94
5.9   Hardware and software buttons   96
5.10   Aspects of semantics: (left) cacao tree seedpod; (right)
       Coca-Cola bottle designed to look like a cacao tree seedpod   97
5.11   Remote control   98
5.12   Calculator and phone   99
5.13   Apple's trashcan (trademark) gives feedback by changing its
       shape when there are objects inside   100
5.14   Remote control. The keyboard flap helps to reduce the com-
       plexity when closed   102
5.15   Fax machine. The complexity of the operation panel is
       reduced by hiding the alphabetic keyboard under a flap   103
5.16   Remote-control button shapes   104
6.1   Dashboard of a car equipped with a car-navigation system   111
6.2   The author in a car equipped with a navigation system
       prototype   111
6.3   Screen remote control for an electronic programme guide   117
6.4   Screen design for night conditions of a car-navigation system   118
6.5   Screen remote control   120
7.1   Overview of types of control: (left) standalone controls;
       (centre) controls to be used together with a display; (right)
       virtual and other controls   133
7.2   Key with a key-click   134
7.3   Key geometry   134
7.4   Recommended key size (left); not recommended key size
       (right)   134
7.5   Force for pushing a key as a function of the downward
       displacement or travel   134
7.6   Foil key   135
7.7   Bistable push-button   135
7.8   Bistable toggle switches   135
7.9   Trigger switches   136
7.10   Continuous control knobs   137
7.11   Control knobs with discrete positions – rotary switches   138
7.12   Thumbwheel (left) and shuttle (right) on a horizontal device
       surface   138
7.13   Shuttle in a combined control on a vertical device surface   139
7.14   Thumbwheels for setting of time and aperture on the elec-
       tronic camera Nikon F-401. Both wheels are locked auto-

matically when turned into their normal position and can be unlocked by pushing the small button between them. (right) Thumbwheel with three positions (up/down/neutral) used for menu selection tasks in the mobile phone Sony CMD-Z1. At the same time, the wheel serves as a monostable key for confirmation   140

7.15   Continuous slider   140

7.16   Discrete sliding switches   141

7.17   Radio buttons (key row with alternating function)   141

7.18   One-shot key with control signal in both states and corresponding state diagram   142

7.19   Key row (pseudo radio buttons) consisting of one-shot keys with alternating control signals and corresponding state diagram   142

7.20   Possibilities to imitate rotary switches or toggle switches by the use of one-shot keys and control signals   143

7.21   Arrow keys with (a) analogue; (b) digital; (c) pseudo-analogue; (d) alternating control signals   144

7.22   Multifunctional keys and corresponding shift keys   145

7.23   Soft keys in the upper line of a calculator keyboard   145

7.24   (Left) Two-dimensional isometric joystick or trackpoint built in to a keyboard; (right) two-dimensional isotonic joystick in a model aeroplane remote control   146

7.25   (Left) Two-dimensional and (right) three-dimensional isometric joystick   146

7.26   Six-dimensional control ('6D-mouse') for steering an industrial robot. For every one of the three axes linear movement and rotation can be applied. The device is build into a remote control with screen and keys   147

7.27   Two-dimensional discrete joystick   148

7.28   Principles of steering with analogue and digital controls   148

7.29   Control error over time A and control force over time B in an acceleration system using an analogue (upper diagram) and a digital (lower diagram) control. It can be seen in this case that better results are achieved with the digital control   149

7.30   Arrow keys, possible arrangements   149

7.31   Trackball   150

7.32   Mouse   150

7.33   Bang and Olufsen audio/video remote control. The glass panel is touch sensitive on the whole surface. An LCD is built into the upper and LEDs into the lower part of the panel   152

7.34   Control unit for industrial applications that has a 10.4 inch TFT (a type of LCD) touch-sensitive screen and an inte-

	grated Pentium processor without fan, and is resistant against moisture, dust, etc.	152
7.35	Apple Newton (trademarks) personal digital assistant	153
7.36	Trackpad build into a laptop computer	154
7.37	Positioning device used with a graphics tablet	156
7.38	Transition from real to virtual controls giving five examples	157
7.39	Summary of standalone controls and their properties	159
7.40	Summary of controls to be used together with a display	160
7.41	Summary of virtual and other controls and their properties	161
8.1	Keyboards for different applications as viewed from the front	162
8.2	Two possible key arrangements for numeric keyboards: (left) telephone; (right) calculator	163
8.3	Key arrangements for alphanumeric keyboard: Sholes, QWERTY or American keyboard; Dvorak, ASK, or simplified American keyboard; linear alphabetic keyboard; random arrangement	164
8.4	Different linear-alphabetic keyboard arrangements for small appliances	165
8.5	(Left) Linear-alphabetic keyboard arrangement; (right) frequency centred keyboard arrangement optimized for the English language	165
8.6	Character frequency for the English language	166
8.7	Input solutions for numbers	169
9.1	Virtual touch screen used for information kiosks in public spaces	176
9.2	Contactless data carriers in form of cards, phicons (physical icons), and sticks for virtually moving software applications or data	179
9.3	Antenna of a car immobilizer using contactless data carriers	180
9.4	Interactive toys: (a) Sony Albo; (b) Microsoft ActiMates Interactive Barney; (c) the MIT Curlybot	181
9.5	An early PalmPilot PDA from 1996 featuring handwriting recognition	182
9.6	A Handspring Visor PDA from 2000. The fact that the UI did not change a lot over four years shows the quality of the original design	182
9.7	The SensAble PHANToM being used at the University of Birmingham to provide force feedback when visual and auditory cues are removed	184
9.8	Braille tactile display element	184
9.9	Auditory display for landing assistance using two superimposed sound signals with different length	185
9.10	Brain activation patterns during different conscious brain activities	186

9.11	Frequency bands of the EEG	187
9.12	EEG signal	187
9.13	Early brain–computer interface used to move a cursor (cross-shaped) for a pointing task at two targets (rectangles)	187
10.1	Spectrogram of the phrase 'F 4'	196
11.1	Birmingham University's wearable computer	211
11.2	Glove interface	212
11.3	Paramedic using first prototype	215
12.1	Information versus data	220
12.2	Analogue displays with a moving pointer (recommended designs)	222
12.3	Scales, pointers and their recommended dimensions	224
12.4	A display housing that makes it impossible to read the display from an inclined viewing angle; B scale and pointer in one plane; C scale with mirror	225
12.5	Error rate when reading the displays for a period of 0.12 seconds. The display element with a mobile scale has the lowest error rate of 0.5 per cent	226
12.6	Displays with a mobile scale	227
12.7	Digital display elements	229
12.8	Three UI elements with backlight. The left element is definitely a binary display. The right element is easy to recognize as a push-button. But what is the middle element?	230
12.9	Consideration of appropriate viewing angle for displays. The viewing angle should be made easy by the form of the display (left and middle) or it should be possible to adapt it by rotating the whole display (right)	231
12.10	Column of 1-bit display elements (LEDs)	232
12.11	Representation of figures on a 7-segment numeric display	233
12.12	Alphanumeric character set for a 5×7 dot-matrix display	234
12.13	Characters of a 5×8 dot-matrix display that differ from those of a 5×7 dot-matrix display	235
12.14	Some characters in a proportional font	235
12.15	Examples of character density at 70, 50 and 30 per cent	238
12.16	Examples of principles of gestalt psychology: (a) and (d) proximity; (b) similarity; (c) and (e) proximity is dominant over similarity; (f) good continuation; (g) good continuation dominates similarity; (h) closeness; (i) good continuation dominates closeness	239
12.17	Desktop and windows of the Apple Macintosh operating system	240
12.18	Icons of the Apple Macintosh UI	242
12.19	Virtual-display elements: virtual analogue display, pie chart, spider-web chart, icon, virtual digital display, dot chart, line chart, curve chart, column chart	243

12.20   Virtual-display elements: grouped column chart, stacked column chart, bar chart, stacked profile chart, deviation bar chart, line chart with amplified differential curve, two refreshing methods for curve chart, filtering of a curve chart   245

12.21   Application-screen examples for measurement devices using different types of virtual display   248

12.22   Examples of action-feedback signs   249

12.23   Elements of the feedback window of the Lotus Notes communication software   249

12.24   Situation-analogue display   251

12.25   Virtual display. By means of optical lenses this display gives the user the impression of viewing a much bigger display at a much greater distance. A display, B mirror, C optical lens   251

13.1   Example of an amplitude-over-time diagram of a piano tone (10 seconds)   254

13.2   Example of a Fourier analysis of the piano tone from Figure 13.1   254

13.3   An equal-loudness contour diagram   256

13.4   Example of a frequency response curve of a small electronic device   260

13.5   Table of notes from the musical scale and their frequencies   262

13.6   Graphical representation of the Nokia 'Grande Valse' ringer   263

14.1   Sony 'Walkman' with electronic buttons on top   269

14.2   Diamond Rio solid-state player   270

14.3   Radio with only three buttons   272

14.4   Real-size image of the Sony MR50 minidisc player   272

14.5   Remote control for the TASO screen-access system. The starting position of the utterance of the textual screen content can be selected using sliders and shuttles   280

15.1   Approach to the use of standards during development of a product   289

15.2   Symbol for power standby   290

17.1   Maslow's hierarchy of needs   307

17.2   A hierarchy of user needs   307

Tables

2.1	Motivation for buying a mobile phone	16
2.2	Preferred method for learning how to use a mobile phone	16
3.1	Different approaches to setting up a team for a research or development project	31
4.1	Input techniques	54
4.2	Processing techniques	55
5.1	Goals for the user interface	78
7.1	Four classes of touch-sensitive input device	151
7.2	Summary of input devices to be used with displays and their properties	158
9.1	Types of gesture	176
9.2	Types of movement sensor	178
10.1	Summary of benefits of speech control	193
10.2	Applications of speech control	193
10.3	Problems associated with speech control	195
10.4	Summary of developments of techniques in speech recognition	199
10.5	Characteristics of user communication	201
10.6	Some myths about speech control	203
10.7	Sources of noise	206
12.1	Symbolism of some colours for analogue displays	225
13.1	Frequency-duration table for the Nokia 'Grande Valse' ringer	264
13.2	Possible software-code table for the Nokia 'Grande Valse' ringer	265
13.3	Example of a set of feedback sounds for a fax machine	266
15.1	International organizations	286
15.2	National organizations	287
15.3	Standards	292
15.4	Standards organizations	293
15.5	Other organizations	294
16.1	Overview of strengths and weaknesses of some testing techniques	297

17.1 Emotions associated with pleasurable and displeasurable products 308

17.2 Properties associated with pleasurable and displeasurable products 308

17.3 Four-pleasure analysis of product requirements for a camera aimed at young women of high socioeconomic status 316

17.4 Four-pleasure analysis of benefits associated with the Apple Macintosh PowerBook 318

18.1 Detailed data on the cultural dimensions in national culture clusters as found by Hofstede 338

18.2 Aesthetic preferences associated with high and low positions on cultural dimensions 339

Contributors

Christopher Baber lectures on human–computer interaction and human-centred systems design at the University of Birmingham. His research interests focus on the use of everyday skills for the design of future technology. He has written books and articles on speech-based interaction with computers and has been researching wearable computers. Current research interests, in addition to speech technology and wearable computers, include haptic interfaces and the role of everyday objects in ubiquitous computing.

Konrad Baumann teaches human factors at the graduate school Fachhochschule Joanneum in Graz, Austria. He has been working as Product Manager with Philips Consumer Communications in Vienna, Austria, being directly involved with the user interface design of Philips fax machines, and before that with Philips Semiconductors. He has a Master's in telematics engineering from Graz Technical University. He is co-author of the book (with H. Lanz) *Mensch-Maschine-Schnittstellen elektronischer Geräte* [Human–machine interfaces of electronic devices], Springer 1998. He provides a course in human factors at Donau University Krems.

Susan Coles is a human factors and interaction designer working in the area of user interface design at Philips Design in Eindhoven. She has worked in the UK in human-factors consultancy, in research in Ireland and the UK, and for Philips Design in Eindhoven and San Francisco. Her experience in user interface and system design since 1982 has covered areas including tourism, military systems, medical informatics and imaging, broadcasting and manufacturing. She went to Surrey (BSc in psychology) and Loughborough (PhD in human factors) Universities.

Patrick W. Jordan is Director and CEO of the Contemporary Trends Institute, marketing and branding consultants to a series of blue-chip companies. Pat is also head of User Research at Symbian and a visiting lecturer at the London College of Fashion. He has authored or edited four books on design and human factors issues, including the best-seller *An*

Introduction to Usability (Taylor and Francis 1998). His latest book is *Designing Pleasurable Products* (Taylor and Francis 2000).

Adrian Martel is a senior interaction designer working for the User Inspired Design Group at the London branch of Satama, a Finnish new-media consultancy. He has a Master's in ergonomics and a BSc in computer systems engineering and over his 10-year professional career has worked in human factors and interface design for companies including British Aerospace in the UK, Philips Design in The Netherlands and Whirlpool in Italy. Recent work includes home-networking concepts, an MP3 player interface, the CARiN car-navigation system and usability studies for audio products. He is a contributing author to *Human Factors in Consumer Product Design and Evaluation*, edited by Neville Stanton, Taylor and Francis, January 1998.

Irene Mavrommati is an interaction designer, project manager and consultant based in Greece. She holds a Master's (RCA) in interactive multimedia, a previous BA and Master's in graphic design, has qualifications in open and distance learning (EAP) and ISO 9000 internal auditing (KEMA). She has worked in the UK, for Philips Design in The Netherlands, for the Computer Technology Institute in Greece, and has her own new-media consulting business. Her work experience covers areas such as distance learning, interactive TV, networked homes, car-navigation systems, and others. She has been involved in several product development and research projects, and contributed to exhibitions, international conferences and articles.

Jan Noyes is Graduate Dean of Science at the University of Bristol, UK. She has published extensively in the area of human factors of interface design including work on keyboards, automatic speech recognition, pen-based technologies, and warning systems within an avionics application. She is a member of the IEE Informatics Executive Committee, and currently Co-Chair of the People In Control Conference. She is a Fellow of the Ergonomics Society having been awarded the prestigious Otto Edholm Medal in 1999 for her contribution to applied research in human–computer interaction. She is co-author of *User-centred Design of Systems*, Springer 1999 and co-editor of *Interface Technology: The Leading Edge*, Research Studies Press 1999.

Leo Poll holds an MS in information technology from the University of Eindhoven and a PhD on 'Visualising graphical user interfaces for blind users'. He was awarded the SNS Bank prize for best PhD thesis in an applied area in 1996. Nowadays, he works as a senior scientist for Philips Research in the UK. His current interests include: mobile user interfaces, context-aware applications, cross-cultural user interfaces and user-centred design processes.

Georg Rakers is a human-factors specialist working as Global Design Manager User-Interface Design Business-to-Business for Philips Design in Eindhoven, The Netherlands. In his 13-year professional career in human factors and interaction design he has been involved in over forty user-interface design projects. Work included interaction design for portable audio, home-entertainment systems, in-car-navigation systems, test and measurement equipment, laboratory equipment, medical systems, manufacturing systems, broadcast systems. He is secretary of SIGCHI.nl.

Othmar Schimmel studied musicology at Utrecht University and the University of Amsterdam with main focus on twentieth century and electronic music, and sonology at the Royal Conservatory, The Hague. After several years of freelance music typesetting for Dutch contemporary composers, music composition and sound engineering, he is now professionally employed as a sound designer for consumer electronic products, user interfaces and digital media. His interests are in music technology, computer music and soundscape composition, and he was involved in projects like Soundscapes Amsterdam (1995), the Soundscapes 'be)for(e' 2000 Festival (1999), and the Soundscape White Lady manifestation (2000).

Bruce Thomas is a human-factors specialist working as User Interface Coordinator for Philips Design in Vienna. He has worked in the UK in a human-factors consultancy, in Germany in a research department at TÜV Rheinland and for Philips Design in The Netherlands and Austria. He is directly involved with the user interface design of a variety of Philips products. He is a co-author (with Luczak *et al.*) of *Ergonomische Gestaltung von Schiffsarbeitsplätzen* [Ergonomic design of workplaces on ships], BMFT 1986 and is co-editor of Jordan *et al.*'s *Usability evaluation in industry*, Taylor and Francis 1996.

Jennifer Weston is a human-factors consultant working at Philips Design in The Netherlands. She has a BS in ergonomics from Loughborough University (UK) and is a registered member of the British Ergonomics Society. At Philips Design she has worked on a variety of projects, both product based and research based, including a major project concerned with the improvement of human-centred processes. She worked for several years at IBM specializing in the human factors of monitors, servers and software. Whilst most of her work has been concerned with the design of consumer products, she also has experience in the fields of human reliability and safety management.

Addresses of contributors

Christopher Baber
Kodak/Royal Academy Educational Technology
School of Electrical & Electronic Engineering
The University of Birmingham
Edgbaston
Birmingham B15 2TT
United Kingdom
c.baber@bham.ac.uk
tel: +44 121 414 3965
wearable computers webpage:
www.bham.ac.uk/manmecheng/ieg/w1.html

Konrad Baumann
Fachhochschule Joanneum
Alte Poststrasse 149
A-8020 Graz
Austria
tel: +43 316 5453 8615
fax: +43 316 5453 8601
konrad.baumann@fh-joanneum.at

Susan Coles
Philips Design
Building HWD-4, Emmasingel 24
PO Box 218, 5600 MD Eindhoven
The Netherlands
tel: +31 40 27 59075
sue.coles@philips.com

Patrick W. Jordan
Contemporary Trends Instute
PO Box 31958
London W2 6YD
United Kingdom
tel: +44 7769 588 920
fax: +44 20 7563 2706
ctrendsi@mail.com

Brenda Laurel
Graduate Faculty, Media Design
Art Center College of Design
Pasadena, California
Experience Design Consultant
tel: +1 408 741 5865
fax: +1 408 741 5458
www.tauzero.com/Brenda_Laurel

Adrian Martel
Satama UK Ltd
The Clove Building
4 Maguire Street
Butler's Wharf
London SE1 2NQ
United Kingdom
tel: +44 70 9211 4039
adrian.martel@satama.com
adrian.martel@iname.com

Irene Mavrommati
Computer Technology Institute
Research Unit 3 – Applied Information Systems
Nafsikas 73 and Nireos Street
26 442, Paralia Proasteiou Patras
Patras, Greece
tel: +30 937 105483
mobile phone: +30 61 430209
mavromati@iname.com

Also: Symbian Limited
Sentinel House
16 Harcourt Street
London W1H 4AD
United Kingdom
tel: +44 20 7563 2938
fax: +44 20 7563 2706
pat.jordan@symbian.com

Jan Noyes
University of Bristol
Department of Experimental Psychology
8 Woodland Road
Bristol BS8 1TN
United Kingdom
tel: +44 117 928 8560
fax: +44 117 928 8588
j.noyes@bristol.ac.uk

Leo Poll
Philips Research Laboratories
Cross Oak Lane
Redhill
Surrey RH1 5HA
United Kingdom
tel: +44 1293 815327
fax: +44 1293 815500
leo.poll@philips.com

Georg Rakers
Philips Design
Building HWD-4, Emmasingel 24
PO Box 218, 5600 MD Eindhoven
The Netherlands
tel: +31 (0) 40 27 59098
fax: +31 (0) 40 27 59071
georg.rakers@philips.com

Othmar Schimmel
Sound Designer
The Netherlands

Bruce Thomas
Philips Design Vienna
Gutheil-Schoder-Gasse 8–10
A-1102 Vienna
Austria
tel: +43 1 60101 5110
bruce.d.thomas@philips.com

Jennifer Weston
Philips Design
Building HWD-4, Emmasingel 24
PO Box 218, 5600 MD Eindhoven
The Netherlands
tel: +31 40 27 59235
jenny.weston@philips.com

Foreword

User Interface Design for Electronic Appliances fills an aching void in the literature of user experience. While pundits have been busy rhapsodizing about the coming *convergence* in interactive media, what has actually been happening is an explosive *divergence* in the form and functions of electronic devices. Convergence, I think, is a myth created to fulfil our wistful desire for everything to be simple and coherent. Instead, we find ourselves in a world where user interface styles seem to be multiplying as rapidly as device types. Our backpacks and belt loops are cluttered with little gadgets that communicate poorly with us, and with each other not at all. The ideal of a life smoothly augmented by technology seems farther away than ever.

This book provides a comprehensive array of explanations, guidelines and techniques that can serve the designer of any new (or 'improved') electronic device, from cellular phones to augmented reality devices. Were this book to become the design manual of choice, we would experience a twofold gain in the quality of user experience: better-informed standards for consistency and ease of use, and increasingly effective customization of devices to specific users, activities and situated contexts. In other words, we can have the best of both paradigms – convergence and divergence – through better design practice.

Baumann and Thomas have resisted the temptation to see user interface design for today's new devices as simple extensions of conventions developed for the personal computer. Indeed, Baumann reaches back to the world of analogue controls for numerous examples of input devices that are universally understood and superbly well suited to their particular functions. Likewise, the contributors to this book do not rely on the received wisdom of the human factors and user interface communities as unquestioned standards for appropriateness. Readers will find refreshing explanations of the realities of human perception, social behaviours and cultural differences along with concrete examples of their relevance to successful design. From brainstorming to team structure to user testing, this book also provides guidelines for managing the human side of design practice.

Finally, *User Interface Design for Electronic Appliances* makes a strong case for adding pleasure to the list of criteria for good user experience.

Pleasure seems to be the supreme goal of many trendy techno-consumer products. But, in recent years designers have too often been tempted to sacrifice thoughtful design practices in pursuit of short-term fads and ephemeral economic gains. It is therefore not surprising that enduring pleasure and extended product lifespans continue to evade our grasp. This balanced and thorough collection of work reawakens our appreciation of a deliberate, multidisciplinary approach to achieving excellence in user-experience design.

Brenda Laurel

Brenda Laurel is the editor of *The Art of Human–Computer Interface Design* (Addison-Wesley, 1990) and author of *Computers as Theatre* (Addison-Wesley, 1991). She currently serves as a member of the graduate faculty in the Media Design Program at the Art Center College of Design in Pasadena, California. She also works as a design consultant in interactive media.

Acknowledgements

This book is based to a large extent on the book by Baumann and Lanz (1998) *Mensch-Maschine-Schnittstellen elektronischer Geräte* [Human–machine interfaces of electronic devices], and we are grateful to Dr Dietrich Merkle and Ms Beverley Ford of Springer-Verlag for the permission to translate and reuse the German text and many pictures from that book.

We are also very grateful to the authors of the individual chapters in this book, without whose generous contributions this work would not have been possible. They provided material under time pressure and despite their own challenging full-time jobs.

In addition, we would like to express our thanks to the following people and organizations: Tony Moore, Taylor & Francis, for support and advice; Philips Design (Eindhoven) and Philips Fax (Vienna) for permission to publish; Prof. Dr Hans Leopold and Dr Robert Röhrer, Technical University of Graz, Austria for the opportunity to write the first text in 1994 leading to this book; Donald A. Norman for helpful advice in the initial phase of this book project in 1998 and for encouraging comments about the book in his seminar speech in Amsterdam in April 2000; Hans Robertus, Gerrit van der Veer, Marja Zuurman and Susan Coles for reviewing an earlier version of Chapter 3 and for helpful comments; Rolf den Otter and Fred Brigham for helpful comments on Chapters 13 and 15, respectively; all co-authors for giving their contributions to Appendix 2, and Georg Rakers and Jennifer Weston for helpful comments on Appendix 2; Edwin van Vianen for helpful comments on several chapters; Harry Kouba for help with processing of figures; Dr Nicole Busch, Siemens (Munich), Dr Martina Manhartsberger, Interface Consult (Vienna) and Aaron Marcus, Aaron Marcus and Associates (Emeryville, CA) for valuable texts and pictures about their projects in human factors; and finally, the Baumann and Thomas families for supporting and excusing our time-consuming activity of writing, editing and managing this project.

Konrad Baumann and Bruce Thomas
Vienna

Part I
Introduction

1 Background

Bruce Thomas

By the time you have started to read this chapter, you will have seen the title of this book *User Interface Design for Electronic Appliances*, and maybe you have read the back cover and the Foreword. You will be aware that this is a book providing guidelines for designing electronic products. So what is different about this book, compared with other books about user interface design or books with human-factors guidelines?

The difference is one of emphasis. Many books on user interface design focus on human–computer interaction; this book focuses on electronic appliances, which may include microprocessors, but do not necessarily have the processing power or display and control flexibility of a PC. Other, more general books with human-factors guidelines have tried to provide all-encompassing guidelines, and as a result are either extremely weighty tomes in which specific information is difficult to find, or the information is so compromised that it is of little help in the search for the solution to a specific problem. Thus, this book contains a collection of material from various sources which is focused on providing help for the non-human-factors specialist involved in the development of small electronic appliances. It contains a collection of simple guidelines, along with explanations of how and why the guidelines should be applied.

The original idea for this book was to provide an English translation of Baumann and Lanz's (Springer 1998) *Mensch-Maschine-Schnittstellen elektronischer Geräte* [Human–machine interfaces of electronic devices]. This book was aimed specifically at electronic engineers and included both human-factors guidelines and guidelines on the design of electronic circuits. The new book should have a more general human-factors focus, leaving out the technical guidelines, but including new material covering recent developments in user interface design techniques and technologies. To this end Konrad Baumann approached me to assist him to find such material.

Our first thought was to translate the original material and to write the new sections ourselves. However, we felt that we did not have sufficient expertise in some of the areas to be covered and that it would be advisable to consult personal acquaintances (mostly colleagues) who were in one way or another more expert in the particular areas we wanted to cover. These people very generously provided such a wealth of material that we felt that it would not do it or them justice to rework it into the book we originally had in mind. Thus, what you have in your hands is an edited work in which the chapters have been provided by a number of people who have worked together at Philips, as well as a few external experts.

Because this book contains material from a number of different authors, there are inevitably some differences in style and some overlaps in the material covered. Nevertheless, as editors we have tried to provide an overall structure which emphasizes the guidelines and makes the material easy to find.

Since many of the authors of this book currently work at Philips, or have done so in the past, you might be excused for thinking that this is a book about the way that products are developed at Philips. This never was our intention and, as editors, we have taken pains to ensure that the guidelines are sufficiently general to be applied in any context. Furthermore, you will find some examples of products from companies other than Philips – some of these were included in the original book, others have been included to illustrate particular points. The views presented here do not necessarily reflect Philips company policy, and the inclusion of product illustrations from other companies does not represent an endorsement of those products by Philips or the editors.

Part I of the book gives some general information about designing electronic devices including some general guidelines by Konrad Baumann (Chapter 2) and the interaction design process by Georg Rakers (Chapter 3).

Irene Mavrommati has provided guidelines on how to design user interfaces, which is presented in three chapters covering creativity techniques (Chapter 4), design principles (Chapter 5, co-written with Adrian Martel), and guidelines for design (Chapter 6).

The core of the book are the guidelines on input and output by Konrad Baumann which were translated from German and updated, concerning the building blocks of user interface design. These form the bulk of Parts III and IV (Controls and Displays, respectively). To these parts new chapters have been added covering advanced interaction techniques (Chapter 9 by Christopher Baber and Konrad Baumann), speech control (Chapter 10 by Christopher Baber and Jan Noyes), wearable computers (Chapter 11 by Christopher Baber), sound design (Chapter 13 by Othmar Schimmel) and tactile displays (Chapter 14 by Leo Poll).

Part V covers other important issues having a profound impact on the design of user interfaces for electronic appliances. These issues include the use of standards (Chapter 15 by Jennifer Weston), usability evaluation (Chapter 16 by Bruce Thomas), pleasure (Chapter 17 by Patrick W. Jordan) and national cultures and design (Chapter 18 by Patrick W. Jordan).

Although we have attempted to provide some general guidelines on the design of user interfaces for electronic devices, we could not possibly provide all material on user interface design or human factors within the pages of this comparatively small book. Therefore the co-authors have provided recommendations on their favourite books in this discipline (see p. 355, collected here as Appendix 2 'Guide to further reading' (by Susan Coles). This is where we started.

2 Introduction

Konrad Baumann

2.1 The challenge of designing a small user interface

This book deals with user interfaces for electronic appliances or devices. These user interfaces are usually characterized by their limitations; for example:

- limited size (if they belong to a hand-held device);
- limited weight (if the appliance should be portable);
- limited sales price (if the appliance is a consumer product);
- limited time to market (often a major challenge with information appliances);
- limited learning time;
- changing users (in case of a public appliance);
- limited acceptable error rate (in safety-critical applications like in production, health care, aviation and automotive applications);
- limited processing power (in case of speech-to-text conversion), and the like.

These appliances do not necessarily have to be small, but many of them are. Others may be big, like an industrial production machine, but nevertheless have a small user interface. By this we mean a user interface having less screen space and/or less or smaller input devices than a personal computer (PC).

Today these appliances are also characterized by dramatically increasing use of microprocessor control and their constantly rising processing power. This fact usually adds a few other challenges for the team that has to specify and develop them; for example:

- feature overflow;
- high menu complexity;
- shortage of input and output devices.

Finally there are some challenges and constraints because of the fact that today's technological products are breaking geographical and social frontiers and that the pace of industry is constantly accelerating:

- lack of applicable standards and guidelines;
- no common terminology;
- no upgradeability;
- low priority and inappropriate process;
- lack of resources;
- unlimited target-user group;
- ignorance of users' needs.

Let us have a closer look at some of these challenges:

Feature overflow   The use of microprocessors makes it possible steadily to increase the number of functions of a device. Nevertheless the user interface (UI) should stay (or become) clear, easy to learn and error tolerant. As every feature or function increases the complexity of a device, feature overflow is clearly our biggest challenge.

Shortage of input and output devices   From reasons of cost and space, the number of keys of an electronic appliance is small compared with a computer or with a mechanical machine. In most cases no analogue input device (e.g. mouse) is available and only a small display.

Lack of standards and guidelines   There are still only a few applicable standards and guidelines for the development of electronic appliances, especially because the task is interdisciplinary (i.e. situated between the domains of hardware development, software development and industrial design).

Unlimited target-user group   The interface should be equally suited for experienced and inexperienced users.

No common terminology   Users and designers have a different way of thinking and a different terminology. Hence problems of communication are possible.

No upgradeability   In the case of a processor-controlled device the computer program is an integral part of the device, hence it must be free of errors and thoroughly checked right from the first version.

Low priority and inappropriate process   In many projects the development initially concentrates on the primary function of the device (e.g. measuring, regulating, data transmission) and the user interface is considered as an add-on to the device, not as an integral part of it. As a consequence the user interface does not get an appropriate development process that contains evaluation, simulation, testing and feedback loops.

Lack of time and money   If the project runs out of time and/or money the user interface usually suffers most. The reasons are the same as described in the previous paragraph.

Ignorance of users' needs   New devices or user interfaces often follow technological or marketing needs – it is not sure whether they help the users to perform specific tasks. This may apply even to UI innovations for electronic devices like screen graphics or speech interaction if applied in an inappropriate way.

2.2   An example: user interfaces of mobile phones

The first generations of mobile telephones differed a lot in their basic features and performance (e.g. size, weight, reception sensitivity, product quality, sound quality, talk time, stand-by time).

However, since these characteristics have run into natural barriers and the appliances from the various manufacturers are thus becoming increasingly similar, the user interfaces of mobile phones have become the most important competitive factor alongside price and brand name (Figure 2.1). As all other aspects of different products become similar, the usability of a device will, in the end, trigger the purchase decision of the customer.

Let us have a closer look at the development of telephone products over time (see Figures 2.2 and 2.3).

In addition to the increased range of functions, the size of the device has decreased by around 90 per cent. So there is limited space available for controls and for the display. Increasing the number of functions will also lead to menu structures too complicated for most users. The operation of such a wide range of functions can only be made possible with a tried and tested user interface. Many manufacturers of consumer products already use sophisticated user interface concepts such as voice-controlled operation, usage of thumb wheels, separation of receiver from control unit and well-thought-through menu structures.

The question is how to start the development of such a concept the right way. How can you determine whether the concept is good before having the possibility to test it? What is the right process for a small manufacturer of a non-consumer product (e.g. a meter produced in small quantity), having only a few man-months available for UI development? How can UIs for existing appliances be improved without changing the entire software or hardware? This book will give you clear solutions and guidelines about how to overcome all these challenges.

(a)

(b)

(c)

Figure 2.1 (a) Siemens mobile (b) Philips mobiles; (c) Sony mobile.

Take call / end call / dial number.

Figure 2.2 List of features and functions of a corded telephone in the 1970s.

Switch telephone on and off / take call / end call / set or enter access code / dial number / redial number / call name from electronic phonebook / save, search, edit and delete entries / set ringer volume / set loudspeaker volume / key lock / signal waiting call / receive waiting call / revert to original call / conference call / display last numbers dialled / display numbers of accepted calls / display numbers of not answered calls / display call duration / set restrictions for outgoing and incoming calls / send or receive short message / e-mail / fax / data / read short message / edit short message / browse Internet / divert calls (all / if busy / if no reply / if not reachable / if fax call / if data call) / program own number and business card / acceptance of call / automatic or with every key / set type and volume of ringer / compose or download ringer melody / vibration alarm / record call / record memo / select telephone provider / display battery status / display signal strength / send DTMF tone / calculator function / diary function / game function / etc.

Figure 2.3 List of features and functions of a mobile telephone in 2000.

2.3 Making a picture of the users

When starting to design a user interface, it is very important to make a clear picture of the users. Most likely, the users will have special characteristics with regard to their educational background, motivation and requirements. Seeger (1992) classifies the users of technical products into seven basic categories (Figure 2.4).

Higher group	Prestige type
	Performance type
Middle group	Innovative type
	Cautious type
Lower group	Aesthetic type
	Traditional type
	Economic type

Figure 2.4 Types of customers of electronic products (Seeger, 1992, redrawn by K. Baumann).

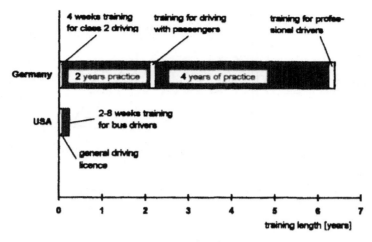

Figure 2.5 Training length for bus drivers in Germany and in the USA (after Seeger 1992, redrawn by K. Baumann).

The more multiple-user types should be satisfied with the same UI, the higher are the demands on the UI itself. If it is possible to restrict the target group for a certain device to a small number of different types, the design of the UI is thus made considerably easier. Furthermore, a classification of the users according to their qualification is very useful. As these can differ greatly, it is normally very difficult to take all users into equal consideration. An electronic meter, for example, can be used by a qualified scientist in a research laboratory. The same device can be used for quality-control purposes in production, with auxiliary staff always taking the same routine measurements.

Figure 2.5 shows an example of nationally differing lengths of training for bus drivers. Two attempts at designing a dashboard according to the qualifications of drivers are shown in Figure 2.6. It is clear that the lower version is more self-explanatory through the usage of symbolic representation while the upper version is more logically and smoothly laid out.

Both versions have the same number of functions. Furthermore, it would be possible to increase the extent of functions of the upper version while the lower dashboard could be combined with a spoken response device.

We should not make the mistake, however, of defining an average user instead of restricting the circle of potential users. As a result the majority of users would not be satisfied with the appliance. In ergonomic product design the general rule is that the dimensions of furniture, doors, etc. must be suitable for all people with body sizes of between the 5-percentile for women and the 95-percentile for men. For these body sizes the integral of Gauss' normal distribution graph (Figure 2.7) includes 95 per cent of the population. Orientation to the average person would result in doors being

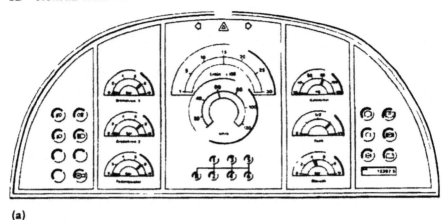

(a)

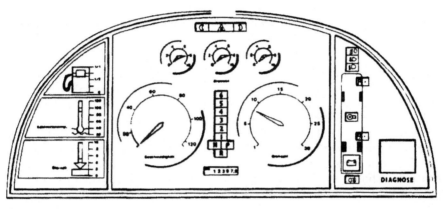

(b)

Figure 2.6 (a) Bus panel for expert bus drivers; (b) Bus panel for less qualified bus drivers (Seeger).

too small for 50 per cent, and armchairs being too high for 50 per cent of the population (Schmidtke 1993).

The application of this rule to the UI means that a user with a 5-percentile qualification must be able to operate the device while a highly-qualified person must not get bored. Impossible? The solution lies in the restriction of the circle of users, either by defining a certain target group or type of user or by determining a certain standard of education.

Guideline
- *Define and limit the circle of potential users as much as possible before beginning the development of an appliance.*

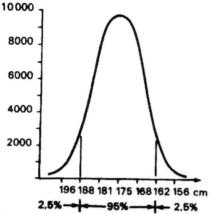

Figure 2.7 Gauss' normal distribution of body size of American soldiers (Grandjean 1979).

To know more about it, refer to the chapters about Interaction design (Chapter 3) and Pleasure with products (Chapter 17) later in this book.

2.4 Communication and ways of thinking

As soon as the circle of users (target group) is defined, the designer, developer or product manager should collect as much information as possible about the prospective users. He/she should acquire their jargon and their views and design the UI correspondingly with respect to the menu options, message texts and symbols used, and to the system model. It is a typical mistake of technicians in software development to confuse their own view of the task with that of the user (Risak 1986).

Guidelines
- *Be sure that the users' subject-specific point of view is taken and their terminology used.*
- *When designing a UI, always have in mind the potential users, their problems and the way in which the appliance can help them solve these problems.*

Technicians are relatively well-educated persons. Their problem-solving competency within the technical domain leads them to the assumption that they would be equally successful in any other domain. However, the very communication with the users often leaves much to be desired. Is that an accident?

Tognazzini (1990) claims that technicians mostly are introverted, intuitive and intellectual character types. He refers to a standardized Myers–Briggs

1	introvert ◄────► extrovert
2	intuitive ◄────► emphatic
3	judging ◄────► perceptive
4	intellectual ◄────► emotional

Figure 2.8 C. G. Jung's basic types of human character (Tognazzini 1990).

personality test made with technicians from Apple, which classifies persons in sixteen groups. The classes are defined as combinations of four binary variables, namely C. G. Jung's basic types (Figure 2.8).

According to Bruce Tognazzini, the creator of the Apple Macintosh user interface, the test is a chance for technicians to understand their own relationship to the users of their products (hardware, software). If variables of their own character are situated at the very borders of their range of values, it is quite natural that communication with and empathizing with extremely different persons requires great efforts (Tognazzini 1990).

The notion of communication is meant in a broad sense here. If somebody is introverted, calm and reserved in contact with other people, this person is not necessarily uncommunicative. Rather such people frequently choose indirect forms of communication, like painting and writing. Even an actor or an actress is relatively safe from spontaneous surprises and in fact this activity is often practised by quiet characters. Technical design and programming belong to the safest kinds of communication because there the information flow is delayed in time and mostly unidirectional. Thus, a technician has much influence on other people's lives, but in general receives only little feedback. Technical development is a kind of unidirectional information flow.

Guidelines
- *All terms used in a UI and user manual (instructions for use) should be taken from common language.*
- *Otherwise they will have to be explained in a terminology understood by the potential users.*

The information flow from the user to the technician should be intensified. The designer obtains valuable insights when watching somebody trying appliances, programs or parts of them.

Herrmann 1988 and Peschanel 1990 define and describe the 'preferred way of thinking' of a person depending on the dominant part or parts of the brain (Figure 2.9). The left half of the brain is dedicated to rational thinking, whereas the right part is responsible for holistic thinking. Furthermore, there

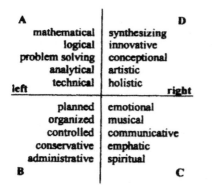 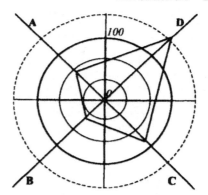

A	D
mathematical	synthesizing
logical	innovative
problem solving	conceptional
analytical	artistic
technical	holistic
left	**right**
planned	emotional
organized	musical
controlled	communicative
conservative	emphatic
administrative	spiritual
B	C

Figure 2.9 (Left) Four parts of the brain according to the HBDI model; (right) An example of a person's preferred way of thinking (Herrmann 1988; Peschanel 1990).

is a horizontal division of the brain into an upper (intellectual) and lower (basic) part. This results in a division into four parts related to logical thinking (upper left part), structural and administrative thinking (lower left part), emphatic and social thinking (lower right part), and creative thinking (upper right part). This approach to classify ways of thinking is called the Herrmann Brain Dominance Instrument (HBDI).

Every person has a different preference or ability to think with each of the four parts of the brain. One or two segments may be dominant for the person, others may contribute less to the person's way of thinking. The way of thinking can be measured via a simple questionnaire with a few pages which takes about half-an-hour to fill in. The questionnaires are evaluated by certified HBDI consultancies. The result of the test consists of four values corresponding to the preferences of the person to think in the four ways described and can be visualized as a spiderweb diagram with a line linking the four values. If a person has equal preference to think in all four directions the resulting diagram will look like a square.

Research has shown that there exist preferred ways of thinking in organizations like companies. This is not only because of the fact that many employees of a company may have the same profession or training, but also to the fact that management prefers to hire people who think similarly to themselves. It is clear that decisions about taking a risk or not also depend on the dominant style of thinking of a person or an organization (Peschanel 1990).

Also it can help to estimate the thinking style of one's boss when presenting a prototype or concept to him or her. The same applies for one's employees, colleagues or customers. When developing a UI or writing a user manual, it is helpful to think about the preferred way of thinking of the prospective users or readers. This does not mean we have to subject all these

Table 2.1 Motivation for buying a mobile phone (Honold 1999).

Country	Motivation for buying a mobile phone	Customer demand for
Germany	Emergencies, time planning, managing delay	Time organizing functions
Italy	Accessory, a must-have	
China	Staying in contact with friends all the time, fostering relationships	
India	Convenience, independence of the infrastructure	

Table 2.2 Preferred method for learning how to use a mobile phone (Honold 1999).

Country	Learning how to use a mobile phone	Customer demand for
Germany	Conventional user manual	Clear and fast overview on all possible functions
Italy	Learning by doing	
China	Sales clerk is most important source of (step-by-step) information about basic functions	Videos to show procedures
India	Whole family meeting, only one person reads user manual, oral and personal information	

people to standardized tests – it is quite easy to estimate the dominant way of thinking of a person or a group of persons if one is familiar with the basics of this theory and knows them well enough. The latter can be achieved by interviews or focus groups and is described in Chapter 4.

A series of focus groups – discussion meetings with users – were carried out by Pia Honold between February and April 1998 in four different countries of the world. The questions raised were: 'Why did you buy a mobile phone?' and 'How did you learn how to use a mobile phone?' A summary of the results is shown in Tables 2.1 and 2.2. As all answers to both questions are fundamentally different for the four countries, it can be concluded that it is definitely a challenge to create 'products for the world' meeting customer needs in very different countries.

Although the automotive industry, for example, is much more mature than the telecommunications industry, one can easily see that customers have different needs and wishes with respect to cars in the United States than in Europe. More details on cultural differences are given in Chapter 18.

2.5   Adding new functions

Processor-controlled UIs were first introduced in products for the upper target group (high-end products). Therefore the consumers were

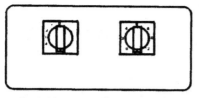 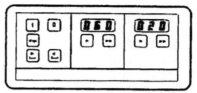

Figure 2.10 User interface variants for a food-manufacturing machine (Seeger 1992).

accustomed to consider appliances with foil keyboard and digital display as modern and efficient. When buying a product they used to prefer them even when modernization implied a loss of ergonomic quality.

Examples for this are the user interface variants of an industrial-food-manufacturing machine in its simple and in its high-tech variant (Figure 2.10). The newer, elaborate variant is not self-explanatory. Operating it requires more time and attention. The readjustment of the parameters, for example, is possible in one direction only. The possible range of values of the parameters is not obvious. At the same time, the efficiency of the machine has essentially not been improved. Here, technical progress has taken a step backwards from an ergonomic point of view.

The paradoxical situation that customers buy the elaborate but worse product rather than the older and simpler one will not last forever. Even for novelty-oriented customers, the so-called early adopters of technology, liquid-crystal displays and menu selection have already lost their attractiveness. For the late adopters, ease of use and the fulfilment of basic functions is essential. In the near future most customers will compare the value of appliances without being blinded by gimmicks as, for example, they are with cars today.

A good example for an up-to-date product having a simple UI not overloaded with features is the microwave oven shown in Figure 2.11. The UI design has been ergonomically optimized considering the size and operational force of the commands, display size and contrast, and a separate start/stop key. This makes the device equally suitable for elderly people and for young computer-literate ones.

Guideline
- *The user interface of a processor-controlled device must fulfil the same ergonomic demands (e.g. creation of an overall picture of the system, explicitness) as a UI having purely analogue input and output elements.*

Simplicity (i.e. the deliberate limitation of features of an appliance strictly to those necessary) has even more impact on the usability of an appliance

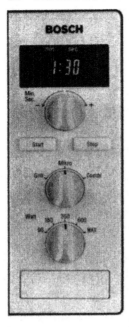

Figure 2.11 User interface of a microwave oven (Siemens).

than the way a given range of features is implemented. Over-featured devices may sell to early adopters (consumers buying new technology); however, only a small range of the features will be used in everyday life.

If a feature is not necessary but unavoidable for sales reasons, it should not have a prominent position in the UI. In other words, secondary features must never disturb the quick access to primary features.

As the range of available and affordable technical features is constantly rising, automation is a way out of an ever-increasing UI complexity (Figure 2.12). In the early years of new technical features, product managers often insist on making the feature visible through the UI – especially in appliances where the feature cannot be seen as a physical part of the product, but is hidden in the electronics. As soon as the feature loses its importance as a sales argument, however, UI designers should insist on making it invisible again. Usually this can be done by means of automation (i.e. some user workload taken over by sensors, electronics and software). Some of the important challenges for product improvement are in the design of these seamless but sophisticated functions. In the end, the whole product intelligence should become seamless and hidden in the device, as stated by Don Norman in *The Invisible Computer* (Norman 1998).

The evolution from technology-driven innovations to customer-driven everyday products where the computer is inside, but out of the user's sight and consciousness, is shown in Figure 2.13.

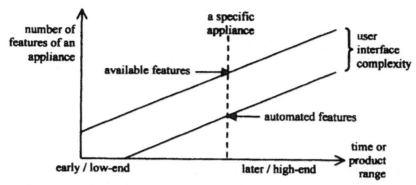

Figure 2.12 Automation should increase over time as quickly as the overall number of features in a specific appliance does. In a product range, high-featured products ideally should not have a higher user interface complexity than low-end products.

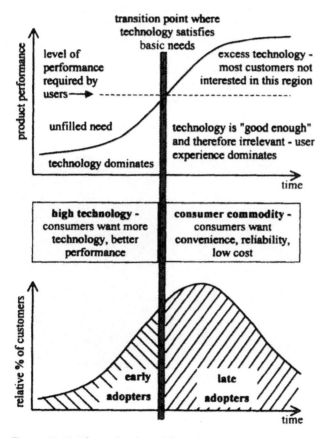

Figure 2.13 The technology life cycle (Norman 1998, redrawn by K. Baumann).

As long as the technology's performance, reliability, and cost fall below customer needs, the market place is dominated by early adopters: those who need the technology and who will pay a high price to get it. But the vast majority of customers are late adopters. They hold off until the technology has proven itself, and then they insist upon convenience, good user experience, and value.

(Norman 1998)

2.6   Ergonomic design as a quality criterion

Some of the proposals for improvement described in this book (such as the optimization of the development process) improve the ergonomic properties of the appliance and at the same time help in saving costs. Many other proposals (such as the use of high-resolution screens or of analogue controls) render an appliance expensive. Hence it is important to show the ergonomic properties of the appliance as a quality feature which is emphasized in the advertisement or mentioned by the salesperson in order to positively influence the buying decision of the customer.

As a matter of fact, it may be possible to attach recognizably ergonomic details to the appliance without thereby creating ergonomic value. In everyday life it is mostly the inconspicuous and unspectacular things that prove good (Figure 2.14). A good user interface in general appears less overloaded and hence less costly than a bad one.

Figure 2.14 Visible simplicity of a small hi-fi UI (JVC).

2.7 Application of analogue controls instead of keys

With the aid of keys the function of nearly all other controls can be imitated. This makes a low-cost production possible but at the same time usually means a loss of ergonomic quality. For instance, a control knob or a slider may be moved arbitrarily fast or repeatedly in both directions or else arbitrarily slowly and precisely. In addition, the user gets tactile feedback about the slider's position without looking at it.

When keys are used instead of such an analogue control, the user must do repeated short key presses or one well-defined longer key press in order to imitate the analogue control's function. Even if the appliance has a well-designed key press repeat function, the process of setting the variable requires much more time, effort and attention if keys are used instead of analogue controls. In general the error rate is higher for keys, too.

More than this, a long row of keys does not provide a characteristic structure (system picture) of the user interface, hence may provoke confusion. The functionality of the appliance may be less easily remembered. In the early days of digital appliances, rows of many identical keys were considered especially 'clear', modern, aesthetic and fashionable. However, this time is over. Nowadays, more producers attempt to give their appliances a unique look and feel by a characteristic selection and arrangement of different controls.

So a UI having at least one control knob, thumb wheel, trackpad, joystick, or slider is a clear trend towards better design for usability. These controls do not have to be analogue in their technical implementation, but they should be analogue in the way they are operated: affording not only a binary variable (on/off, pressed/released) to be input by the user, but also input of a continuous (analogue) value corresponding to our analogue world. If for space or cost reasons such a control is not available for implementation, at least one pseudo-analogue control should be used (e.g. a rotary control combining the functionality of three monostable keys for 'up/down/confirmation' for menu selection in a mobile phone UI).

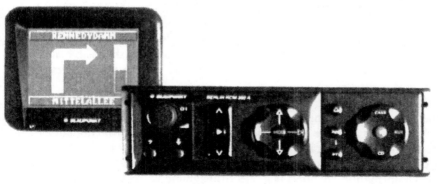

Figure 2.15 UI of a car-navigation system (Blaupunkt).

Figure 2.16 (Left) Combined controls on a car hi-fi (Sony) (right) and on a home hi-fi (Bang and Olufsen).

Shortage of space and ergonomic improvement need not contradict each other. Various controls may be combined in such a way that they adapt well to the anatomy of the user's hand and are less error prone than the number of keys needed for the same task. Also, they are well suited for blind operation which makes them a preferred solution for appliances used in cars. The higher cost and design effort is worthwhile because an original UI is created.

Find more information about all sorts of applicable controls in Part III.

Guideline
- *For each UI use some analogue or pseudo-analogue control or display elements for important functions which give an indication of the purpose of the appliance and its use, and will be remembered more easily.*

2.8   What do user interfaces involve?

A *user interface* is that part of an electronic appliance which serves the information exchange between user and appliance. It consists of controls, displays and an intrinsic structure. If human and machine are considered as a whole, one talks about a *human–machine system*.

The *intrinsic structure* consists mostly of hardware and software, thus of electronic circuits and computer programs. It connects the visible elements of the UI to the (invisible) rest of the appliance. For the user, however, the interface is the system.

A *display element* is a device which transfers information from the machine to the human.

A *control* is a device which transfers information from the human user to the machine. In both cases the information is given in the form of *variables*.

Displays and controls may generally be visual, tactile or acoustic. Even other information channels (e.g. electromagnetic waves) may be used.

The totality of values of these variables is called the *state* of the control or

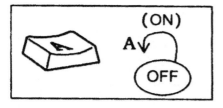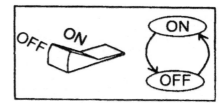

Figure 2.17 Examples of state diagrams.

display. The current state must be accessible (e.g. visible) for the user at any time. This feedback may take place via the position of the adjusting part itself, or via a separate display belonging to the control, or via another display of the same user interface. In the case of a control the adjustment, storage and feedback (display) of the variable are done by the same single mechanical part.

A variable is either *analogue* (continuous, infinitely adjustable) or *digital* (discrete, gradually adjustable). It may be stored as a mechanical or as an electrical magnitude. Several variables combined in a single control are also called a multidimensional variable.

A state of a variable is called *stable* if the variable retains this state after operating the control. Otherwise it is called unstable or transient. A variable with exactly one stable state is called *monostable*, a variable with exactly two stable states is called *bistable*.

For discrete variables there is the possibility of representing their functioning clearly by a *state diagram* (Figure 2.17). A circle means a stable state and an arrow a possible transition from one state to another. In each circle the corresponding value of the variable is inscribed. A transient state is given in parentheses.

Each state diagram may correspond to several possible controls (devices). In particular each state diagram may be realized by a mechanical, electronic or virtual control. For a *mechanical* control, a definite position of the adjusting part corresponds to each state of the variable. For an *electronic* control, in contrast, the state is deposited in a memory, and a separate display informs the user.

A *virtual* control exists as a whole in an electronic memory, thus it has no physical parts. The virtual adjusting part is presented on a display at the UI if required. Non-virtual controls are called *real*. If all three sorts of controls are represented by the same state diagram their ways of functioning are identical too. For this reason real and virtual controls are evaluated according to the same criteria in what follows.

Variables with very many or infinitely many states are called *continuous* (analogue). Their behaviour is not given by a state diagram but by a characteristic. Figure 2.18 shows an example of a block diagram, a state diagram and a characteristic.

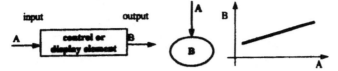

Figure 2.18 Block diagram, state diagram, characteristic.

A *display* is a part of a UI serving the information transfer from machine to person. In analogy with a control, every display has a one or more dimensional variable which is continuous or discrete and which is always in a well-defined state.

There is no strict separation between controls and displays. Some displays are part of a control and vice versa. A touch-screen is control and display in one. The meaning of the terms is, however, always clear in the given context.

> *For the user, the interface is the system.*

Figures 2.19–2.32 are examples showing the appropriate use of analogue and digital elements in human–machine systems.

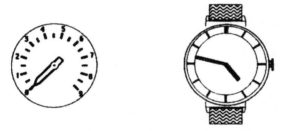

Figure 2.19 Analogue display: for quick rough reading of a variable, for discerning changes in the value.

Figure 2.20 Digital display: for precise reading of a variable.

Figure 2.21 Analogue control: for quick adjustment of a value, gives tactile feedback about the status.

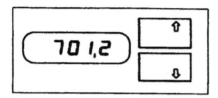

Figure 2.22 Digital control + digital display + keypress repetition function: moderately good for precise adjustment of a value.

Figure 2.23 Analogue control + digital display: better for quick and precise adjustment of a value.

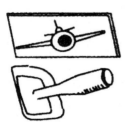

Figure 2.24 Analogue control + analogue variable: for control tasks (e.g. vehicle or plane steering).

Figure 2.25 Digital control + analogue variable: for regulation tasks with strongly integrating behaviour of the variable.

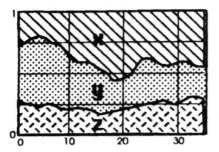

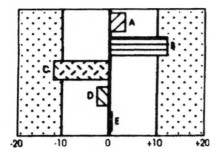

Figure 2.26 Analogue representation of several variables in a diagram: for quick overview and comparison of values.

Figure 2.27 Digital control + digital binary display: for possible automatic mode (i.e. setting of the variable by the appliance).

Figure 2.28 Analogue control + digital binary control: for on-screen cursor movement, drag-and-drop, pointing and clicking.

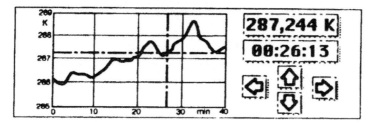

Figure 2.29 Analogue display + digital display: for flexibility in different contexts of use.

Figure 2.30 Analogue display + digital display: for maximal flexibility in reading a value.

Figure 2.31 Symbol: for quick grasping of information (© Apple Computer Inc.).

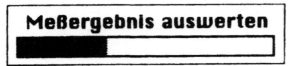

Figure 2.32 Analogue activity sign (progress bar): for processes of known (or estimatable) duration.

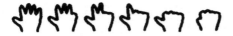

Figure 2.33 Animation as an activity sign for processes of unknown duration.

3 Interaction design process

Georg Rakers

3.1 Introduction

In this chapter I am going to address the interaction design process. The emphasis will be on the *analysis and concept phases,* on how to get the right information and accordingly design the appropriate solution.

In the past many handbooks have addressed *how* to design user interfaces (UIs).[1] The focus of the content in those textbooks was on principles, techniques and methods for designing human–computer interaction. Specifically, guidelines to realize optimal *screen* designs for workstations, personal computers and mainframes.[2] Fewer publications have explicitly addressed the interaction design *process* or have provided guidelines about *how to achieve* high-quality UIs.[3] In this chapter I am going to outline an approach, or method, to do so.

The approach described is based on my personal experience in the electronics industry, at Philips Design, and on teaching and supervising design students, at the University of Wuppertal, Germany.[4]

3.2 My personal inspiration

Before discussing anything else, I would like to start with what in the past has inspired me most, and is still inspiring me: '2.033 Die Form ist die Möglichkeit der Struktur' (Wittgenstein 1921). My understanding, and translation, of that assertion is *form is the possibility of structure.* Form is the possibility of structure to express itself in form. For me, in my daily

---

1 See, for example, Eberts 1994; Galitz 1993, 1997; Helander 1988; Laurel 1990; Mayhew 1992; Shneiderman 1987; Sutcliffe 1995; Thimbleby 1990.
2 See, for example, Galitz 1993, 1997 and Mayhew 1992.
3 See, for example, Dix *et al.* 1993; Henderson 1991; Mayhew 1999; Meister 1987; Newman and Lamming 1995; Preece *et al.* 1994; Redmond-Pyle and Moore 1995; Ritchie and List 1996.
4 For an example of this see Rakers 1998, 1999, 2000; Rakers and Pieters 1989; Rakers and Wittkämper 1995.

profession, that statement of Wittgenstein means: first we have to figure out what the structure is (how it should be built up), then we give that structure a form (such as shape, size, proportions, colour and texture); which is, hopefully, the most relevant form for that particular structure. For me, interaction design is about designing structure by juggling with purposes and functionality, and we arrive at purposes and their associated functionality by studying the user's roles, goals, responsibilities and tasks.

3.3   New freedom, neue Freiheit, nuova libertà

'Screen pollution', bombarding the user with heaps of irrelevant information, is considered to be a bad design principle. The key point here is not the screen pollution, but how relevant is what is on the screen to the application and to the user? It is my personal belief that we, as designers, have the responsibility not only to protect the user from useless information bombardment, but above all to prevent our techno-society from being bombarded with nonsensical gadgets and applications. There is already enough techno-nonsense. Instead of pushing new technological enclosures onto the user, it is our ethical and social responsibility to provide new freedom, neue Freiheit, nuova libertà. Not more of these new, emerging, options, possibilities, thingies, but *new freedom*.

3.4   A truly multi-disciplinary approach

Designing interactive devices or systems is a multi-disciplinary activity. Nowadays it is impossible for one discipline to do it all. In the project team, all design aspects need to be covered by the relevant and related design disciplines. This is to ban any hobbyism. No hobbyists in another discipline's field of professional expertise (e.g. a software engineer who also 'does' the graphics). It also needs to be a *truly* multi-disciplinary team. This means that all disciplines collaborate through all phases of the design process, in parallel, and not as a cascading waterfall of solutions passed on. And if your commissioner remarks that such an approach is way too costly, this is what you could answer: 'If you list all the design activities which need to be done, and you plot them in time (project planning), you will see the dependencies and parallel activities, but most of all you see the 'costs in time' (activity × hours/days × person(s) × hourly tariff)'. If one person is carrying out all the activities, the cost will probably stay the same, but throughput time will most likely be highly increased. If a team of the same discipline carries out the activities, both cost and time will probably stay as plotted. However, if a multi-disciplinary team carries out the activities, cost and time can be expected to stay as plotted, but the quality of the end result will be much higher. At the same costs, you can get much better quality (Table 3.1).

Table 3.1 Different approaches to setting up a team for a research or development project.

	Not recommended		Recommended approach
	Person saving	*Discipline saving*	
Persons	1	Many	Many
Disciplines	1	Few	Many
Cost	Constant	Constant	Constant
Speed	Low	High	High
Quality	Low	Low	High

Working together in multi-disciplinary teams also has some potential pitfalls. The problems that might occur include the following:

Culture The team members will have different educational backgrounds. It is essential that each member accepts and respects the other design perspectives and approaches. The main aim here is to ban any hobbyism.

Jargon Apart from specific terminology, people with different educational backgrounds may use the same words, but they can mean something completely different. 'Concept', for example, is such a nice word. Where for one discipline 'coming up with concepts' means showing some preliminary ideas, proposals or possible design directions, for another discipline it might as well mean engineered solutions which can be brought to market next week. And this conflict of expectations is not something you like to find out at the end meeting; it is something you need to check upfront. Jargon can trigger wrong expectations, misunderstanding and frustration. At the start, when the project team is installed, drafting a glossary of terms and distributing it among the team members may not be the most exciting thing to do; it can, however, be quite effective in making everybody aware. And if you do so, don't forget to send a copy to your commissioner.

Experience Not all team members may have the same level of experience. This can also be a source of erroneous expectations and frustration. More coaching and coordination may be needed than foreseen.

Specification technique This is a very important aspect, and there are several angles to it. The first angle relates to the common language in the team. Ideas, problems, solutions need to be communicated and taken further, but how? Text, flowcharts and program code may be too abstract for visual designers, where sketchy drawings may not be specific enough for engineers. (And there is confusion and misunderstanding all around us, isn't there? I have been in a project where we drew in one picture the steps to be taken in an operating procedure. Click here, then this will open, select that and this will happen. To indicate the sequence big black arrows were drawn. Click here, big black arrow, then this opens, select this, big black arrow, and this will happen. Simple, we thought, and easy to communicate to the software engineers, because it is visual! I will always remember the comment of one of them: 'It is impossible to program these arrows into the user interface.') Another aspect relates to realization. Every discipline wants to have the design implemented 'as designed', without room for any interpretation. So which task-based interaction specification technique provides the best format for realizing this? Storyboards, for example, can nicely show events in time and context, to demonstrate the main principles, but they certainly are a lousy format for specifying detailed interaction. Another important aspect relates to traceability. A specification is not the end product; it is a means for getting there. And any design solution, in whatever format or form, in whatever phase of the project, needs to be based on requirements (new freedom). The key question here is: Which specification technique is most useful for easily tracing back currently proposed design solutions to agreed requirements?

To the above described potential pitfalls there exist no standard solutions. It does pay off, however, to invest in experienced teams; that is, teams which are aware of these pitfalls, that have proved to work well together and who can work out a solution that suits them, there in that particular project, at that particular point in time.

3.5   Overall design process

The overall interaction design process does not exist. It may depend on how you prefer to work, or on what the business objectives of your company are, or on what your customer wants you to do, and it might even be based on what your boss tells you to do. The overall process does not exist. It could be a waterfall life-cycle model of *requirements specification, design, construction, testing and installation* (Redmond-Pyle and Moore, 1995). It could be another waterfall model of *user identification, user requirements specifications, interface design and application testing* (Wood 1998). Or it could be a process which contains an *orientation, analysis, concept, specification, realization and evaluation phase*, and a formal project closure (i.e. Philips Design's *high-design process*). Whatever the process is, it is probably characterized by the statement that 'interactive system design is likely to involve not one process but several concurrent processes' (Newman and Lamming 1995). 'Concurrent' I think should be read as both 'Dixieland music' (= orchestrated) and 'they-all-play-together-at-once' (= total chaos). And I sincerely hope that you are not too often in projects of the latter kind.

Figure 3.1 Tri-partition of the interaction design process.

However, in its basic form I think the overall interaction design process is simply characterized by think, draw, make, with the associated design phases analysis, design and realization. In this chapter this basic *tri-partition* is assumed (Figure 3.1). As said in the Introduction, the emphasis will be on getting the right information and designing a solution that fits. Furthermore, it is assumed here that end-users are involved at all stages of whatever process you are following.

3.6 Forget about the user

It is now 30 years since Hansen (1971) stated *'know the user'* and 13 years since Shneiderman (1987) wanted us to go out there and *'fight for the user'*. Since then, user-centred and task-based systems design has become a standard approach in industry. Apple Computer Inc., with their evangelistic *constant user focus*, was one of the first companies advocating this approach. It may be time, however, to rethink what is meant by know the user, know the task. Maybe it is time to forget about the user.

ISO 9241-11 defines users by *user types, knowledge and skills*, and *personal attributes*. This, however, may be too superficial. Because, guided by these definitions, you may not only have found out that it takes quite some time to collect all that information. Now you may also be puzzled by the question, raised by the significant finding that nobody in your sample wears spectacles, how can this contribute to the quality of the design solution? Looking at users as such, describing them with a job title, or as 'the primary user of the system', or in terms of experience and attributes, may simply not be at the right level of granularity. Or even more important: it may simply not be *inspiring* enough to bridge the gap between analysis and design, between getting the right information and doing something appropriate with it.

So we have to go deeper. And another way of looking at users is to describe users in terms of the roles they have or play (Johnson 1989; Johnson and Johnson 1991; van der Veer 1998, 1999, 2000; van der Veer *et al.* 1996a, 1996b). Van der Veer *et al.* consider roles to be a categorization of people that perform certain subsets of tasks. Roles are either allocated to people by free choice or by the organization, which is the human structure in the community of practice. Also, the time structure in which roles are allocated to people involves categorization. Roles may be performed temporarily, can be negotiated, accepted or refused. A role may be equal to a single physical person, a single physical person may also be the owner of several, and different, roles, or several persons have the same role. For example, one single person at the reception desk of a hospital may have the role of the receptionist (with the two tasks of receiving and scheduling patients) and the role of the telephonist (with informing and communicating tasks), but also the role of the transcriber (with the task of typing out the dictation and voice recordings of the medical staff).

Task-oriented or task-based work system design is strongly promoted by ISO 9241. In ISO 9241-11 a task is defined as *the activities required to achieve a goal* (and two small notes accompany this definition: (1) these activities can be physical or cognitive; (2) job responsibilities can determine goals and tasks). In ISO 9241-11 a goal is defined as *an intended outcome*. And for ISO 9241-11 our design life is quite easy: '*When specifying or measuring usability, the following is needed: a description of the intended goals*'. But that is easier said than done. The question is how do we arrive at those tasks? How do we spot these so badly needed goals? And a related question is: How do we know that we have not omitted some important goals and tasks?

One quite useful approach is to combine the ideas about roles of van der Veer *et al.*, the ISO 9241-11 definition of a goal and that small note about job responsibilities. To start with, just *forget about the user*. (Also, temporarily, forget about the tasks. Don't perform a task analysis yet.) Go one level of granularity deeper, focus on the roles, goals and responsibilities people have. Users have roles, these roles are associated with responsibilities, these responsibilities request actions to be taken, goals to be reached, and boundaries and limits to be set and adhered to. But how will we as designers ever be able to detect all these goals, roles and responsibilities? Well, here is a trick (Figure 3.2).

Humans have or believe in values, which are the drivers of their interest in

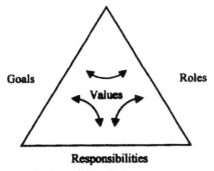

Figure 3.2 Mirroring perspectives for determining goals, roles and responsibilities.

things. These beliefs or ideals (values) are the intrinsic motivation for playing or having roles, wanting to reach goals and taking or having certain responsibilities. These roles, goals and responsibilities can be considered to be *mirroring perspectives* of each other, or viewpoints, as visualized in Figure 3.2. This means: roles include tasks, goals point to tasks (tasks point to goals) and responsibilities determine tasks. So, for example, if you don't know the goals, ask users for their roles. If you don't know the roles and the goals, ask people for their responsibilities. The basic idea is, making an inventory of one side of the triangle will give a perspective on the other two sides. This way you can also cross-check. Do the goals match up with the responsibilities? Are there identified roles, but no goals yet? Let's be quite clear: this will not be a guarantee for not omitting anything, but the result will be richer, more complete. A description of the roles pins down the user. A description of the actions to be taken to reach goals reveals the user's domain tasks (where domain is defined as the area of application). In this current Section 3.6 (Forget about the user), I wrote *user*. But that is not what I meant. It should be read as '*human*' or '*person*', because 'user' assumes a solution. It assumes that someone *has* to *use* at least *something*. But maybe this time a better solution is that we don't have to use anything. So forget about the user, and also get away from the solution-dependent task descriptions. How to get to the person's domain tasks is the subject of the next section.

3.7 Task-Situation Model 0

Knowing the roles, goals and responsibilities first is needed to get to the domain tasks. It is essential to get to these domain tasks, because domain tasks are solution-independent activities. Domain tasks are what people really want or need to do. It is essential to understand those tasks. And it is essential to get all the design solutions removed from them. In the 'analysis phase' therefore, the process of '*know the user, know the tasks*' needs to be filled with (a) getting to know the roles, goals, responsibilities, plus (b) getting to know the design solution-independent domain tasks. (In the 'analysis phase' other concurrent processes may be visual-trend analysis, social-trend analysis, market-trend analysis or technology-trend analysis, but those are outside the scope of this chapter.)

So what is the matter with solutions? The problem with solutions is that they are problems. Future problems. The *desktop metaphor* might once have been the perfect answer to the question how to support on a monitor screen professional users in an *office* environment, but the folders, windows and trashcan metaphor have nothing to do with sailing, broadcasting, being ill or having your car mended in a garage. Still, for some reason, there appears to be a strong urgency to get this *office* metaphor implemented in virtually every thinkable domain. But is this omnipresent office metaphor really required to support other, non-office, domain tasks? And how about 'new freedom'? Can we develop a new, better solution? Actually, how *can* we

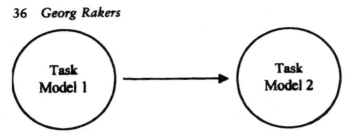

Figure 3.3 Transition from the current situation to the next situation.

generate new solutions? What is the mechanism behind designing well-tuned, neat solutions?

Van der Veer *et al.* have developed an approach for modelling complex systems (van der Veer 1998, 1999, 2000; van der Veer *et al.* 1995; van der Veer *et al.* 1996b; van der Veer and Mariani 1997). Simplified (and this is my simplification of their approach): van der Veer *et al.* make a distinction between what they call Task Model 1 and Task Model 2 (Figure 3.3). Task Model 1 is the current situation which needs to be described and understood in order to specify and model the, not yet existing, next situation – Task Model 2.

In essence, studying Task Model 1 in order to be better able to model Task Model 2 is indeed key to what we as designers have to do. A potential pitfall, however, of the approach of van der Veer *et al.* is that the whole development of Task Model 2 might lose itself in refinements of Task Model 1. An example of the goal, or desire, to keep something – which is in the mind – visible, retrievable and accessible independent of person and time is carving in a stone. Going from a quill pen to a ballpoint is a breakthrough and innovation, 'new freedom'. Refining printer functionality by adding more options to a print window is not. Maybe nowadays we need small, mobile, portable electronic devices which can spray their multimedia content on any surface, including paper.

What I think is missing in the approach of van der Veer *et al.* is the specification of the domain tasks in a **design-solution-independent** way. What is missing is *Task Model 0*. We need to understand Task Model 1 to specify Task Model 0, in order to design a solution for Task Model 2. To have my terminology in line with van der Veer *et al.*'s terminology, I have previously termed Task-Situation Model 0 'Task Model 0' (see Rakers 1998, 1999, 2000). However, because I need this 'task model' aspect in Section 3.8 (Design requirements specification), for dividing 'look and feel' up in several concept-design aspects, here in Section 3.7, I will introduce what I now think is a better term for Task Model 0: *Task-Situation Model 0*. In Figure 3.4 I have accordingly renamed the other two models of van der Veer: Task-Situation Model 1 and Task-Situation Model 2.

In order to design a solution for Task-Situation Model 2 we need to understand and specify Task-Situation Model 0, which consists of the

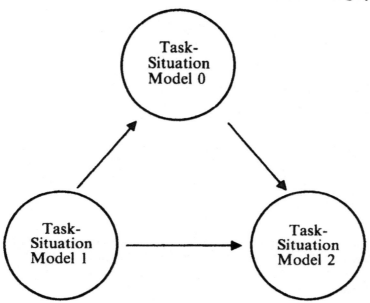

Figure 3.4 Transition with the domain situation.

solution-independent domain tasks. Given the data in Figure 3.4, the high-level questions to be asked are:

1. *What is the current situation?* (= current solution dependent, T-S Model 1);
2. *What is the domain situation?* (= solution independent, T-S Model 0);
3. *What is required for the next situation?* (= future, next solution dependent);
4. *Is this a solution that fits the requirements?* (= future, next solution dependent, T-S Model 2).

Knowing the roles, goals and responsibilities, and by asking what the domain's 'task-situation' is, the domain tasks can now be mapped out. Given goal *a*, the associated domain task is *x*, or associated domain tasks are *x, y, z*, etc. We don't need the domain tasks for the sake of having the domain tasks, it is not a goal in itself, we need these tasks to come up with a design solution. We need to know the work *patterns* (as solution-independent descriptions); which is how these tasks are organized. The focus therefore shall be on understanding and specifying the *work flow(s), task flow(s)* and *event flow(s)*, on mapping out the structure, interrelationships and dependencies. As described in Section 3.6, a description of the actions to be taken and the goals to be reached reveal the

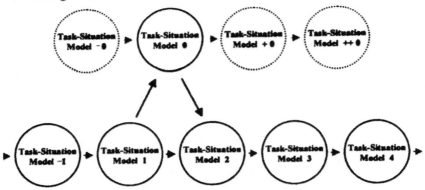

Figure 3.5 Full transition story.

domain tasks. The specification of the limits, boundaries and overlap between tasks reveals the domain's 'task-situation' structure. And this task-based, goal-oriented structure, filled with functionality, shows the logical clustering/structuring of the UI. It shows what people expect to be together at the same time – its location and dynamics.

To be explicit, Task-Situation Model 0 is not equal to what other disciplines may call 'business model' or 'domain model'. Task-Situation Model 0 captures and specifies all those domain tasks which need to be supported and reflected in the UI. That's the crux. A 'business' or 'domain' model may in addition also specify '*stakeholders*' who are affected by the system to be designed, but *who are not interacting – directly or indirectly – with the system*. Purchasers, planners or parents are some examples. There are also users of the system who are not the target audience – the primary users – but who do interact with it. If and when you should consider or exclude these users depends on the nature and goals of your project. Examples of these user types are sales representatives and demonstrators at exhibition stands (at exhibition stands demonstrators are the primary users, but they are usually not the primary users, if you know what I mean).

Task-Situation Model 0 is an abstraction of the current situation in order to be able to project the next situation. Theoretically, Task-Situation Model 0 can be assumed to be stable or staying the same over time. In Figure 3.5 this is depicted by having the model 'hovering' and 'drifting' along 'in the sky' from the previously derived Task-Situation Model (−0), to the current one (0), to the Task-Situation Model to be derived next time (+0), etc. For some domains, like the medical one, this may be true. For others, like the broadcast industry, this might not be true. However, understanding how Task-Situation Model 0 changes over time may give you a competitive edge, and might keep you in business, or even steer that business, because your analysis and understanding of the domain is more profound. But it is not only about T-S models 1, 0 and 2. There is more to it. Apart from understanding roles, goals, responsibilities and tasks, it is also quite useful to study the past or the future (which is beyond Task-Situation Model 2). In Figure 3.5 the full story is depicted. In that figure:

Task-Situation Model −1 = the previous generation of the product or system;
Task-Situation Model 1 = the current situation;
Task-Situation Model 2 = the to-be-designed next situation;
Task-Situation Model 0 = the target domain tasks;
Task-Situation Model 3 = future scenarios;
Task-Situation Model 4 = science fiction.

By knowing and understanding Task-Situation Model 0, studying past solutions (e.g. Task-Situation Models −1, −2, etc.) may also provide you with ideas for new solutions. And looking into the future is also a reflection on Task-Situation Model 0. It certainly is not only about 'visualizing' the future. Implicit in these ideas is that the framework, as depicted in Figure 3.5, is also useful for analysing all types of systems and devices, including competitor equipment. Where did they come from? What did they have (Model −1)? What is there now (Model 1)? What is their understanding or focus (Model 0)?

3.8 Design-requirements specification

If the ultimate aim is to provide people with 'new freedom' to really enrich their quality of life, in order to be able to grow, to develop into free spirits, then new solutions need to be based on profoundly thought-about starting points. These starting points for design, or *design requirements*, are based on the analysis of both Task-Situation Model 1 and Task-Situation Model 0. Agreed upon by all parties involved, the 'design-requirements specification' is not just a list of things to be matched by the design, it is the contract between the outcome of the 'analysis phase' and the outcome of the 'concept and realization phases'.

I find it quite useful to have the following *Table of Contents for a Design Requirements Specification* (definitions and descriptions are given following the list):

a. user profiles;
b. user requirements;
c. description of domain tasks;
d. domain-task requirements;
e. description of working environment;
f. environment requirements;
g. UI design requirements
    (g1) technical model,
    (g2) social-technical organization model,

(g3) navigation model,
(g4) task model,
(g5) presentation model,
(g6) instruction model.

Here a–f are the solution-independent parts, and g is the solution-oriented part.

User profiles Specification of user types, roles, goals and responsibilities.

Domain tasks A general overview of all domain tasks; preferably visually to show work flow(s), task flow(s), event flow(s), task breakdowns, structure, interrelationships, dependencies. Domain tasks can be divided into primary tasks, secondary tasks, supporting tasks. Primary domain tasks are the primary concern of the user. These tasks are domain specific and usually solution independent. (For example, a surgeon in an operating theatre wants to know what is wrong with the patient – diagnosis – and what is the best way to treat the patient – intervention. Both are examples of time- and solution-independent domain tasks. What Philips or Siemens or General Electric have to offer are current solutions.) Secondary domain tasks are tasks which are critical to and contribute to the successful execution of the primary tasks. Depending on the domain, preparation can be such a task. Supporting tasks – or tertiary tasks – are tasks which you actually don't want but get for free because you are using a product or system. Learning and cleaning are examples of such tasks. Repairing can be another example; but if you have a garage and your profession is car repair, then it is a primary task.

I never agreed with the term 'look and feel' or the like. But over the years I have learned that designing a concept involves several aspects, which are all equally important. 'Look and feel' does not cover these aspects at all. Not even when all 'these looks and feels' are transformed into 'usability' or 'quality of use', and 'visual identity'. So, when designing a concept six aspects need to be addressed explicitly. However, this partition into six aspects is purely a theoretical one, to be used by the designers of that system. For the end-user there is only one system – the whole. The six aspects that are explicitly to be addressed and designed are:

Technical model This specifies all technical requirements needed to set out foreseeable design restrictions and possibilities.

Socio-technical organization model This defines aspects of the UI which are necessary to support system usage in an organizational work context; supporting relations between people affected by and/or using the system in the same 'organization'; supporting the organization's goals (e.g. structural flexibility). How people are affected or will be affected is shown in the interrelationships and dependencies of Task-Situation Model 0.

Navigation model This defines the structure of the UI (statics). This structure is a domain-dependent, task-oriented, logical clustering of the required functionality. Logical here does not mean hierarchical and branching, but what is logical for the end-user. What is logical for the end-user is shown in Task-Situation Model 0. The 'navigation model' defines where specific parts of the required functionality are located in the UI.

Task model This defines how the user has to go through the UI to get to a particular part of the functionality (dynamics). Parts of this might already be shown in Task-Situation Model 0.

Presentation model This specifies how the information appears visually in the hardware and software of the UI and what the visual behaviour of each individual UI element will be.

Instruction model This defines the conditions under which the type, structure and format of user guidance is given to the user, while learning and using a system.

Above I have used the word 'model' to indicate that each aspect is based on a rationale.

As it is crucial to have the foreseen end-users involved at the beginning of the 'analysis phase', it is essential to involve them at the end. When the draft version of the 'design-requirements specification' is ready, show and share it not only with your commissioner and project team members, but also let your end-users review the draft. The question is simple: 'We have interviewed you, observed you, and studied your working environment. Here is a summary of our understanding. *Are we right?* Please comment.'

3.9 Concept modelling: form follows structure

We no longer live in the days of 'fitting the task to the man', but in the days of matching the to-be-designed user experience with the user's understanding. Concept modelling is about matching the to-be-designed 'conceptual model' with the user's 'mental model' and the domain's Task-Situation Model 0. But most of all it is about *form follows structure*.

As described in the sections above, describing roles pins the user down. A description of the actions to be taken to reach goals reveals the domain tasks. The specification of the limits, boundaries and overlap between tasks reveals the domain's 'task situation' structure. And this task-based, goal-oriented structure, filled with relevant functionality, shows the

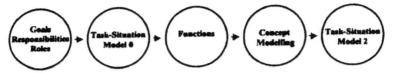

Figure 3.6 Key stages in the interaction design process.

direction for the logical clustering or structuring of the UI to be designed. In Figure 3.6 the succession of key stages in the interaction design process and the switch from 'analysis to concept phase' are shown.

Concept design is about imposing a meaningful structure on the world, meaningful as perceived by our brain. There are two sides here: as designers we can create a structure, but as humans we have to perceive this as a structure and as being *meaningful*. Concept modelling is about matching the to-be-designed conceptual model with the user's task-situation and mental models. And that is what *interaction design* is all about:

(a) knowing and understanding the target domain's work patterns at the work-flow, task-flow and event-flow levels
(b) knowing and understanding the user's knowledge of the (professional) world, the user's task-situation knowledge (= (a), or a subset of (a)), and the user's device knowledge (= mental model); and
(c) translating this into a matching and meaningful 'conceptual model' (divided into the six aspects as described in the previous section), which together appears as a whole to the user, as the UI structure, interaction and visual behaviour.

Task-Situation Model 0 was explained in Section 3.7 and the six aspects explicitly to be addressed in a 'conceptual model' were described above in Section 3.8. But what is the meaning and purpose of a mental model? Actually, what *is* a mental model?

Talking about *mental models* has gone out of fashion today, the peak was between 1983 and 1995; but the notion that users have an internal 'something' about the external world is quite useful. From a design perspective, the key here is to match exter·al arrangements with internal knowledge structures, experiences with understanding. A Task-Situation Model 0 can be inferred, abstracted from concrete observations and a 'conceptual model' is what we design. But how can we 'see' and influence a user's mental model? In 1949 in his book '*The Concept of Mind*' the philosopher Gilbert Ryle made a distinction between *knowing that, knowing how*, and *a ready frame of mind*. 'Knowing that' is, according to Ryle, learned by drill, and we learn 'how' by practice. Ryle states that 'he is in a "ready" frame of mind, for he both does what he does with readiness to do just that in just this situation and is ready to do some of whatever else he may be called on to do'. Thus, Ryle (1949) distinguishes between (a) 'knowing that', (b) 'knowing how' and (c) 'a ready frame of mind', as the combining context.

The mental-model paradigm first emerged in the field of human factors of complex and slow dynamic systems. In the field of human factors of highly complex systems, such as control rooms in the process industry, the concept of operators possessing a mental model of the dynamic system under control has been around for nearly four decades (e.g. Cohen and Ferrell 1969; Conant and Ashby 1970; Moray 1987, 1992; Rouse and Morris 1986; Smallwood 1967; Umbers 1979; Veldhuyzen and Stassen 1977). Here, the human regulator is assumed to use an *internal model* (also called decision model or optimal control model) to assess the current state of the *external* system, and to infer and predict future system states. The structure of an operator's mental model is considered to capture two basic properties of the system under control: a part containing static knowledge and a part containing dynamic knowledge (for reviews see Rasmussen 1986; Rouse and Morris 1986; Umbers 1979; Wilson and Rutherford 1989). Landeweerd (1979), for example, states that the structure of an operator's internal representation of the system under control comprises (a) a mental image and (b) a mental model. The mental image refers to the structure of the process (*what is located where*); the mental model refers to the functioning of the process (*what leads to what*). The content of these static and dynamic components consists of (a) how-it-works knowledge (facts about the system; both facts about what is located where, and what leads to what) and (b) how-to-use knowledge, procedural knowledge, a set of explicit control strategies to manipulate process dynamics (cf. Roth *et al.* 1986).

The mental-model concept emerged in the early 1980s in the cognitive and computer sciences. In cognitive science, Norman (1983) made a distinction between (internal) mental models and (external) conceptual models. The term mental model is reserved for 'what people have really in their heads and what guides their use of things'. Conceptual models, however, are 'devised as tools for the understanding or teaching of physical systems'. Moran (1981) divided the structure of the user's conceptual knowledge into (a) knowledge that organizes how the system works and (b) knowledge of how it can be used to accomplish tasks. Kieras (1987) divides between (a) how-it-works knowledge (facts about the system) and (b) how-to-use knowledge (procedural knowledge). Both views bear close similarities to those described above of mental models of complex dynamic systems. Polson and Kieras (1984), however, identified (a) how-it-works knowledge (facts about the system's structure and functioning), (b) how-to-do-it knowledge (using, controlling the system) and (c) job-situation knowledge. Thus, knowledge of the task, or knowledge of the context in which knowledge of the system needs to be applied, plays a *third* part (Rouse and Morris 1986). Johnson-Laird (1983) stated, regarding, the structure of a mental model, that mental models can be considered to be structural analogues of the world. '*Their structures are analogous to the*

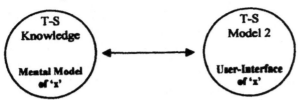

Figure 3.7 The user interface in the task-situation model.

structure of the corresponding state of affairs in the world, as we perceive or conceive it' (Johnson-Laird 1980, 1983). Rouse and Morris (1986) also made a similar observation: 'If a perfect mental model can be assumed, one need only perform an engineering analysis of the system of interest to identify the model. In a sense, there is only one choice.'

Based on Ryle's (1949) 'knowing that', 'knowing how' and 'a ready frame of mind' concepts and evidence from human factors and computer and cognitive sciences, also Johnson and Rakers (1990), Rakers (1987, 1992), Rakers *et al.* (1988), Rakers and Pieters (1989), Rakers *et al.* (1990) and Tennyson (1990) all made an explicit distinction between three types of knowledge: (a) *knowing what* there is in a system or product (declarative knowledge), (b) *knowing how* the system or product functions and should be operated (procedural knowledge) and (c) *knowing when and why* 'what' and 'how' should be applied (the contextual, or criteria knowledge about the conditions of use). For example, the ingredients for a course or meal are the 'what'; the washing, cutting and frying are the 'how'; and 'first wash it, then cut it in five equal pieces, then fry all pieces in olive oil' is the 'when' (the 'why' five equal pieces is about the olive oil, frying time and the moment of wanting to serve the course).

Figure 3.7 depicts the UI of system or device '*x*' as a subpart or sub-structure of Task-Situation Model 2. Knowing device '*x*' is part of the user's 'task-situation knowledge'. Designing a user interface and its interaction means providing it with a structure which is logical in the context of the domain of application.

Figure 3.8 shows that designing UIs for electronic devices, systems or networks means that the 'navigation, task, and socio-technical organization models' are structured in such a way that it is obvious for the user what there is, how to use it and when it can be used. Most mental muddles come from illogical locations (clustering) of functions (what and where), inconsistent interaction procedures (how) and absence of feed-forward information (when).

Form follows structure Designing interactive systems means: first we have to figure out what the structure is (how it should be built up), then we give that structure a form (such as shape, size, proportions,

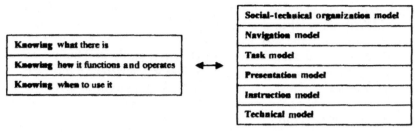

	Social-technical organization model
Knowing what there is	Navigation model
Knowing how it functions and operates	Task model
Knowing when to use it	Presentation model
	Instruction model
	Technical model

Figure 3.8 Structure of models.

colour and texture). The refinements in the process from Task-Situation Model 0 via functionality to concept modelling now are (see also Figure 3.6):

- cluster the required tasks;
- allocate the required functionality to the tasks;
- cluster the functionality per task;
- separate clusters of functionality;
- design a solution for how to get from cluster to cluster (= select and design the appropriate navigation structure);
- select and design the appropriate interaction mechanism(s);
- select and design the appropriate visual presentation and behaviour;
- select and design the appropriate structure, form and format for user explanation.

There are two *major pitfalls* in the 'concept phase', and both relate to structure. The first one is that you have structured the UI according to the book. The 'navigation model' you have designed is a perfect textbook example. However, for end-users it is not so logical. How come? I have been in projects where 'all' disciplines were involved and all did their utmost with their knowledge and skills, but when the concepts were shown to real end-users, however, we still had not got it right. I stress once again that, throughout the process (in the 'concept phase' as well), the potential end-users need to have the final say. We design not for ourselves, we design for them.

The second potential pitfall is also about getting the right structure in the UI, but is subtler. It is about structure mapping and about how 'true' Task-Situation Model 0 is. Let's assume that in Task-Situation Model 0 we got everything right. We captured all tasks, did a proper task breakdown and figured out all the interdependencies and interrelationships, and above all we checked this with real end-users. And, yes!, they approved: that's how it is. This does not mean that the structure of the tasks in Task-Situation Model 0 should automatically be mapped 1-to-1 into the structure of the interface. Often, just mapping this interrelationship 1-to-1 into the interface is exactly what you need to do. But a UI is part of a technology and that technology may get in the way when you literally *copy* the structure of the tasks in Task-Situation Model 0 into the UI. Also, partly or wholly restructuring may provide 'new freedom', not because it is different but because it forces you to think about the 'trueness' of Task-Situation Model 0.

A few more words about mental models are necessary regarding the 1-to-1 mapping of structure. If Johnson-Laird is followed, a mental model is a structural analogue and analogous to the structure of the corresponding state of affairs in the world, then there arises a practical problem with respect to the cognitive demand when users carrying out complex tasks with highly complex, dynamic systems (Moray 1992; see also Grant 1992). Moray (1987, 1992) therefore suggested 'that a model can be regarded as a homomorph, rather than an isomorph, of the real system, ['homomorph' should be read as 'similar' and 'isomorph' as 'identical']. Homomorphs provide a reasonable way to represent a system that is too complex, in all its details, for it to be understood. The mental model is probably a set of quasi-independent subsystems into which the total system can be 'decomposed'. However, 'a homomorph is isomorphic to a reduced version of the original system' (Moray 1987). If this is true, and I think it is true, then there is an even bigger demand to design UIs which are structurally sound.

3.10 Realization

In the 'realization phase' real working parts of the functionality can be tested, but getting users involved should be done in all phases of the process. Although often called user testing, or end-user evaluation, this activity has nothing to do with testing users. We test system behaviour with real end-users involved, not user performance. Since Gould and Lewis (1983) and Gould (1988) made their pleas for early and continuous user involvement and iterative design, this has been adopted in many textbooks (see e.g. Preece *et al.* 1994; Mayhew 1999). So here, I take this as a given. In this section I want to address some other issues of the 'realization phase'.

The first issue is about the 'design-requirements specification', acting as the compass to keep the project on course. Whatever the costs of the 'analysis and concept phases', in the 'realization phase' we are talking real money. And real money brings real life. The proposals designed in the 'concept phase' will have to be checked against the requirements as agreed in the 'design-requirements specification'. The same goes for the chosen, and now to be implemented final concept. But because we are talking real life, all kinds of funny excuses may now arise arguing not to do so (e.g. 'yes, you are right, let's check it, but that takes time, you know, so let's check it after the first version has hit the market'). In the 'realization phase', the 'design-requirements specification' is not only the contract or a tool for traceability, it is *a compass* to keep the project on course.

The second issue is about the prototype. I make a distinction between a simulation and a prototype. Whatever form a simulation takes, in the end it is something that you will throw away. A prototype is the first working version (proto-type) of what in the end will become the real thing. For some reason working versions of real things are hard to change. And this has often more to do with attitude than with technical limitations ('yes, you are right, but not now, let's add/change/delete that in version 2'). The only solution for this is, if possible, to stay on top of it, check what the other disciplines are doing and increase the number of update or progress meetings.

Another issue is about *chaos*. As the *realization phase* means actual realization, in this phase all the disciplines involved are very busy with 'detailed design' or 'implementation'. Although you might expect the opposite: realization is a production line where you just knock out the solutions. It is not. The potential pitfall of the 'realization phase' is 'they-all-play-together-at-once'. As the German proverb says: *the devil is in the details*, here in this phase the details *are* the devils. If not constantly checked, the disciplines may absorb themselves in really working out the details and so lose sight of the overall picture. And it is not only those in the engineering disciplines who might take a short cut, because this way it works better (for the system). All disciplines may suffer from this. As mentioned above, stay on top of it; realization doesn't mean conveyor belt or production line.

Finally, when everything is shipped to market, you may feel happy (and exhausted). The real aim is to have a customer who will be delighted.

Part II
User interface design

4   Creativity techniques

Irene Mavrommati

4.1   Introduction

This chapter looks at the process for the design of interactive user interfaces (UIs) and outlines the creativity techniques that can be used to facilitate this design at each stage.

The first section looks at techniques and tools for creativity and concept development. We will examine why they are needed and how using them together in a UI design process gives an integrated result that is much more than the sum of each in isolation.

A few words about the design process

Traditional design must communicate as well as seduce.
In interactive media, design has a third imperative: it must facilitate use.

> (Paraphrase of a quote from the site of the San Francisco-based
> web-design company 415 Productions, Inc.)

Defining the big picture

The first thing that should be done at the start of any project involving the design of interactive media is to define 'the big picture'.

One part of this is to establish the priorities of the client in terms of the balance between the interrelated factors of development time, cost and quality. Doing this at the beginning helps to avoid any misunderstandings later on.

Then it is important to have a clear idea of the requirements for the user experience of the UI in terms of:

- brand identity;
- style (aesthetics and 'personality');
- the target audience (their existing knowledge and style preferences);
- our message to the users;
- our aim (what we want to accomplish with respect to the users).

In addition to these explicit issues there are a number of implicit issues we have to refer to – the usual UI interaction issues which are involved in ensuring a positive experience for a user interacting with the UI and through it navigating within a virtual space. The interaction should be easy and comfortable and the virtual space should be simple enough to be easily memorized and consistent with the things with which the user is already familiar. More than this it should somehow be pleasurable to use. We will see more of these issues later (see also Chapter 17 'Pleasure with products').

A 'needs analysis' document can play a key part in establishing answers to all the above issues and facilitating our process. If the job involves a client and a development team then the communication and agreement of these issues with these groups whether by a paper document or through e-mail is very important. The document facilitates the agreement of the needs issues by being something which is easy for everybody to discuss. Once finalized the document provides a means of reference and benchmarking which can be referred to later on by everybody involved.

Getting started: Concept creation

As a second step we – with a team of two or more people – can create a number of ideas which fulfil the requirements in the 'needs analysis' document. Some techniques are described in Section 4.2 which can enhance the creativity of these sessions.

Getting on: Concept development with extended mapping

The structure of the interface then needs to be defined and it is helpful to explain this by analogy with the architect's work in designing a building.

Consider the architect's plans showing the framework and layout that defines the building's structure and some of the details of how various functions can be performed in various places. The layout in the plans will combine matter (walls) with empty space – which, in design, are called 'forms' and 'anti-forms' – so that he can both optimize the use of space and express something of the nature, personality and context of the building. Some rooms will be more frequented than others, or used by different people, perhaps for different tasks. Less frequently-used rooms may tend to be more difficult to reach, or further away, and somebody would need more effort and time to get there. The architect's plan expresses the philosophy behind how he expects the building to be used.

In the case of an interactive UI we are talking about the structure of an information space and the movement possible within it. While this may not be as tangible as a wall in a building, it is equally capable of confusing, amazing or moving us with emotions.

It is in this phase that information architecture happens, and the plans are laid down according to which the building will be constructed. Mapping the structure of the application in flowcharts or other tools will be explained in Section 4.3.

Taking it further: Concept development with interactive prototypes

While flowcharts or other diagrams are invaluable for making a map of navigation, they cannot express the dynamic aspects of how the user experience may be implemented (e.g. in terms of animation and sound). Something else is clearly needed for 'fleshing out' the detail of the execution of the concept.

The tools that would be appropriate for this depend on the nature of the interactive application and the complexity of the concept. If the multimedia is destined for a proprietary platform which will require some time spent on production then it is useful for us to make an interactive prototype for the software team to consult during development. Alternatively if we are also making the interactive application through it a CD-ROM, website or 'info-kiosk' then we can select appropriate software tools for each.

Turning points: Evaluation

The final element in the design process is evaluation. This is often thought to be something which happens at the end of the design process; if it were done at this point its usefulness would be significantly limited. The best relationship between design and evaluation reflects what is perhaps the oldest design process – that used by a craftsman in making one product at a time. The craftsman would frequently pause to reflect on his work, consulting with the client and trying some aspect of it. This continuing cycle of design–evaluation–modification is now known as iterative design.

But what is being tested? At the beginning this might be basic storyboards, progressing through interactive demos to perhaps a real system at the end. Each time the fidelity is increasing, but the flexibility of it to be changed is decreasing.

Evaluation is a whole topic by itself and is covered in Chapter 16, as well as in several other books we recommend (see Appendix 2 'Guide to further reading').

4.2 Getting started: Concept creation and creativity techniques

In this section we will be looking at techniques and tools for creativity and concept development.

The first phase of interface design is to understand the problem and to create a number of possible solutions to it. A good way of doing this is, if possible, to get a team of two or more people together to have a creative session which will produce a variety of ideas. Even if you cannot get a group

Table 4.1 Input techniques.

Related products review	Other similar systems and techniques are reviewed by spending some time in playing with them and experiencing them. A balanced view is taken rather than viewing them purely in functional terms at one extreme or as a work of art at the other.
Benchmarking key features	Key parameters of the existing and new products are compared with the main competitors.
Mood boards	Collages are made of photographs (either from magazines or photographed for the occasion) which 'set the scene' for the new product.
Trend boards/matrices	These are collages of clippings which are used during the creative process to represent visually the direction of trends for personal values, beliefs and lifestyle for the target users and technology.
Visual mapping (matrix)	Like a trend board a visual mapping is a collage, but the clippings are arranged according to how they relate to scales of independent parameters applied to each axis. The axes tend to relate to 'emotional' qualities which can be applied in the design.
Matrix with key features	As with a visual mapping a matrix contains images of relatively similar products positioned according to two axes. The difference is that here the axis parameters used are more 'rational', such as the levels of performance, cost and features. This helps in visualizing the features for the product being designed.

together, some of the techniques we will be discussing can still be used. The creative session can last for anything from a few hours to several days.

During this session a number of techniques (Tables 4.1 and 4.2) can be used to enhance creativity in the beginning of the design phase; we will be meeting these in more detail a bit later on.

Before the workshop

Before starting it is important that the aims of the workshop and what we want to achieve at the end of it (which is related to our initial analysis) are clearly communicated to and understood by everyone involved.

It is also good practice, particularly in important projects, for the organizers to plan the workshop phase in advance, not just to keep on schedule but also to optimize the results, keep on-track with the workshop objectives and get the most out of the sessions.

Table 4.2 Processing techniques.

Brainstorming	Freely and uninhibitedly expressing ideas. The result is a rich collection of new and often daring concepts of features of a system, that are noted and presented thus triggering in turn new ideas. Cross-fertilization of ideas is encouraged.
Snowballing	A technique starting with a group of two, then four and augmenting the number, while grouping, filtering and merging the concepts. Controlled cross-fertilization of ideas is facilitated.
Role playing	Playing out scenarios of user behaviour help define new features of a system.
Discussion/questions and answers	A group familiar with the subject of the brainstorm can discuss the possible features of the product. One way of stimulating detailed discussion in certain directions is to use a prepared list of questions.

Figure 4.1 Workshop.

At the workshop

Once the workshop (Figure 4.1) begins an informal environment should be created within which everyone will feel at ease to release their creativity. One part of this is that everyone should be familiar with each other and often this means that a few minutes are spent with people briefly introducing themselves.

The workshop objectives should then be reviewed and the background knowledge of participants brought 'up to speed' with presentations of related projects and material, facts about trends, the target-audience profile, the position in the market place and the competition. Creativity techniques are then used to stimulate a number of concepts and ideas, which are just collected until the end of the creative session. It is important not to judge the concepts before this as doing so can block the creativity of the team members and some good ideas may never see the light.

Towards the end of the session ideas can be grouped and combined into a small number of concepts which cover the main ideas. The final shortlist can be made by filtering out those concepts which, when evaluated against a list of goals and limitations, least satisfy the selection criteria.

The creation phase finishes with a single concept being chosen for development from this shortlist.

Materials that may be needed in preparation for such a creative session are:

- white sheets of paper (in abundance);
- Post-it notes;
- large sheets of white paper (board paper);
- markers, pens and pencils of various colours.

Some creativity tools such as moodboards and matrices give their best effect by being on show throughout the session.

Let's now look at the creativity techniques in a bit more detail.

Basic rules

Here are a number of basic rules for creative sessions:

- create as many different ideas as possible;
- do not censor them at the beginning;
- make associations, think spontaneously and document these thoughts in quick notes;
- try to avoid behaving as superiors and disciples in a creative session – everyone's ideas are of equal value.

Input techniques: Getting in the mood to begin with

In this section we review a number of techniques which get participants in the mood for the creative session and stimulate ideas about the functions and interface of the new product; these will be developed later as a more concrete concept.

Related-products review

Participants should be given time to play informally with and experience other similar systems and techniques so that they can get familiar with them. It is important for them to try to get a balanced perspective of all

Figure 4.2 Keywords.

attributes of the systems – neither just in functional terms at one extreme nor seeing them as a work of art at the other.

Benchmarking key features

Benchmarking can take many forms. Here the participants in the creative session can be divided into groups, each of which will look at one type of related products.

A short time is set aside where group members note down the key features of their selected products on 'stickies' (Post-it pads) or other small pieces of paper. After this the whole creative session reconvenes and the key features are written up on a board (Figure 4.2). A representative of each group talks through them, giving their group's perspective and ideas about new features. Finally the whole team selects the ones thought appropriate for the new product.

Moodboards (to keep focused on brand identity and brand values)

Moodboards are collages on large boards of images, loose words, magazine clippings, fabrics and objects that are related to the desired products' look and feel, emotional aspects and brand identity.

In compiling these images and showing them during the brainstorming session this, again, helps to set the right mood for the 'look and feel' of the product or interface being developed and stimulates imagination from the emotional and functional perspectives in balance.

It is also a good idea to supplement any moodboards with a number of related magazines which participants can browse, refer to and use for inspiration.

Trend boards/matrices

Another important part of the creative process is an understanding of where the trends for personal values, beliefs and lifestyle for the target users and technology are heading as this helps with the identification of possible gaps in the market that could be addressed with the design or product concept. A trend expert, futurologist or sociologist may supply this trend information.

Like moodboards and visual mappings the trends are ideally presented visually throughout the creative session (Figure 4.3). They can be clustered on a trend matrix: A grid where aspects run along one axis and the corresponding trend categories – short term, emerging, long term, certain, uncertain, regional or global, etc. – run along the other.

Concepts generated can later be categorized according to the trends – or gaps in the market created by those trends – for which they provide a solution.

Visual mapping

Another stimulus during creative sessions and the design process are visual mappings which help visualize the positioning of the product in the market and aid comparison with other products (Figure 4.4).

Like a trend board a visual mapping is a collage, but the clippings are arranged according to how they relate to scales of independent parameters applied to each axis. The axes tend to relate to 'emotional' qualities which can be applied in the design (e.g. 'conservative to serious', 'playful to sentimental' and 'male to female').

Here a square or rectangular board is created where the horizontal and vertical axes each represent a 'look-and-feel' or character attribute of the desired product or the brand. The axes to be chosen for the mapping should be the ones most important for the particular product or service under development. Some examples of possible axis attributes range from:

- conservative to innovative;
- traditional to contemporary (modern);
- individual to group, novice to expert;
- younger to older;
- emotional to rational;
- masculine to feminine;
- handcrafted to hi-tech;
- simple to confusing;

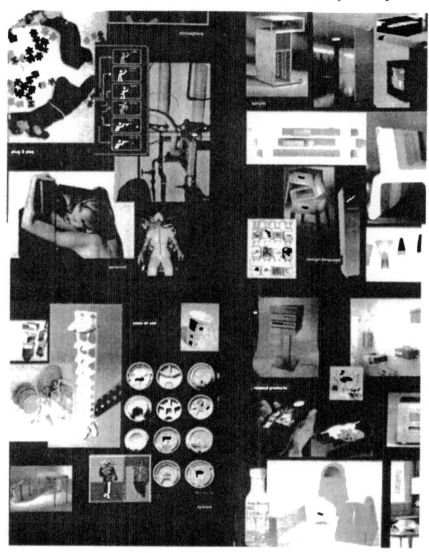

Figure 4.3 Trend board.

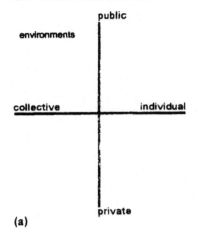

(a)

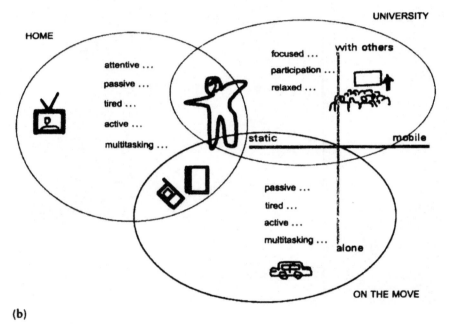

(b)

Figure 4.4 (a) Axes; (b) visual mapping.

- pleasurable (satisfactory) to disappointing;
- intriguing to dull;
- alternative to mainstream;
- low-priced to high priced;
- following a trend – where similar products go to – or alternatively perhaps setting a trend.

Small images are then fixed to the board in a position corresponding to how it is judged to relate to the parameters on the two axes. As this is done participants can discuss why they think the positioning is appropriate.

Matrix with key features

As with visual mapping this matrix contains images of products arranged according to two axes, but this time the axis parameters are relatively 'rational' qualities. Products are arranged along the axis according to the level of performance, cost or features represented. Such an approach helps visualization of features for the product being designed.

One particular type of visual mapping is the 'key feature matrix' where the axes are:

- horizontal – ranging from traditional/old fashioned/standard on the left to modern/new features/innovative on the right;
- vertical – importance ranging from 'optional', through 'nice to have' to 'essential'.

Consumer profiles/customer profiles

It is also important to focus on the users of the product.

The first thing to clarify for everyone is who the direct customer (the client) and target audience are. Sometimes the direct customer is an intermediary to the final client – the target audience that will use our product – but the goals and expectations of the former may be focused differently to the needs and wishes of the latter. The design should strive to reach a balance between the interests of the two.

The next step is to map the profile that the target audience – our users – would be expected to have. This helps everyone to imagine what their context of use and needs would be, and which functions would be useful in fulfilling them.

Some of the profile elements that might be appropriate to be defined for this target group are:

- sex (male or female);
- education;
- age;
- character;
- marital status;
- lifestyle;
- possible profession;
- possible aims in life;
- possible beliefs;

- hobbies;
- type of house;
- transportation;
- other appliances used;
- amount of travelling (with distance and means of transport);
- benefits sought.

Processing techniques

These techniques can be used in the creative session to take the concepts further. More than one of them can be used in combination to give more variation and produce more diverse results (Kokkos 1998).

Brainstorming

Brainstorming is the examination of a subject from many different view-points and participants are encouraged freely and spontaneously to express their ideas (Figure 4.5). They must express every idea or proposal that comes to mind without judging it against any criteria (even when the idea may seem unrealistic or too futuristic). There should be no criticism during the generation phase and all the ideas are kept as notes (or yellow Post-it notes) that are all stuck onto a board or wall and then grouped in clusters.

In this process a high level of interest and participation, cooperation and team spirit is encouraged between the participants, and the creativity of all is used. Many approaches and solutions may be generated, and by exchanging

Figure 4.5 Brainstorming meeting.

points of view and concepts the members of the whole team will be stimulated with new ideas.

One way to start with brainstorming is free association with symbols or keywords, starting by describing the general look and feel of the product: its texture, what it can be associated with, its smell and taste, the sound of it. This relaxes the team, and can then be taken further into more concrete parts of the interface.

Snowballing

Snowballing is the exchange of opinions around a subject in a way that is a little different to normal discussion. First the team decides which subject has to be discussed and what the objective of the snowballing will be. Then each person individually examines the subject for 5 minutes or so and keeps notes or sketches of ideas.

Each participant then compares notes with another for 10 minutes, identifying common elements and differences in the two approaches, exchanging ideas and enriching the approach or solution. The concepts are then combined into one unified complementary concept. The above procedure is then repeated in groups of four people.

Finally the combined results are presented to the whole team, there is a synthesis of the results and conclusions are made about the appropriateness of the solutions.

This process doesn't take as much time as discussion, partly because it keeps participants focused within the subject.

Role playing

For role playing the idea is to think of a story (or 'scenario') about a typical situation of use within the context of the product (Figure 4.6). There could be one or more roles that play a part in this story, possibly resulting in differences or conflicts in the outcome.

The beginning of role-playing then is the setting of the scene, and the distribution of roles for the participants to play. Each role may be played by more than one person because different people may have different reactions and needs within the same situation.

It is useful for the scenario to refer to the actions of the users such as walking, reaching, glancing as well as the distance involved and the daily habits and routines that they may be a part of. These really give an idea of the Human–Machine Interaction (HMI) issues involved. For example:

- in the morning Mrs B wakes up, turns off the alarm clock and gets out of bed to make some coffee ... while her husband turns on the coffee-maker ...:

Figure 4.6 Role playing.

- Mr P is standing on the train platform with his mobile phone, holding lots of bags. He wants to switch his phone on but doesn't have a spare hand to . . .;
- Miss K decides to call her friend who is thereby interrupted by the phone; now she has to answer it . . . .

These stories reveal problems and conflicts that may give us HMI-based ideas on what controls and functions would be needed to resolve them.

All the stories or conflicts and their possible HMI solutions can be noted down on the fly as keywords on yellow Post-it notes, for example, which are collected later for reference. Video recordings or notes are also useful for doing this. Additional notes can be kept about things like the feelings of those involved while using the system, what this means for their day-to-day experiences, the dynamics in communication and interaction with the device and the relationship of the device with its environment.

This process leads to an expression of sentiments and an understanding of the participants through the simulation of the living experience. It helps to highlight the various situations and conditions of usage, the perceptions of people about the product or system and their attitude and behaviour towards it. This kind of knowledge is invaluable in defining requirements and priorities for the system interface.

Discussion/questions and answers

When the team is relatively knowledgeable about the subject of the brain-storm they can then discuss it, either together or split into groups. This discussion should include the elements and possible features of the product and notes should be made to record the main points, conclusions and keywords.

Questions and answers can be used to help stimulate a discussion related to the subject for which detailed understanding is needed and concepts will be created. These questions may require a yes/no answer or further analysis based on reasoning. The team can answer these questions freely without following any predefined order. To make them more appealing they can be supplemented with illustrations, sketches, audio or video.

4.3   Getting on: Concept development with extended mapping

It is now time for the development work to start on the chosen concept. The development and communication tools used in this phase include storyboards, prototyping and flowcharts. The principal technique used in combination with these is of role-playing through various scenarios of use.

This section examines why these tools are needed and how they comple-ment each other in a process providing much more together than if they were used separately.

Storyboards

Storyboards are a great communication tool both to novices within the project team and to others outside it. While storyboards were initially used as a way of defining a sequence of events in linear media (e.g. in films or comics) they can be applied to interactive media by relating a possible scenario of use where the interactivity of the user is fixed as part of the story. In this way storyboards can be used to define the concept and basic functions of an interface.

Obviously, though, as a storyboard cannot present the full detailed scope of the interactivity of the interface and interrelation of elements within it there is still a complementary role to be played by flowcharts, for example. In this way storyboards may help in the initial definition of flowcharts and continue to be used as a parallel definition tool.

The best starting point is to identify a number of common situations with the product that will have to be dealt with by the interface and to begin making draft storyboards where each situation is expressed as an imaginary scenario. This usually helps to reveal many details of how the interface should behave that may not previously have been anticipated.

1 Mr John, an estate agent gets a new digital camera...

2 He takes snapshots of his house...

3 Then he takes out the plug-in memory, that carries the images

4 ...and he puts it in the reader attached to his computer...

Figure 4.7 Sample storyboard.

Typical questions to be answered within each storyboard are:

- What is the profile of the user?
- What situation are they in?
- What do they need?
- What does the user do to the product in question?
- What are the results from the product that solve his initial problem?

In its simplest form a storyboard can simply be a textual scenario of use where a description is given in a narrative style of an imaginary situation with actions, problems and context. This helps to set the scene for the product concept. A storyboard can evolve from the textual description to take a more visual form where a sequence of key actions is illustrated. A storyboard in this form looks like a cartoon strip (Figure 4.7).

The visual storyboard itself can be made by:

- separating key actions of the scenario into a sequence of drawings or sketches (not the user actions in detail, just the main steps in order to illustrate the point);
- pasting them in sequence onto a large board;
- writing a brief caption below or to the side of each picture describing what is happening.

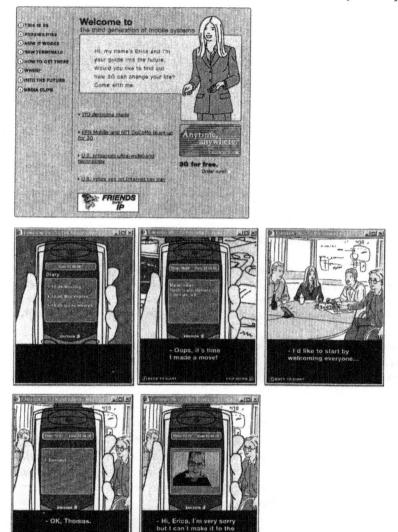

Figure 4.8 Screen shots taken from an animated storyboard (Ericsson).

The storyboard can also be made as a sequence of printed sheets or web pages, which can be distributed, accessed and referred to by team members (Figure 4.8).

Apart from the static storyboards described so far it is also possible to make animated ones where the basic structure remains the same but each static picture is replaced by a short animation or piece of video, often with an accompanying narration. These can achieve a more dramatic effect, but, because they take more time to produce, tend to be used to communicate an already crystallized project idea in a quick, understandable and impressive way. They are not generally used to specify extra context of use through alternative user scenarios.

Flowcharts

Why use flowcharts?

The purpose of a flowchart is to map the territory, flow and levels of use of an application or product.

A flowchart is important at the beginning of the creation of an interactive application because it:

- gives everyone in the team a common understanding and reference for discussion;
- shows the full extent of the application, defining the structure within which the interface and its various facets can be constructed, edited and rationalized. There may also be a hierarchical aspect where smaller units within this structure may themselves be represented by flowcharts;
- promotes understanding of the consequences of changes in the overall structure;
- shows how the interface parts interlink, visualizing the consistency within the whole application.

Making flowcharts requires logical groups or clusters of activities and modes to be defined. This is difficult and needs careful consideration by the whole project team. It is almost inevitable that it will involve many disagreements, compromises and trade-offs because of the different priorities and perspectives of the team members. The result, however, is an agreement that will prevent problems later on.

What does a flowchart look like?

We are all familiar with flowcharts in different forms: We have encountered tree diagrams such as family trees, or process description in schematic forms of flowcharts. They can be seen as draft sketches, as sheets with Post-it notes, or as a print of any size, created by flowchart software programs.

Flowcharts are diagrams where most of the screens or modes of the interface are each represented as a block. The block may contain sketches, images or plain text. Each block is named or numbered with a code to avoid ambiguity – a reference to 'A1', 'A2' or 'B1' is much better than 'the green screen with the bar on top'. Arrows between the blocks show the conditions under which a transition is made.

The lower level behaviour of elements within each screen/mode is not usually described in the flowchart itself unless essential – it might then be marked in as a reminder. More usually, however, such low level behaviours are explained in an additional dedicated appendix or flowchart.

Flowcharts can have several forms and the one to use depends on the kind of application, what is to be explained and the stage of the development process. These various types will each be described now.

How to make a flowchart

There are several ways of making a flowchart, ranging from draft at one extreme to final and polished at the other, depending on what it is needed for and the phase of the project.

At the beginning of a project – when work is moving quickly and spontaneous changes to the intended structure occur very often – it is best to use draft flowcharts with images or sketches that can be quickly made and positioned/repositioned on a large sheet of paper. Lines connecting the flowchart blocks can be quickly drawn by hand, or also somehow fixed onto the paper (e.g. using string). Sometimes in relatively simple projects it is possible to write directly on the board itself.

Here is a basic sequence for making a flowchart:

- Make a set of Post-it notes, each representing a screen in the user interface of the product with descriptive text or a roughly drawn sketch of it.
- Once you have notes corresponding to all of the screens you think you need, put them in sequence on a sheet of easel paper from the first to the last.
- Identify the most important and often used screens, and position them first in the hierarchy at the top of the flowchart. Realign the activities to fit under the appropriate group. Do the same for the second screen rated by frequency and importance, and then the third, and so on.
- Then identify the logical steps that link these screens with user actions (buttons pressed, rotaries turned, etc.). Try to start visualizing in your drafts the relationship with the hardware parts of the HMI, so that the on-screen UI stays consistent with the hardware interaction part.
- Paste an image of the hardware control next to each frame representing a UI screen as a reminder.
- Use arrows to connect the activities and decisions.
- Identify the major steps of the process that the user should do in order to perform a task in the UI (with the hardware and software components of it). Try to improve the logical sequence of these actions.
- Go through several scenarios of use using the flowchart, triple-

checking consistency and ease of use both mentally and in detail. Rearrange the flowchart squares and the lines in the process and do not be afraid to create new draft sketches to add to the existing ones or to replace others.

- Realign activities where appropriate for optimizing the interaction, or to bring the main steps of the process into line.

You may, of course, draw these flowchart screens by hand or use Post-it notes or pre-drawn elements to combine them. Whichever way you do it flowcharting is a great method for pre-designing your interaction through the product or user interface and for helping others to understand how it will work when it is completed.

Later in the project, when the basic structure is laid out and is more crystallized, a software flowcharting tool can be used for communication purposes.

Text flowcharts/technical flowcharts

Text flowcharts provide an outline of the context whereby every element of the UI is described textually in a box, sometimes also schematically. In this way the context of an application can be mapped out and initially separated into screens. Later the text can be replaced with diagrams or sketches of the context of each screen to help with the screen design.

Text flowcharts include the full content and a description of the functionality of every main screen, but also give full freedom to any experienced on-screen designer who may take over later to translate it into whatever layout he/she finds appropriate.

Programmers and system designers use flowcharts too, but in a slightly different form. They include diamond shape blocks representing the decision paths of the program (e.g. for yes/no decisions) (Figure 4.9).

If the flowchart is being used to communicate to people with a technical, (e.g. programming background), these diamond shapes can be included in crucial parts of the program (related to states of the system or previous user actions) to make them more understandable.

Flowcharts from the user perspective

User-perspective flowcharts normally appear later in a project, gradually replacing the draft sketches with the actual work in progress of the screen design. Elements normally used in storyboards can be added (e.g. small textual descriptions under the screen picture or 'block' of behaviour of multiple screen elements, animations, sounds and colour changes).

These flowcharts can be used to check the UI, usually before a demo is made. Aspects under scrutiny are the interrelationships of screens, gaps, inconsistencies and repetitions within the screen designs from the user's

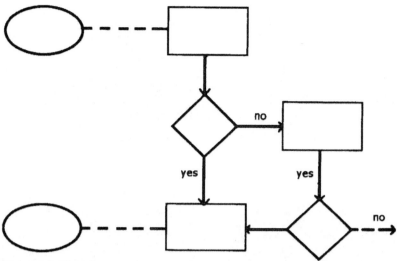

Figure 4.9 Flowchart.

point of view. Once shortcomings are revealed in this way, logical solutions can appear more easily as well.

Sometimes heated discussions can take place when such a flowchart is discussed within a team during the development stage; this is normal since people in a team have different perception, as well as different taste, preferences and sometimes even come from different professional fields.

Detailed overview flowcharts

A big structural overview flowchart with detailed subdialogue flowcharts is always useful (Figure 4.10). It may itself consist of several flowcharts, one or more being at a more abstract level (the main screens only) with other more detailed flowcharts for each of the main flowchart's subparts. If it is fully detailed the structural flowchart may be bound as a booklet; otherwise as a single big sheet it may become unwieldy and cluttered.

Other types

Any other form of flowchart that may explain complex interaction systems while perhaps not having a tree structure is also welcome, including three-dimensional or mobile structures (Figure 4.11). There is usually a lot of scope for an imaginative but clear portrayal of the project structure.

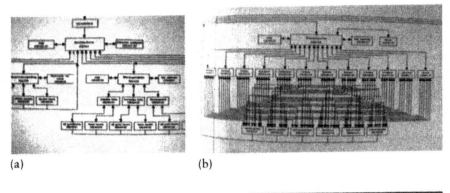

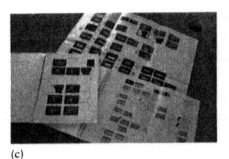

(a)   (b)

(c)   (d)

Figure 4.10 Detailed flowcharts.

4.4   Prototyping – taking it further: Concept development with interactive prototypes

The interactive prototype is a tool – in this case a designer's tool – for further development of the application or product. It is also often used as a communication tool for showing perhaps-hesitant developers the added value of certain options over others, convincing them that the extra effort needed to implement them in the final product is worth it in the final result. Finally it can be used as an evaluation tool.

Static representations like flowcharts and pictures of screens are important for defining modality and transitions within an interface. However, they are not enough to allow full definition and understanding of a system or the experience of using it, especially for a highly complex one. It is important to represent the whole interaction dynamically to give insight on aspects such as:

- the requirements on the user in terms of mental workload and attention;
- its readability in relation to interaction (i.e. clicking/touching/distance);

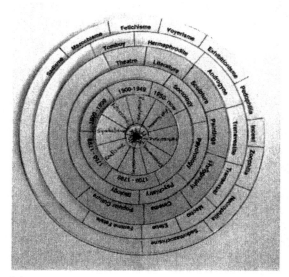

(a)

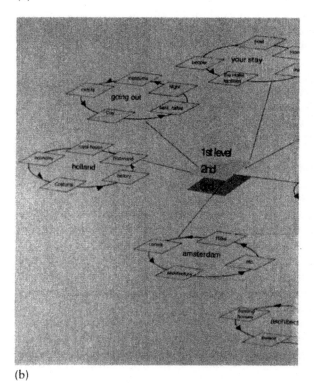

(b)

Figure 4.11 Other types of flowchart.

- the effect of sounds, animations and colours;
- the logical (or not) sequence of the screens;
- other sensorial stimuli.

The experience as a whole is more than the sum of these separate parts, and in bringing the integrated experience to life the demo performs a complementary role to the flowcharts in the design process. The demo enables focused examination of the design and detailed discussion about aspects that would otherwise have to be imagined.

The demo is usually a computer mock-up representation of the product. When possible it is advisable to have a physical mock-up whereby it works with what appears to be real hardware, but sometimes it is too costly and time consuming to get such work in progress. Mocking-up the system as an on-screen demo is usually much more flexible and it is easier to iterate, change things on the fly and develop it along with changes in the design than it is with the hardware mock-up of the real system.

The fidelity of this demo may vary from a simple draft showing just the main functionality (the things presented in the storyboard) to a realistic interactive presentation. Usually it is made with software tools such as Macromedia Director, HTML or Visual Basic and in most cases this is different to what the system-development team will use for final implementation. It is not usually worth making the demo too realistic in the small details because:

- the system engineers will have to recreate the system for real with their own tools – demo 'code' is not generally reusable by them;
- as the level of realism required in a demo increases the time taken in making a demo rises at a much greater rate.

Of course, a balance should be made if the demo will be used for testing later. It should have the necessary functions completed in an interactive way, as far as are required for user evaluation.

While selecting what to include in prototyping, a simplification is made of the system, cutting down the complexity by selecting certain functions that we want to display, test and evaluate.

Horizontal prototypes (the simplest of which can be the non-interactive scenarios) reduce the level of functionality (the depth in interaction) while showing the total UI in the first level, thus showing all the features in extent, but not their in-depth functionality.

Vertical prototypes, on the other hand, concentrate on displaying the whole functionality and features of a selected, critical area of the system.

Demonstrators of scenarios of use are a completely flat version of prototypes, made easier and quicker to do by reducing the level of functionality as well as the features. They are normally used in the initial steps of the

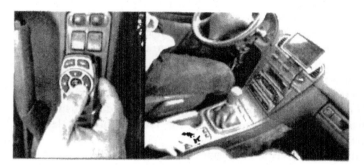

Figure 4.12 Prototype of a car-navigation system (Philips Carin, now available under the VDO Dayton brand name).

development process as a development tool, or as quick mock- ups for evaluation purposes.

Both the flowchart and demo should be reviewed for gaps in the structure or weaknesses in the user experience. Both are then updated accordingly.

In developments where system engineers make the real system, using the demo and flowcharts as a blueprint, the prototype – when it becomes available – can also be used for expert appraisals and usability testing (Figure 4.12).

Additional models – demonstrator scenarios

Additional models – such as visual prototypes on paper – can be used together with the demo or prototype to get a more complete understanding of the system. These can also be tested with users. Animated scenarios can be based on the storyboards mentioned above. They can be very simple prototypes reducing the horizontal level of functionality of the system or device, as well as the number of features. They display visually a selected scenario of use judged as critical or of added value to the user. They are generally cheaper to design and implement, since they are not interactive but mostly linear, and give a good and understandable impression in user testing sessions. These scenarios are a way of getting quick and frequent feedback from the users, but not testing the functionality of the system in detail – which is done with more detailed prototypes.

The simplest way of developing and evaluating a system is in scenarios implemented as paper mock-ups. Either storyboards or paper or clay three-dimensional models that help us visualize the system in a more concrete way.

Controls and input devices

The position and structure of buttons on the input device or front panel is also important in defining the on-screen interaction of a system. A TV-based interface depending on a remote control, for example, should have a visual connection between the locations, colours and shapes of the control buttons with their representations on the screen.

For this reason an important part of concept development is having a sample of the system's input device(s) available to play a part in playing out scenarios. If it is not technically possible to control the demo using these hardware controls then a representation of the latter should at least be visible.

The best solution, however, is to design the whole UI together (i.e. the hardware-based and the screen-based parts of it). This is possible if all the involved disciplines work closely together from the beginning of the project.

Indicative tools for demo prototyping

The tools to be used depend on the expertise of the team, the detail required within the demo prototype and the nature of the product being designed. Tools most often used in interactive design are:

- web authoring packages (e.g. Macromedia Dreamweaver or Flash, or Microsoft Frontpage);
- interactivity software (chosen from a variety of authoring platforms that exist [e.g. Macromedia Director, Microsoft Visual Basic, Authorware], or any programming language [e.g. Delphi, Java, etc.]);
- static and moving image-processing programs (e.g. Adobe Photoshop, Premiere or AfterEffects, CorelDraw and various three-dimensional and drawing packages that exist);
- audio packages to create sounds for the needs of the demo.

5   Design principles

Adrian Martel and Irene Mavrommati

5.1   Overview of goals and principles

In the last chapter we looked at the process, tools and techniques which define the framework within which the interface is designed. Simply applying these is, of course, no guarantee that an interface will be any good; consequently this chapter looks at some usability goals and the principles of good interface design, which are used to achieve them.

Before we go any further, though, a word about the judgement of quality: How can a good design be recognized? The fundamental issue here is the design has to be made for, experienced in and judged from the right context. It is meant for particular types of people (who may share a particular perspective and experience which may be different to that of the developers) in particular situations (the occasions and places where the product is experienced in terms of the amount of light, sound levels, amount of space, viewing distance and social situation, etc.). Judging a good design, then, can only really be done from that same perspecive and for this reason it is important to involve target users from the beginning.

Furthermore it is unlikely that a set of quantitative measurements will suffice as a form of judgement. Many issues can only really be assessed by observation of a target user with the product in a representative situation.

Let's move now on to the goals and principles themselves. The fundamental goals for the user interface are outlined in Table 5.1.

The rest of the chapter is dedicated to explaining the principles mentioned in that table.

5.2   Suitability

The first major issue is the suitability of the product for doing its job. This largely comes down to looking at the target users in context with certain tasks to do and goals to achieve, and seeing how appropriate the main aspects of the interface are for this. The following list of considerations need to be considered:

Table 5.1 Goals for the user interface.

Usability goal	Principles used
1. To be suitable and appropriate for its purpose through:	Suitability
– the way the implementation of the product facilitates the user in physically handling it (size, access to openings, portability)	Suitability
– the scope of the controls that are provided to access its various functions	Suitability
– the implementation of the controls and other points for which access is needed (physical access, size, position and immediacy)	Suitability/Fitts' law
– the visibility of any indicator lights or display or audible sounds made	Legibility/immediacy
2. To be easy to learn and understand, building on what the target users already know, to give a simple, clear and coherent impression of:	Self-descriptiveness, learnability, familiarity, assumed knowledge
– What is it for?	Semantic language
– What is it doing right now?	Feedback, understandability/ relevancy
– What can the user do with it and its controls? What shouldn't they do?	Affordances, feed-forward constraints
– How are functions, controls and displays related together?	Prioritization, grouping, consistency
– How can these controls be used to achieve the objectives or change the system's status?	Transparency
– How can the design avoid and correct any unintended situations?	Error tolerance
3. Be satisfying and pleasurable to use	Satisfaction

- the functions;
- the information given both in terms of *what* is given (relevancy) and *how* (legibility);
- the controls used;
- other physical attributes.

Let's talk about these in more detail.

Appropriate functions for the tasks and goals

The first aspect is how appropriate the functions provided are for achieving the users' goals and matching the functions available to the user with their expectations. This involves finding a balance between functional overcomplexity or overkill which may confuse normal people, and oversimplification to the point where they aren't able to control aspects which are important to them. In a nutshell the keywords may be 'relevant' and 'simple' but 'enough'. At one extreme, then, the functions available to the user should be relevant to achieving their tasks and goals rather than being overcomplicated and divided into more functional groups, subgroups or control aspects than they will normally use. Giving too many possibilities for control would only confuse the user, make simple functions seem (or be) more complex than they really are and give them more to have to learn. In essence the principle 'less is more' should be applied here.

This can perhaps be illustrated with an example, though a rather extreme one.

Example   *Imagine that we're going to make a dimmer for a light – the user would expect a single control to make it brighter or dimmer. We would be overcomplicating this if we instead gave separate controls for two technical parameters contributing to brightness. Furthermore, adding a control to adjust flicker would just be distracting and irrelevant (and would always be set to 'no flicker').*

At the other extreme it would be equally excessive if the user had insufficient choice in the main functions as a result of key aspects being automated. Automation tends to work when use is very well defined, but there may be situations when the user wants something different from normal. If, for example, a central-heating system always turned itself on and off at certain hours without a manual override, an owner wouldn't be able to control it manually in unusual situations like being at home on a very cold day.

One way to limit the functions provided to those necessary is to use the majority proportion rule whereby the system should aim to please the majority of the users (say 80 per cent). The functions that a minority (say 20 per cent) of the users would want should also be included but only if it is possible simply and easily to accommodate them, otherwise they should probably be skipped to avoid complicating things for everyone else.

Appropriate controls

The way in which the functions are implemented as controls in terms of location and form should also be appropriate.

The most important or frequently used functions should be operable most easily while those which are less important can be slightly less convenient (Fitts' law). Those which should not be operated accidentally should be

(a)

(b)

Figure 5.1 Examples of frequently used controls on portable appliances: (a) CD
player buttons; (b) portable cassette player.

implemented in such a way that while not difficult to use they can only be
used deliberately.

For example, the most used controls on portable appliances should
naturally fall under the fingers and be used in a way consistent with the
fact that the product is being gripped at the same time. One instance of this
might be a portable CD player (Figure 5.1) where the most used function is
the play button – this should be large and located close to the place where
the fingers tend to hold the product.

Controls, which should not be used accidentally, should be well away
from areas where the hand is normally found. For example, you can see
how the power button on the mobile phone in Figure 5.2 is by the aerial; see
also the power button on the Kidcom appliance shown in Figure 5.17.

Figure 5.2 Example of a power button on a portable appliance (Nokia).

This also applies to the number of steps taken to access functions. The most frequently used or important functions in a complex product may be given instant access by pressing a large, convenient button ('hard key') (e.g. the 'call/answer' button on the mobile phone). Less frequently used fuctions may be accessed in a less direct way:

- smaller buttons hidden under a cover (e.g. the numeric keypad in the phone above);
- by holding down a button (e.g. in portable CD players which have no dedicated scan/search buttons this function is accessed by holding down the next and previous buttons);
- through a soft key – a hard button whose purpose changes according to a graphic shown in a nearby display (e.g. the two buttons just under the mobile phone display or the buttons on cash machines);
- through a menu selection (e.g. settings on a TV or a mobile phone). The more important the function the higher and earlier it should be in the hierarchy and the less key-presses should be needed to get to it. Menus can be used to access large hierarchies of functions; though as the display only provides a small window on where the user is in this hierarchy it can be easy to get lost or not know where to find the function.

The type of control is also important. On most consumer electronics products the following guidelines apply:

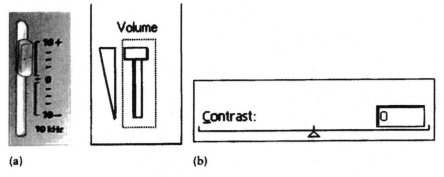

(a)   (b)

Figure 5.3 Analogue sliders [(b): Apple].

- functions which are analogue and can be varied on a 'continuous' scale (i.e. the light dimmer mentioned earlier or a volume control) are best implemented using analogue or continuous controls (e.g. knobs or sliders [see Figure 5.3]);
- functions, which require a single 'discrete' action, are best implemented with buttons (e.g. press once to 'play' now). 'Virtual buttons' on a touch screen are also a possibility here; though as pressing the screen doesn't have the same responsive feel ('tactile feedback' see p. 156) they take second place to real buttons;
- functions which require a selection between many 'discrete' settings are best implemented with sliders, knobs or thumbwheels which click between a number of positions, or with a button being used to change ('toggle') between options indicated close-by in a display.

A more extensive discussion of controls is given in Part III.

For multimedia products where a pointing device is available – most typically a mouse or trackball and its on-screen representation, the cursor – there are many other interaction possibilities:

- selection between a number of exclusive possibilities can be done by clicking on one of a set of circular 'radio buttons' or icons. The selected radio button or icon will take on a highlighted appearance and all others will take on a normal unhighlighted appearance;
- a function which can be selected without being activated is selected with one click (see the last point) and activated using a double-click (i.e. quickly clicking twice);
- a number of options which can be selected in combination can be implemented using check boxes. Each click on a box toggles its state between selected and unselected;
- an action can be initiated by clicking on a button or hyperlink;

- a movable screen element can be moved or categorized by dragging it somewhere (e.g. values can be set within a range by dragging a slider);
- extra information ('feed forward' – see p. 93) about a screen element can be brought up when the cursor is touched against it or rolled over it;
- the specification of a relationship between the cursor or selected moving elements and other elements or screen locations can be facilitated by making the former 'magnetically' attracted to the latter when it gets close.

Despite the opportunities presented by pointing devices (such as a mouse) they're not always convenient in the home, and the most common input devices for domestic products – apart from the controls on the device itself – are remote controls, either wireless (like most remote controls these days) or wired (as with many games consoles). The future is bringing other possibilities though – sound recognition, eye tracking and gesture recognition are often mentioned.

Appropriate and relevant information

It is then important to keep all the information provided relevant – remember that it is largely from the information provided that people learn about the product and what its situation is. This makes it important that it is easy to take in, is clear and to the point. Giving information that people rarely need only increases the clutter and apparent complexity of the interface, demands more attention from the user (becoming more demanding and tiring) and gives people more to learn.

Some of the factors affecting the appropriateness of information are:

- the frequency of use. Important information should be quick to access;
- its significance within the interaction. The key questions are how knowing this information affects the user's tasks and the user's ability to reach his final goals. There is a possibility of overkill if a lavish presentation (e.g. three dimensional) is made about something that would be much better presented more simply;
- whether interaction is passive ('lean back', 'pull') or active ('lean forward', 'push');
- its importance. The more important the information is the clearer it should be expressed.

Appropriately written text: Legibility and immediacy

Even appropriate information is still useless if it cannot be read and so the issue of legibility and immediacy is important. Clearer and more legible information will tend to:

- have an appropriate size of typeface/font. This is usually larger rather than smaller, but shouldn't be so large that the amount of information shown is overly constrained;
- use proportional rather than non-proportional fonts;
- contrast more strongly against its background. Contrast is made up of luminance contrast – which is the contrast of light against dark – and chrominance contrast – which is that based just on different hues. It is more important to make a big luminance contrast (light–dark) because our eyes perceive it more strongly, and lower light situations or colour-blindness can greatly reduce the perception of chrominance contrast (between hues) even further;
- be well organized in terms of the positioning of screen elements;
- be less cluttered by surrounding graphics (not too 'busy');
- contain not too much distracting and tiring movements or animations.

A consistent theme in this chapter is attention to the target users and their context. It should also be considered that people in different cultures may have different thresholds of acceptability for clutter and the amount of animation; some cultures consider pleasant and contemporary what others may consider annoying and excessive.

Finally information will have to be more legible:

- if older people are among the target user group (and improving the legibility so that they can use it will improve it for everyone – a basic principle of 'design for all');
- if the product will be used under poor lighting or environmental conditions.

Other physical attributes which should be appropriate

One final but big area of appropriateness, for small consumer products especially, is that of their physical attributes, and here anthropometry plays a key part. Anthropometry is perhaps the best known part of ergonomics and looks at the varying size of parts of the body for different people. It is important to make sure that the size of physical aspects of the product – which are supposed to be holdable, grippable, pressable – are appropriate for the more critical size of hands or strength to be used and the space needed by the user to use them:

- hand-held products should be easily holdable in one hand for the majority of target users (i.e. should be designed for and tested with target users with hands towards the lower extreme in size).

- the physical strength needed to use the device should be no more than the weakest target users are able to give;
- any openings such as battery compartments or connector bays should be an appropriate size, not just for the batteries or connectors which fit into them but also for the space needed for access by the fingers of the users (e.g. putting in or taking out batteries or gripping connectors should not be too fiddly for users with large hands).
- buttons should be large enough to be easily pressed and far enough apart that the wrong one is not pressed by accident;
- knobs or handles should be easily grippable by the fingers or hand as appropriate;
- if something is to be gripped there should be adequate gripping area for the fingers and a surface texture that is not too smooth or slippery.

5.3 Self-explanatory interfaces: explicitness

When somebody picks up a product for the first time they have to build up an idea of how it works and how they can use it. All this knowledge – which we refer to as a mental model – has to be communicated to them through the design of the product and through any additional instructions that are provided. Relying on the instructions too much is generally a bad idea because many people will not read the instructions, either because they are not interested or, in the case of a product used by several people, because the instructions may not be available. The challenge is, then, to give this knowledge to the user through the design.

This section looks at the various factors that need to be combined to make a self-explanatory interface; experience together with new patterns and relationships need to be incorporated into a simple, clear and coherent user experience. This does not necessarily mean that the user experience should be overly uniform as that might reduce the pleasure aspect which we will see later.

Familiarity, assumed knowledge and common sense

The starting point is what the target user already knows.

As people vary in age, profession and cultural and social background their expectations, level of experience and they way they have become used to relating to and using products will also vary. Something that may seem relatively straightforward to a computer-literate user may not be easy or even comprehensible to an 80-year-old or a child. At the same time, there are many things children can use fairly routinely nowadays that older people would have to think carefully about. Remember also that as people get older their memory and speed may not be all it used to be.

All the users of the product will begin their relationship with it by expecting things to work in a way that is already familiar to them. For this reason it is crucial that the development team understands the most basic level of knowledge that it can assume that all members of the target group will share. Users may refer to this assumed basic knowledge as simply 'common sense'. Everybody knows, for example, that if you throw something up into the air it will come down again as that is a piece of knowledge we all share. However, as the knowledge gets more particular the boundaries of 'common sense' begin to blur because some people may have experience of something while others might not. Some knowledge therefore turns out to be less 'common' than expected, and this can really be a surprise because once you know something it is very difficult to imagine not knowing it.

Beware, then, of treating any knowledge as 'common sense' when it is actually something above the experience of many target users. Particularly dangerous, perhaps, is the situation of the development team trying to imagine what users will consider 'common sense' and how they will behave. The fact is that the development team probably knows far more about the product than regular users ever will and when speculation is in the air the only way to be sure is to go out and test it with some real target users. It is for this reason that user evaluation phases should be planned as iterative parts of the design process. Should this not be possible then at least consult people who have about the same knowledge as the target users or who can genuinely represent them in some way.

So, again, the product should basically only need knowledge known to be shared by the target users in order to be used. That's not to say that everything within the interface has to *be* absolutely familiar, just that it should be based on known elements such that it *seems* inherently familiar.

Building conceptual models onto basic knowledge

It is almost certain that the user will have to learn something about the system they are using and that it will not be immediately familiar, especially if the product being designed is somewhat innovative (e.g. relating to new media). The act of learning is really that of building a mental model of what the product is and how it works so that the user will know how to approach any task which involves it.

Mostly this mental model will be pieced together by the user through a combination of familiarity (where they recognize things they have seen before) and deduction from clues provided within the 'system image' by the developers. The 'system image' is the 'user experience' – literally what the user perceives of the product in terms of its design (visual clues), instructions (written and graphic explanations) and how the product behaves when the user interacts with it. It is the only communication medium through which the developers can impart their own understanding of the system to the user and this is why the right information has to be made explicit. A

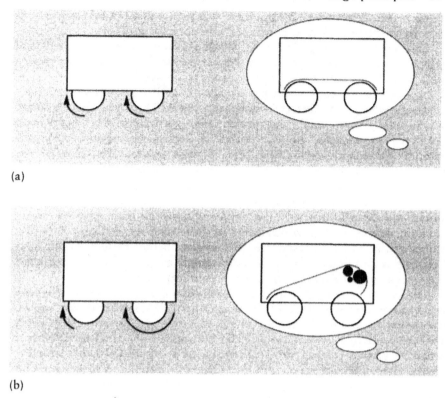

Figure 5.4 Conceptual models: (a) wheels attached by an axle; (b) gear-wheel arrangement.

certain amount of trial and error may be involved by the user in the learning process and the mental model they build up will be the simplest mechanism that explains what they perceive. (In many ways this is related to the area of perceptual psychology known as gestalt.)

Imagine, for example, that I have a box which has a wheel on each side – we will refer to these as wheels A and B. If I notice that turning wheel A makes wheel B turn in sympathy then the simplest explanation of the contents of the box is that it contains an axle which connects the two wheels (Figure 5.4).

Imagine that I now notice that they're not turning at the same speed – if I turn wheel A then wheel B turns more quickly. Then I would imagine that perhaps there is a gear-wheel arrangement inside the box as this is the simplest explanation that I can think of. In this way I modify my mental model of the system. Now it could be that the contents of the box are much more complicated than that – there's a sensor connected to wheel A and a motor which drives wheel B – but there's no reason that I should suspect this as long as the behaviour is consistent with something simpler.

This starts to make it clear how people get confused about systems and relationships; they put together the clues as they understand them to make a simple explanation ... which just happens to be wrong.

Here is a real example. Some years ago one of the authors worked for a company which manufactured microwave ovens. The ovens make a sound when they are cooking, and people soon associated this sound with whatever was making the microwaves. A logical assumption, perhaps – certainly a simple explanation. But if they opened the door of the microwave and the sound continued they might get alarmed that they were being exposed to microwaves. Actually the relationship they assumed between the microwaves and the sound was incorrect – the sound was actually that of a fan removing moisture that operated during cooking; microwaves are *always* shut off if the door is open, even if there is still moisture to be removed. There's no reason that people should have known this and this is a fairly typical example of the kind of 'folk science' that people use to explain things which turn out not to be true.

What we as system developers can learn from this is that the user has the best chance of constructing a mental model which is similar to the one we developers already have if the product's 'system image' presents clues which are as meaningful, relevant, simple and as honest as possible In some ways it is like the plot of a good film – if there are too many subplots and complexities to remember it is easy to get lost. One rule of thumb is that if you can describe the basic purpose of the product and its most important features in just a few sentences then you have done well. An interface that is simple to explain is easy to learn and the more exceptions and quirks you have to add to the explanation ('oh and another thing ... don't forget ...') the more difficulty the user will have.

As people's perceptions play such a big role in interpretation of the 'system image' into a 'user model' this is something that has to be checked with target users at least once in the process. Details do, apparently, matter and lack of attention to these can cause much misunderstanding, often in ways which are rather suprising to the developers (who, as we saw earlier, 'know too much').

Guideline
• *Pay attention to the details.*

Most of this section talks about making interfaces 'self-descriptive', 'self-explanatory' or 'explicit'. Whatever the term used the principles contributing to it – affordances, transparency, consistency, feedback and all the rest we will describe shortly – are still the same. Applying them, though, should not go to the other extreme of making the interface dull, patronizing or uninteresting as these can also make the details of the system image less memorable.

Use and abuse of analogies and metaphors

Earlier on we talked about how people built up their mental model of a system by using their own experience as a foundation. We also talked about how people learning about a system try to piece together the clues presented to them to make the simplest possible interpretation of aspects of the system image. For much of the time these principles of familiarity and simplicity work together at a low level of complexity (e.g. 'when that light is on it means that the function it represents is active'). When, however, a more complex part of the system image is understood by being related to something complex with which we are readily familiar we find ourselves talking about analogies and metaphors.

An analogy is when two compared things are similar in one key sense that aids understanding, even though they may be dissimilar in many others. A folder icon on a computer screen, for example, is a visual analogy of a cardboard folder and its similar appearance deliberately reflects that it shares the attribute of collecting together a number of documents even though a real folder contains papers and a virtual one contains electronic files (Figure 5.5). Whenever something is described as being 'like' something then an analogy is being made.

Guideline
- *Whenever something is described as being 'like' something then an analogy is being made.*

A metaphor is also a comparison to aid understanding, but of a more extended nature. Interface metaphors are off-the-shelf mental models of the use of real and familiar objects which are selected and transformed

Tap here if you want to change the folder you are viewing.

Figure 5.5 Virtual folders (Apple Newton).

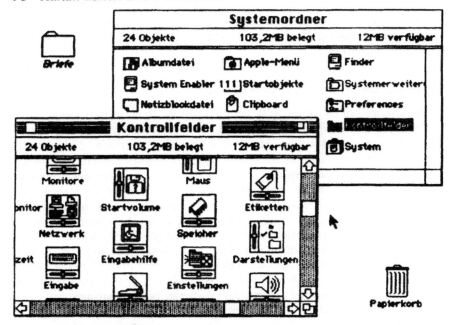

Figure 5.6 Desktop (Apple).

(or perhaps 'distorted') by designers to accommodate new meaning and functions in a different context – that of the system.

One of the most famous examples of this is the famous 'desktop metaphor' used on personal computers which originated at Xerox PARC and Apple (Figure 5.6). It is analogous only in the sense that objects can be spread around it, but in being used as a metaphor you are able to treat it in a similar way.

Another related example in computing is the 'windows' metaphor which seems to give a sometimes restricted view of objects (e.g. files) beyond. Unlike the windows of a building the frame can be resized to give you a better view – and this gives a hint of where metaphors start to break down.

First, the similarity with a familiar concept only extends so far. While the similarities that exist can be used as an aid to understanding sometimes complex situations, the strong imagery can get in the way as soon as the interface element to which the metaphor is being applied stops behaving exactly like the more familiar object. It can then end up slowing down the user's learning curve because objects behave differently than expected and may even have hidden unexpected functions. At this point the level of confusion through the shortcomings of the metaphor can become greater than the gain in understanding through using it, and that's perhaps the time to stop using it.

At the the other end of the scale, staying loyally consistent with a metaphor to maximize familiarity can also reduce the potency of the user experience. Many metaphors involve the idea that a virtual world of some sort is generated on screen as a Graphical User Interface (GUI) and that the user can interact with objects – or rather bundles of functionality – within it. An extreme form may be the implementation of a three-dimensional virtual world within which functions or content can be found by navigating around. While this may seem very impressive, easy and familiar at first, this implementation can ultimately prove impractical both because it soaks up so much processor power and because moving around without short cuts can take much time and effort, actually adding more steps to a function. In this way the issue of the 'appropriateness' of the implementation comes up again – in this case the use of the three-dimensional metaphor may be overkill. A more satisfying experience might be generated by instead implementing the system image with a more basic model, more appropriate interaction with the functions and richer animations (through redistribution of processing power).

Then there is the effect on expert users who may view the familiar and friendly environment which is so helpful to novice users taking their first steps as rather patronizing and limiting. For these reasons the use of metaphors can be a controversial subject between interaction designers and they should only be applied very carefully. (The physical equivalents, product semantics, can share some of the same problems as we will see on p. 96.) Usually a better alternative is simply to find a good and consistent system model ('user experience') which is more true to itself and the remainder of this section looks at some of the major aspects which contribute to this.

Guideline
- *Metaphors should be applied carefully, otherwise a good and consistent mental model can be the alternative.*

Learnability

Our discussion so far in building mental models is aimed at making the interface easily learnable and that also means making sure that the user is aware of what they can potentially do with it and how. While a proportion of 'power users' may work their way through the manual and explore the functionality in depth most users will not do this naturally and so may be unaware that some features may be particulary useful to them. Sometimes, therefore, some extra 'learning support' can be useful but this has to be done in the right way.

While some entertainment products have a 'demo mode' which gives an

Figure 5.7 Elements providing help in the user interface (Apple Macintosh).

overview of the main features, this often doesn't tell the user how to actually use the feature; it is more a piece of salesmanship. Some computer packages such as Microsoft Word offer 'tips' when you start them up and which can be deselected when they become annoying; this hints at the danger in trying too hard to make products learnable. For example, the 'wizards' found within some software packages are optimized for inexperienced users and lead them by the hand in accomplishing their tasks; for them this may be satisfactory because they do, at least, achieve what they want while learning. However, users with more experience (possibly the same user as they become more accomplished) may find a wizard frustrating and patronizing because, compared with a more 'open' interface configuration with key short cuts or pull-down menus, the wizard gives them more limited control and takes longer to do what they want (Figure 5.7).

The trick, then, is to provide an interface which contains all the power and short cuts which experienced users will demand, while offering an extra 'teaching' level on top which helps inexperienced users and which can be removed (either completely or for known tasks only) when help is no longer needed. This is, in fact, how wizards are usually implemented – only inexperienced users need select them.

This illustrates the difference between simple 'ease of learning' and the greater area of 'ease of use' in general. A system that's *easy to learn* (as it has a strong mental model) is not necessarily *easy to use* as it might become limiting, while one that's *easy to use* tends to be *easier to learn* as its model is visible and easily explored. In this way the rest of this section on 'explicit interfaces' will concentrate on ease of use and then hopefully ease of learning by association.

As a final note, once people have invested some time learning how to use a system this influences their concept of 'familiarity' and they will tend to resist changes to it (e.g. in future versions of the product, even if in fact the change would make it easier). This is often seen with changes to the interface of computer applications where suddenly people cannot find functions in their usual locations any more. Another often-used example of that is the QWERTY keyboard, which was designed in the days of mechanical technology. The base of learned experience is strong enough to resist rearrangements of the keys, such as the Dvorak version which would actually be quicker to use after a learning period. For more details refer to Chapter 8 on keyboards.

Affordances, feed-forward and constraints

One of the themes so far has been the importance of the product expressing what its status is and how it can be used, and the use of affordances and feed-forward are key tools here.

The affordances of an object are whatever the user perceives they can do with it. A cylinder which can be easily gripped between the fingers and is attached only by a central axis seems to afford turning – and is probably a knob if it is on a control panel. Similarly a small flat surface that has a crack around its outer edge seems to afford pushing and a user may try pressing it if a button might be an appropriate way to proceed. A large surface with a crack around the edge may be perceived as being a panel which can be opened, and the designer can also hint at the way of opening it through perceived affordances. If the panel is to be opened by pushing and sliding the surface at certain points, then the designer can emphasize the affordance of sliding by putting ridges or textures there which would offer a finger or thumb good grip against slipping on the surface. If, on the other hand, the panel is just to be pushed in at various points to open it, then the pushing affordance can be expressed more than the sliding one by making surface features which appear to make pushing comfortable while not giving much grip to lateral sliding (e.g. small 'dished' areas). These are cases of the perceived affordances matching the designer's intentions for the objects.

The design of objects can mean, though, that the affordances perceived are misleading. The classic example used by Norman (1988) is that suprisingly often door handles which can only be pulled also look as if they can be pushed. It is amazing also how often people react to the design of a handle before they read any message next to it saying 'pull' or 'push'. It is also possible that the affordances intended by the designer may not be perceived at all, as in some of the following examples:

- some furniture has a door which is opened by pressing at a certain place on an apparently featureless panel. This is also sometimes true of the battery compartments on portable electronic products;
- some refrigerators cleverly conceal the place where the door can be

Figure 5.8 Inserting a floppy disk.

gripped to pull it open, and the user only knows which side to pull by looking for the well-concealed hinges;

- some electronic products have a knob which can also be pressed (e.g. to confirm a selection made with the knob) and if the end of the knob is rounded the affordance of pressing may not be perceived by users;
- a product tested recently by one of the authors had several large buttons which turned out to be just sections of panel – nonetheless the affordance of pressing was there and so the mistake was made.

In this way the design of parts of a product should signal how they are intended to be used in a way which is perceived with the right affordances. This is also usually possible while retaining the aesthetic qualities sought by the designer; the hinting need only be subtle as described earlier.

At the same time users can be discouraged or constrained from perhaps negative or otherwise unintended actions by hiding perceived affordances that might lead to these being done. For example, a floppy disk for a computer has the perceived affordance of being able to go into the disk-drive slot, and this is coupled with a constraint – the disk and drive are designed so that the disk can only go in one way, the right way (Figure 5.8).

Note though that on this computer the constraint was not complete – the floppy disk could be put into the CD drive by mistake because the CD door was just a large hinged flap. Newer CD drives are designed in a way where this cannot be done by accident. In this way constraints are one factor which try to prevent errors, as we will see on p. 106.

Constraints also provide a potential solution to the problem of providing childproofing without putting older or weaker people at a disadvantage with solutions relying on physical strength or dexterity.

So we have seen that the hinting of affordances to the user is a rather subtle language through which the product (and designer) expresses what the user can do. A much more direct and explicit kind of expression is the use of 'feed-forward' to show the user what they can do, which functions are active or available and their purpose. This can take many forms; for example:

- a label like 'press here' on a battery compartment cover;
- the physical form, illumination, captions and symbols of dedicated buttons ('hard keys');
- the changing labels and graphics which appear in or close to recon-figurable buttons ('soft keys') as on cash machines (ATMs [Automatic Teller Machines]);
- the changing graphics, messages, prompts and animations on screen-based interfaces. On a screen-based product it is quite comon for highlighting to be used to show the currently selected item(s) or function(s), and at the other extreme for items and functions that aren't available to be faded out or have graphics hidden. The cursor can also be changed to reflect something about the screen elements over which it is passing (e.g. a cursor might normally be an arrow, changing to a finger when a screen element can be clicked on). Feed-forward can also be expressed by making things that can be clicked on change their appearance as they are moved over ('roll-overs').

Within all of these areas attention getters (like flashing lights) may be used to attract the user towards a problem or a suggested action.

Of course virtual controls also have perceived affordances (e.g. something may look as if it can be clicked on or dragged with the cursor). In this way the words which are underlined and differently coloured on web pages are usually hyperlinks.

Such virtual objects sometimes use a real-world object as a metaphor to express how they may be used, both in the way they appear and in the way they can be interacted with, sometimes using sound, movement or animation to support this (e.g. a virtual 'screen' button may look like a hardware one, complete with three-dimensional 'look and feel', and even the 'click' sound that the hardware would make as feedback sound when clicked on ['pressed']). As was said on p. 91, though, care needs to be taken in this area because the metaphor should only be adopted to the extent that it is true and honest for the virtual object too. As an example, rotary controls implemented on screens do not usually work very well as it is difficult to see how operation of them relates to use of the cursor, a very unreal situation compared with the way in which a real knob is used (Figure 5.9).

A final kind of feed-forward (or arguably feedback) which can be provided by more complex products is guidance and support in helping the user understand the system; obviously this is especially useful for novice users and there should be a way of activating or deactivating it,

Figure 5.9 Hardware and software buttons.

perhaps via a help menu. Any information provided should be informative, easy to read and relevant.

Semantic language

In the last section we said that the affordances of an object perceived by the user give them clues as to what they *can* do with something – this can be moved sideways, that can be pressed, or pushed – and from these the user can work out how it is intended to be used. A stronger instant impression of an object's *purpose*, the underlying significance or meaning behind it and how it should be used to achieve it, can be emphasized through the use of semantics in its design language.

Semantics can certainly be applied at the fairly low level we were tending to use with affordances – the level of individual controls or objects, which you manipulate. More than this, though, they can be applied at the higher level of visualizing the purpose of a whole product (e.g. making a remote control look like a remote control and not a calculator). In doing this the semantic language can make instinctive references – perhaps exaggerating key functional aspects such as the remote control's pointability and the transmission of the infra-red beam. The semantics can also tap into historical references to what similar products looked like earlier in their evolution (e.g. the basic configuration of a telephone hasn't changed that much in all the time they've been around and they are still recognizable because of some 'telephone-ness' quality). Here, of course, we have to make sure that such references are familiar to the target user group, a part of their 'assumed knowledge'.

Some aspects of semantics are also commonly applied to corporate identities and the brand language of products. The story goes that the glass bottles used for packaging Coca-Cola were designed to look like a cacao tree seedpod (Figure 5.10) so that they would be more distinctive to the touch. Even if this subtlety is lost on most people (who do not know what a cacao tree seedpod looks like) the new plastic bottles at least have a similar shape to the older glass bottle and this in itself is a semantic reference.

Let's illustrate semantic language at the product level by comparing three electronic devices – remote controls, mobile phones and calculators – which

Figure 5.10 Aspects of semantics: (left) cacao tree seedpod; (right) Coca-Cola bottle designed to look like a cacao tree seedpod.

in some ways have similar affordances. Each has approximately the same size, can usually fit in the hand, may have numeric buttons and may have a display. Despite these similarities they are used in quite different ways and we generally manage not to confuse them at a glance, partly because of the semantics that have been applied which perhaps exaggerate key functional aspects.

For calculators the key aspects are that:

- the usual location is sitting on a desk, so they are generally flat;
- the main use is the input and output of numbers, which accounts for the numeric keys and display being most prominent.

For remote controls the key aspects are that:

- they are usually held in the hand, so they are made a shape that's easy to grip;
- they are pointed at the product being controlled, so they have a clearly defined dark translucent area on one end from which the control beam transmits;

Figure 5.11 Remote control (Bang and Olufsen.

- the most frequent uses are for controlling volume and quickly selecting between channels (e.g. TV), tracks (e.g. CD) or menu settings (e.g. DVD, TV settings), and so up/down buttons for volume and a means of quick selection are most prominent (Figure 5.11).

For mobile phones the key aspects are that:

- they are usually held in the hand, so like remote controls should be easy to grip;
- ... but are also kept in pockets, so they are as small as possible;
- ... and are used for talking and listening, so an earpiece and microphone are prominent;
- the most frequently used functions are calling and hanging up, selection from a phone-book list or dialling through numeric keys, so these are prominent.

Sometimes we may be unaware of how much we have got used to the differences. Figure 5.12 shows a calculator on the left and a phone on the right. But what is wrong with the picture?

The layout of numeric buttons on the calculator is not as it should be – the '1' should be at the bottom left, not the top left where it is for a phone or remote control. This difference is not normally obvious to us because we are used to dealing with them in different ways. As long as the semantic language is clear and as long as someone does not try combining a phone with a calculator then this should not be a problem.

This brings us to the downside of applying product semantics to convey what a product is for. Semantics are, in some ways, like physical metaphors and the same problems apply – there is a conflict when the semantics applied to the product or the controls do not quite match up with its real role or purpose. This has become a real issue when, for example, we get

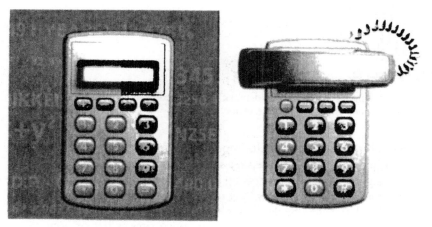

Figure 5.12 Calculator and phone.

convergence of products. If we did ever combine a calculator *and* a phone what should it look like? More than that, how should the keyboard be arranged?

For this reason the application of semantics is less 'trendy' than it has been, and the current emphasis is on the design of products and the aspects within them as a clear, coherent and consistent user experience within itself.

Feedback

The basic principle of feedback is that the user should always be kept informed about the state of the system, where they are within any navigational structure such as a menu system, what actions they have taken and whether these actions have been successful. Feedback information given should be kept constantly and automatically up to date.

The system status can be conveyed to the user using indicator lights, changing screen graphics or different cursors, depending on the product type.

The feedback information should be shown where it would be expected to be seen, and certainly within the field of view of anything else which may occupy the user's attention at the time (e.g. the controls being used, to which the feedback may relate). It can be done in quite a subtle way and should generally not distract users from what they are doing unless a change in their behaviour is appropriate.

When the system moves into a state which has consequences for the user's behaviour – particularly in forcing them to wait (e.g. the system is busy) then clear and immediate feedback which gets the user's attention is important. Animation is one good way of doing this as people are naturally attracted to

Figure 5.13 Apple's trashcan (trademark) gives feedback by changing its shape when there are objects inside.

movement. If the state may last for more than a few seconds there should be an animation, icon, image or message shown explaining the basic reason for the delay (e.g. the system is processing information or connecting to another device) and an indication of how long users will have to wait before they can continue. The latter may be done graphically (usually with a progress bar of some sort) or textually (by indicating how much time is left or how much still has to be done [e.g. when copying files]), or in both ways. If the period of time is so long that the users may turn their attention to something else then it is beneficial if some extra audible indication (a brief sound) can be given when the waiting time is over.

Obvious and immediate feedback should also be given for every user action so that the users know that it has been registered by the system. This should be shown in a way which is coherent with the rest of the interface. If possible, even more than one of the following senses should be used for simultaneous feedback:

- visual: a changing appearance, animation, highlight or other nearby graphic;
- audible: a sound being made;
- tactile: a feeling of positive movement in a hardware control, for example a 'click' feeling when a button is pressed or when a knob with discrete positions is turned.

Feedback in response to an external input device (such as pressing a button on a remote control) should also express something more about the controlling device used (e.g. it is quite common to have a signal with a 'radiating' graphic indicating that a control signal from a remote control is received).

The following recommendations can be made about the feedback time (Woodson 1987; Gilmore 1989):

- for a key-press, the recommended feedback time is less than 0.1 seconds;
- for the results of an action (e.g. one that involves lengthy processing or data transmission) feedback should be given within 3 seconds;

- for starting or rebooting a computer system, a delay of up to 1 minute is acceptable;
- it should always stay the same for a specific task. A constant time between every action and the corresponding feedback is even more important than how long this time actually is (Spinas 1983);
- there is also a minimum recommended speed for scrolling. It must be possible to perform scrolling slowly enough to keep control over it and stop it at any given point;
- if a delay is longer than the recommended time, the system must show that it is still working by a constantly updated feedback message. Preferably the remaining waiting time is displayed.

Guideline
- *For every single operation the appliance should give obvious and immediate feedback to the user.*

Prioritization

The device being designed will probably have a number of functions, but some of these will be more important than others. Of these the more important or frequently done functions should be made easier to do, a principle also known as Fitts' law. This can be implemented in many ways with important functions:

- being accessed by one-touch operations (e.g. a dedicated button) while less important ones might be hidden away in lower levels of a menu structure where several button pushes are needed to reach them. Medium-level functions may be given 'soft-keys' which are hard buttons that are reconfigurable according to context;
- being in prime locations (e.g. the front of a control panel, the top of a TV or the centre of a remote control where they fall under the user's fingers) while less used ones may be further away or even under a flap or cover (Figures 5.14, 5.15);
- being bigger and easier to press or grip, while less used ones may be smaller and have less of a profile above the surrounding surface.

These priority principles also apply to how easily it is to pick out and read information.

Understandability

Information conveyed to the users should be understandable to them.

Figure 5.14 Remote control (Bang and Olufsen). The keyboard flap helps to reduce the complexity when closed.

The idea of understandability is that the meaning of all menu terms, symbols and icons should be *obvious* and unambiguous to the users – yet another example of having to accommodate the level of knowledge or familiarity of target users. The general rule is that the design should try to stick to terms, symbols and icons that are known to be familiar and meaningful to target users (perhaps through experience with other products of the same sort). It is useful to maintain an up-to-date glossary of these which can be called on when necessary. If something novel (e.g. new symbols or terms) are to be used then their understandability should be checked in the right context with some target users.

Terminology used should not contain unfamiliar jargon, and dialogues, explanation and labelling should be concise, brief and constructive in tone without being too patronizing. A final note is that in formulating labels it helps if the keywords (usually the noun) is put first.

Grouping

Grouping is a way of implying relationships between controls, displays or visual information. One common way of achieving it is by giving related elements one or more visual attributes that they consistently share – typically colour, shape or proximity (e.g. screen buttons with the same function should appear in the same part of the screen, and have the same general appearance). Another example might be that the relationship

Figure 5.15 Fax machine (Philips). The complexity of the operation panel is reduced by hiding the alphabetic keyboard under a flap.

between an external input device and elements controlled in an on-screen interface might be accentuated by giving them similar colouring and shapes. For this to work it means that things with a different function should not share the attribute.

Another way to achieve grouping is to use a principle identified by a group of German psychologists working on an area of perceptual psychology called gestalt. The principle is that the mind tries to explain what it sees as the simplest combination of objects – something we saw earlier when we were talking about mental models. In practice the grouping effect can be achieved by having related elements close together with a line drawn around them or by making them all contribute to a single simple shape. The remote controls shown in Figure 5.16 are typical in using oval shapes to combine separate up and down buttons (left and right) and round shapes to combine four direction buttons (centre and right). One effect of this kind of grouping

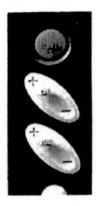

Figure 5.16 Remote-control button shapes (all Philips).

is to make the panel look much simpler than otherwise as there are less shapes to be perceived. See Chapters 7 and 12 for more details.

Consistency

Earlier we were talking about how the user builds up a mental model of a system when learning about it, and this is most easily done if the system is consistent. There are two main types of consistency that we will describe here.

The first kind of consistency concerns stereotypes and other conceptual ideas already familiar to the (target) user. This should mean that there is the minumum to learn, and this is related to assumed knowledge.

Some stereotypes are symbolic. You would expect blue to mean cold where red means hot, that moving a slider to the right increases it and possibly that moving a slider up increases it as well (though this stereotype is not as strong as the horizontal one). Dials that move to the right or up are also showing increasing values. Just as the hands of a clock move clockwise as time passes, dials should also show an increasing value clockwise. The graphics can also be consistent with symbols which are widely known.

There should also be consistency with familiar models of use. For example, if you buy a new calculator it should be easy to use the main functions without instructions because the experience of using it is the same as other calculators you may have used before, at least for major functions.

Interface elements such as controls and displays should also relate together in a way consistent with the simplest model based on the afford-ances implied. If a list is shown with one item highlighted, then in the simplest model the 'subject' of the operation is the highlight and this is perceived to have the affordance of moving vertically whereas the list is perceived as fixed. (This is the most common situation, though these affordances do, of course, depend on the way the list and highlight are represented.) The action of any control buttons should be consistent with this perceived affordance and so pressing the 'up' button would be expected to move the highlight to the item above. If the list moved up and the highlight stayed where it is then this would be inconsistent with the model. Suprisingly often it happens that products are inconsistent with the perceived affordances while achieving a consistent result (e.g. when the highlight needs ultimately to stay in the same relative position). In such a situation pressing 'up' will make the highlight stay where it is while the list moves down and the result is that, indeed, the item above is now selected. The user can overcome any initial surprise and learn that this is the way it works, but the model is more strongly reinforced if the highlight first moves up to the item above, and then appears to pull it, with the list attached, down to the highlight's original position. This is more consistent with the expected simple model.

The second kind of consistency is that *within* the system model and its representation so that what the user does have to learn is minimized. There should just be a few rules and concepts rather than many dispersed or quirky ideas and exceptions. A particular situation or task should always be perceived to have the same 'user experience', no matter what the context. At a low level this means that there should be consistency in the representation of similar or related elements (e.g. in structure or behaviour) and this makes consistency closely related to the aspect of grouping. Things that look the same should behave the same, and conversely things which do not behave the same should not look the same. This also means that physical controls should be consistent with any representation of them or effect they have on the display.

At a higher level this consistency should be applied to the experience of tasks (e.g. tasks like the selection of an item or items, or navigation in a menu structure, should always seem the same no matter what the context).

In the end, though, consistency should not be taken to a point where the interface looks dull and overly uniform as this compromises the visual pleasure aspect which we will talk about shortly.

Transparency

If a control is transparent this means that the effects of actions are predictable. Before the users press a button, for example, they should have a fairly clear idea of what will happen, and of course this is supported by a consistent operation.

At a higher level the whole UI can be seen as transparent when the structure and the representation of the controls, tasks and information within it seem so natural to the users that they almost do not notice the interface itself. If it is not getting in the way of what they want to do the users can devote their attention to thinking about their goals rather than how to achieve them. Making a transparent UI means that users do not have to search for the means to do what they want – ideally everything they need should be right there, clearly visible when they need it. This is closely related to 'appropriateness' discussed earlier and the whole issue of explicitness because it should seem to take no time to learn how to use it.

Error tolerance

Two types of errors can be differentiated:

1. Things which are legitimate from the user's point of view but which the system cannot cope with; and
2. Times when the user genuinely makes a mistake (although in some ways the mistaken action was valid).

The former of these tend to happen when the user does something that the design does not allow for and this says more about the limitations of the system design than the errors of the users – in such a case the concept of it being an 'error' becomes rather artificial. The trick for the system and its designers is to anticipate what people will try to do, to try to hint at affordances which lead people to do things that will work and to constrain people from doing things which will not. This is better than thinking in terms of stopping people from making errors as such, which can lead to an interface being more restrictive than it needs to be. The other principles of explicitness such as consistency and transparency are important here. Genuine user mistakes tend to happen if the users:

1. Change their mind about something they have done; or
2. Have to specify input according to rigidly defined rules (and can therefore make a mistake).

The solution for the first – when people change their mind – can only be implemented in some kinds of product, particularly the computer-based ones. For these the system may be able to remember what the user did and provide some kind of 'undo' function.

For the second the preferred behaviour of the system it to try to pre-empt a mistake and prevent it before it actually happens ('prevention is better than cure'). In some types of product and interface the provision of input by a user is separate to the processing of it (when errors might arise) and it may be possible for users to check their input and correct potential error situations before they happen. In such a situation the system might constantly monitor input and give discrete feedback of when it is not sure that users meant to specify what they did. A perfect example of this is later versions of Microsoft Word, in which text will be underlined in green if there may be a grammatical error or red if there is a spelling mistake, and this happens before a spell checker is activated.

If the error or mistake does really happen and the system cannot automatically fix it itself, then any prompt that it gives the user should explain what has happened and preferably let them know what they should do to remedy the situation. It is important that this be done in a way that is constructive rather than accusing, easy to understand (relative to their assumed knowledge) and not patronizing. Errors shouldn't be used as a way to teach a user who is trying – and failing – to perform a task. This can be very unsatisfying.

The means of recovering from the 'error' situation should be easy, preferably in a way which impacts least on the user's task (e.g. a 'cancel' or 'undo' button) or, if the worst comes to the worst, a reset button (perhaps a physical button in screen-based interfaces).

5.4 Satisfaction

The aspects that we have discussed so far have been mostly related to the relatively objective and scientific aspect of how efficiently the system communicates itself and can be used. There is, however, a much more subjective side to the user experience which is related to satisfaction:

- how interesting or intellectually stimulating it is (e.g. we find some applications with a narrative stimulating, that provide an intellectual challenge at certain steps);
- whether or not it is pleasing to the senses (sight, sound and touch).

At its best, satisfaction can have the 'wow!' effect which will get it invaluable word-of-mouth promotion, positive reviews in the Press and gives it the kind of reaction that makes it an impulsive purchase. It can also provide extra motivation in use which should make it easier to learn.

So what is it that actually contributes to satisfaction? In many ways it is in the details. Some aspects are relatively cheap (e.g. rich graphics, animations and sounds) while others can be quite complex and expensive (e.g. certain materials, production methods and underlying technologies).

Therefore one of the factors to be determined at the beginning of design is to what extent the 'wow' effect is to be included in the design based on the anticipated market position. The next thing to consider is how this can be achieved (e.g. with animation, fancy graphics, pleasing textures or elegant hardware). The final consideration is to dedicate the extra time and expense to the project that this level of quality will involve.

For more information about pleasure with products refer to Chapter 17.

6 Design of on-screen user interfaces

Irene Mavrommati

This chapter gives guidelines on what to consider when designing an on-screen interface. It is split into two sections: (A) visual design guidelines and (B) interaction guidelines.

One way in which you can test the on-screen interface you are designing is by imagining some common situations and then working through what you, as the user, would be expected to do and what it would do. As part of this make sure you bear in mind where it might be used, in which context and by whom. Also try to view it from the distance you expect people to be normally looking at it, and be rather critical. Doing this will help you get a perspective on the size of letters, use of colour, contrast, graphics shapes and so on, which would be appropriate for making it clear and readable.

There are two basic categories of electronic devices, each with different interface variations.

The first category is of electronic devices that are self-contained in that they do not rely on other devices; they may or may not have a screen or display (e.g. mobile phones, personal digital assistants [PDAs], clock radios, digital cameras and so on). The interface issues such as consistency and feedback, as mentioned in the previous chapter, are relatively straightforward because only one single appliance is involved.

The second category is of systems that are made up of more than one single device (e.g. have a separate screen or remote control). Examples here might be a car navigation or entertainment system with controls on the steering wheel, or an electronic programme guide, videoplayer or TV which all use remote controls. For such devices special additional attention has to be given to making sure that there is consistency in appearance and interaction between the devices and that they complement each other. If, for example, the system consists of a set-top box with embedded applications or services and a remote control then the interface designer will have to consider how to make control across all the devices and services of the system consistent.

A   VISUAL DESIGN GUIDELINES

6.1   Designing on-screen graphics

The interface you will be designing may use a number of screens according to the application. Make sure you know absolutely all of the constraints and problems of the screen you are designing for. Additionally try to arrange tests of the actual system and sample screens before you take a certain design direction. Start with screen tests even from the first draft screen designs.

Keep in mind that screens may vary in:

- size;
- resolution;
- safety area (the gap around the edge within which graphics cannot be shown);
- colours available, ranging from monochrome screens to coloured screens (e.g. liquid-crystal display (LCD) matrix screens). You will have to determine which colours are shown optimally;
- contrast, brightness, luminosity;
- readability;
- the way light is reflected;
- 'colour life' (the length of time for which colours are well represented by the screen). This can vary from colour to colour according to the screens (e.g. in some electro-luminescent displays blue has a shorter than average colour life as it tends to fade before the other colours);
- underlying technology (e.g. LCD screens, electro-luminescent displays, cathode ray tube (CRT) for TV and computer monitors);
- cost;
- the platform being used (e.g. PAL or NTSC for television) which has an effect on resolution and colour (PAL and NTSC are different television standards in North America and Europe);
- the screen ratio (e.g. 4:3 or 16:9 for television);
- cultural differences (see Chapter 18).

The last two affect the decision of which screen to use, while the others affect how the screen design will be done. Chapter 12 discusses some more specific aspects of screens and visual displays.

Screens may also differ in the way they are used, in terms of

- viewing or reading distance (e.g. a projection is seen from much further away than a monitor);
- environment lighting (are there reflections off the screen which make it more difficult to read? How does the brightness of the screen compare

with that of the environment? A bright screen seen at night, for example, is uncomfortable to read because of the glare).

The viewing distance can influence the size selected for the text font.

Guideline
- *The recommended font size h for a given reading distance d is:*
  $$h = 0.0052d$$

Reflections off the screen can reduce the contrast between the text and the graphics making it more difficult to read and this can also happen if the graphics are 'too dark' for the environment. If, however, the screen and graphics are too bright for the environment then this 'glare' can make the screen uncomfortable to look at and will affect the ability of the viewer's eyes to adjust to lower ambient-light level. This may be a problem in some situations (e.g. when using a car-navigation system at night where the driver's ability to see outside clearly is vital).

The immediacy of the information presented and the time taken to understand it is also very important, especially in contexts where the user's attention is shared between it and other tasks such as in our car-navigation system example. One key factor in this is the positioning of the display which, in a car for example, may normally be in the driver's peripheral vision; the driver may need to move his eyes or even his head to read it properly. Another factor is the amount of attention the user can give to the display (and indeed the system) at any one time (e.g. a car display should distract the user's attention away from the road for an absolute maximum of 2 seconds). These factors have consequences for the readability, colour, luminance and contrast of the screen and information and graphics shown on it.

Guideline
- *Plan and conduct regular screen tests*

There is a huge diversity of screen or display types and ways of showing information on them. Car stereos, in-car navigation systems, mobile phones, digital cameras and hi-fis, for example, all have displays with on-screen interfaces that are very different to each other and very specific to the function of that product (Figure 6.1 and 6.2).

It is particularly important to remember the constraints of the target screen or display and its viewing conditions when the interface is being

Figure 6.1 Dashboard of a car equipped with a car-navigation system (Philips Carin now available under the VDO Dayton brand name).

Figure 6.2 The author in a car equipped with a navigation system prototype.

designed on a computer. The computer monitor is viewed close-up, has good colour quality and high resolution while a mobile phone display, for example, may be a low-resolution 'black and white' LCD. This also applies when you review the designs on printed paper – bear in mind that some interfaces which appear dull and rough on paper still look interesting on screen. Conversely some interfaces that look good on computer or paper look overdone or dull when tested on the actual screen. Do not make a final decision until you see the interface on the actual screen; the design may look more three dimensional than you expect, warmer in terms of colour or generally different.

The refresh rate of the target display may also be different to that of the monitor – on many devices the screen is updated much more slowly and this may restrict the use of animation. In addition, the amount of flicker influences legibility and this becomes significant for displays such as TVs.

Also think about the conditions of use: the high contrast levels within the screen graphics which are suitable for bright-light viewing may be uncomfortable in darker environments where low contrast levels are adequate and more appropriate.

So, the underlying message is during the design process you should keep in mind the screen you are designing for and its environment. Have frequent screen tests to help you build up an understanding about which fonts, colour palettes and graphics work best in the target situation.

6.2   Visual perception and visual attention points

There are several tricks of visual perception that can be used to aid reading and to help make certain screen elements stand out. First, our eyes tend to get attracted towards points of attention or contrast on the screen; for example:

- corners of the screen;
- angles in lines – either actual ones or ones perceived from extensions of lines – (e.g. where axes meet in a T shape or in an arrow);
- tilted lines tend to create more stress and contrast and attract attention while the horizontal lines have a more calming and harmonious effect;
- objects with complex shapes;
- moving objects;
- brighter or contrasting colours;
- size (big objects rather than small ones).

Our eyes tend to rest in:

- points of harmony;
- horizontal lines;
- smooth colours;
- white spaces and margins.

For example, large brightly coloured screen graphics will attract more attention than small, dimly coloured ones.

The visually attractive areas and aspects should contain information or navigation elements such as buttons of higher priority. One great example of this is the location of the 'start' button of Microsoft Windows at the bottom left corner of the screen. Smaller, more static, calmer representations or buttons should be used for details which are intended to be read more slowly and carefully.

Guideline
- *Prioritize buttons in an on-screen interface using screen-design elements which attract the eye (e.g. colour, size or animation).*

Design with points of contrast and harmony could help readability as well as expression.

6.3 Reading the screen

A significant factor affecting how a two-dimensional screen of information is scanned for highlights is cultural differences in reading which are fixed from an early age. In Western cultures (e.g. Europe and the USA) reading is done from left to right and top to bottom, while others (e.g. Arabic or Asian cultures) may read from right to left, or even from bottom to top.

One common aspect in all cultures, at least, is that the screen corners and sides attract the eye and these locations are therefore good for generic navigation buttons. The eye is also attracted to screen features according to their colour contrast, shape, size, location on the screen, animation or movement. Actual reading of the screen tends to start from elements which are 'attractive' in this way though the flow of reading and the priority of what is 'read' depends on the total screen design.

There are various other things that tend to attract our attention:

- complex shapes (e.g. a hand) rather than simpler geometric ones such as a circle or square;
- screen elements which, if they were made into a three-dimensional object, would be physically unstable or unbalanced. This certainly applies to diagonal lines, for example;
- conflict points – where lines meet at a corner, or extensions of lines of shapes.

Guidelines
- *Use locations close to edges of the usable screen for standard navigation buttons throughout the application.*
- *Provide enough empty space in the screen-design.*

It would be a mistake to design screens which are too compact with too much information squeezed into them. Empty space provides a resting point for the eye and helps us to separate different groups of information so that a structure can be more readily seen and the text within it read faster.

Example   For TV screens there are several television-broadcasting conventions. The main two are NTSC and PAL. NTSC (used in North America and Japan) uses 525 lines. When designing for NTSC, measuring in pixels, the screen size is 360×240, and the safety area within is 320×210.

6.4   Use of a grid

One way of optimizing the readability of a screen is to base the layout on a geometric grid. This helps both in reducing the number of corners and conflict points in the layout and, by using the grid as a common starting point, facilitating consistency in screens (Figure 6.3). Text, the main buttons, standard elements and white spaces should all be aligned with the grid. When creating the grid it is crucial to take into account the safety area around the display within which nothing can be shown except the background. Any text or buttons that extend into this area may be cropped. The size of the safety area depends upon the type of screen used.

Guidelines
- *Start your design by using a grid.*
- *Provide safety areas and empty spaces.*

6.5   Use of colour

Colour can be used to effortlessly express mood and emotion and the way of doing this has its roots in language, symbolism, tradition and superstition.

At a basic level colourful things are more compelling and eye-catching than neutral things, not just on the screen but also in nature. When colour is applied strategically it can distinguish important from unimportant things, signify tasks, and show structure through colour grouping, colour selection and so on.

There's also a link between age and the hues of colour found most attractive. Babies look around for brighter colours with a preference for yellows and reds, hues of a relatively long wavelength. With maturity most people develop a greater liking for blues and greens, hues of shorter wavelength.

Colour and perception

The way a colour is perceived is affected by its hue (lightness or darkness) and saturation (how vivid the colour is).

The basic rule is that if you want a screen element to be noticed it should be strongly coloured; the more saturated it is the more it will stand out, with yellow seeming to be the most immediate colour. However, this guideline considers screen elements in isolation, so when set against a background they will stand out more the more contrast they make against the background. Contrast between dark and light colours or between weak and strong ones is collectively known as luminance and this has a greater effect than contrast that is due just to a change in hue (chrominance). Chrominance contrast is also dangerous as it often depends on the viewer not being colour-blind.

More than this lighter colours or long-wavelength colours, like reds and yellows, are naturally better for foreground or highlighted features as they appear closer to the viewer. Similarly dark colours or shorter-wavelength colours, like blues, are better for backgrounds as they seem further away. The reason these hues seem to stand out or sink into the screen is because these colours are perceived at different depths within the retina.

At the same time, care has to be taken with reds and yellows because they sometimes do not look good on screens as a result of a kind of 'bleeding' effect into adjacent colours. It looks almost as if they are somehow detached from the rest of the screen design.

So the selection of colours for the screen elements and background depends on many factors. In the end the only way to be sure that they work is to test them by showing them on the 'target' screen.

Cultural symbolism of colour

Several conventions for the way colour is used in nature and everyday life (e.g. for signal colours in traffic signs, road signs, etc.) can also be applied on the screen where there is a conceptual connection. For example, people recognize from traffic lights that a green colour may mean 'allowed' or 'pass' while red may be used for 'forbidden' or 'do not pass'. Similarly a flashing colour may mean 'pay attention' or 'press here now'. Yellow or red backgrounds for physical signs may be used for warnings, and in the same way a yellow background may be used for error screens. (Error screens may also contain more angular shapes than other screens because angles give a kind of visual tension.)

At the same time, care should be taken because culture plays a large part in the symbolic association of colour (e.g. green must not be used for Muslim cultures, since it has religious associations, pink may be associated with erotic subjects in Europe, yellow has a similar association in China).

Guidelines
- *Take the symbolism of different colours into account.*
- *Do not leave it till the last minute to test the colour design on the actual screen.*

In Asia more complex aesthetics and brighter colours are more acceptable than in Europe and indeed visual complexity is a part of everyday life in cities like Hong Kong, as well as of Asian art. In Europe the same aesthetics may be seen as busy or loud and calmer designs are preferred.

6.6   Visibility

Colour-blindness

Colour-blindness among men has a frequency of about 8 per cent (predominately red/green colour-blindness, Birren 1978), while among women it is less than 0.5 per cent. Colour-blind people tend to describe red, dark yellow or green as brown and therefore highlighting of important screen elements in these colours or tints may go unnoticed. For this reason there should also be adequate luminance contrast between screen elements as described earlier. One good trick for ensuring this is to make sure that your screen design is still readable enough when shown as greyscale, adapting the contrast or brightness of colours as necessary.

Guidelines
- *Design for all viewers, including the mature and the colour-blind.*
- *Test and fine-tune contrast by converting your design into greyscale.*
- *Make sure you know the range of light intensity and design accordingly (day screens, night screens, backlight, etc.)*

Older age

With age the eye lenses harden and lose their flexibility and clarity, letting less light into the eye and most people over the age of 45 to 50 are affected in some way by age-related eye problems like presbyopia. Much older people need increased brightness and contrast within the screen in order to compensate for these problems. Designing on-screen interfaces for older people also implies designing clearer shapes and may also mean using a larger size of text.

Consequently if mature people are expected to be using the product you are designing, try to have a small test with some people representing your

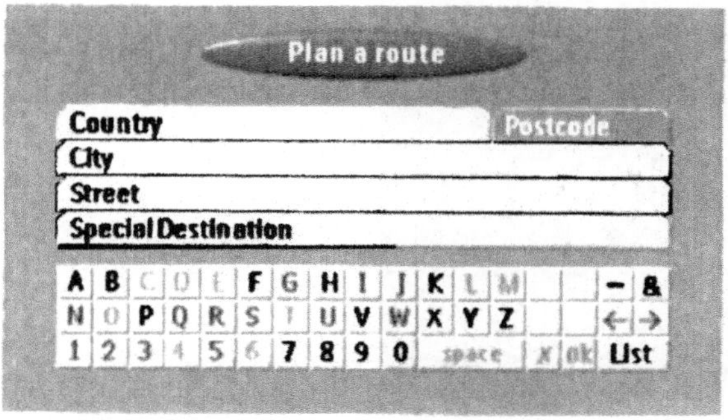

Figure 6.3 Screen remote control for an electronic programme guide (Philips).

target age group to verify the screen colours and the overall contrast and readability.

Lighting conditions

Visibility may differ according to ambient light levels. In strong lighting conditions dark text on a light-coloured background with high contrast levels may read best, while in lower light levels the reverse (dark background and light coloured text with less contrast) may be better. One of the reasons for this is that a screen tends to be most comfortable to read when its overall luminance level does not exceed that of the surroundings (i.e. avoiding glare), while a dark screen in a light environment may show reflections too easily.

In most cases you will be designing using a luminous monitor seen in normal room lighting conditions and so the lighting conditions of the 'target' display in its environment may be overlooked (Figure 6.4). For this reason make sure you are able to regularly test (or even design if possible) the screen graphics using the 'target' screen in a range of lighting conditions representative of those where the product will be used.

Keep in mind that for example a night screen for a car navigation system seen in a dark environment will look different if it is designed while there is plenty of white space around it (the white border or background of a monitor in ample diffused lighting). Therefore when you try to design night screens in day conditions the contrast and colour saturation may appear differently because your eyes are not adjusted to the lower light levels. This would result in difference in readability with eyestrain being caused in a night situation (Figure 6.5).

Figure 6.4 Screen design for night conditions of a car navigation system (Philips/ VDO).

6.7   Screen elements

Organization

On-screen organization, grouping and consistent style help people understand and use the on-screen application faster.

Be explicit about the way you define:

- status;
- menus;
- buttons;
- highlights/lowlights/active and inactive fields.

Define how each element of the interface works and what it signifies so that inconsistencies are avoided. Try to write it down on paper as a simple description, so that you or others can refer to it at later stages and avoid inconsistencies.

Keep a glossary of terminology; this will help to avoid difference of terminology used to describe the same thing in different parts of the application.

Make sure that things that act the same are visually consistent in the design. But allow for extra differentiation within the style to increase the ability of your users to recognize the things that act differently.

Fix at certain locations the screen elements that will enable people to navigate, and give them the same look; they should be easy to find as they help people maintain a sense of orientation.

Group the graphic elements of the screen (by using colour, shape and proximity), according to their similar functionality. It helps quick recognition of the screen elements and makes it easier to read, orientate and recognize the element's functions.

Last but not least, pay attention to detail; take time to refine the design and tweak it as much as you need to achieve the best balance of screen elements. Keep doing this until you feel there is an optimal balance accommodating user needs, aesthetics and the requirements of the application. You can be helped in this by regular screen tests and by people giving their own evaluations.

Guideline
- *Plan and take time for iteration on the design in order to achieve an optimal balance, accommodating user needs, aesthetic concerns and the application requirements.*

Prioritize the screen information and controls so as to make the most important or frequently accessed functions easiest to find (Fitts' law).

The most important information can be made to stand out by using a different colour code to the rest of the interface. In multimedia graphic interfaces, differentiation can also be achieved using different shapes, movements or behaviour for elements.

You can organize information by making the more important content stand out using colour (both using the strength of the colour and contrast), larger size, underlining or framing, and by using different shapes to the rest of the interface. You can also organize the information using its position on the screen. Remember that the time taken to see something is in direct proportion to its size and viewing distance and position the highest priority functions according to this, using bigger buttons and arranging them closer to the corners or edges of the display.

Buttons and their shape: Balance between hardware and software design parts

Buttons and controls should have consistency throughout the application, both in terms of looks as well as behaviour. Their shape may be designed according to their functional meaning. Groups of buttons should have distinctive size and shape to differentiate them from other groups. Buttons may come in different flavours. Some may appear a bit dimmed while others may have small features on them to promote grip. At the same time, though, it is important not to overdo the decorative effect of buttons. As was said earlier the more important or frequently used buttons should be larger.

For applications that use a separate input device (e.g. a remote control or a knob) the device and the screen interface should complement each other so that the relationship between the two is clear. If possible the layout and appearance of the physical buttons and on-screen controls should be com-

Figure 6.5 Screen remote control (Philips BATE).

plementary and share a common look and feel, including the way that motion of physical controls is represented on-screen (Figure 6.6).

Highlights and lowlights, active and inactive fields

The buttons or controls on the screen can have different graphic treatments to indicate their state. Common states and their representations are:

- *active:* the 'normal' graphic appearance of the button indicating that it can be selected and that something would then happen;
- *inactive:* a dimmed image or text indicating that at that moment it is not possible to interact with this element and that nothing will happen if it is selected. Such images have also been referred to earlier as 'lowlights';
- *disappearing:* this is a more extreme form of inactivity. When this button or control is not relevant to the application context and it is expected to stay inactive for a long period it is simply not shown;
- *highlighted:* the button, control or field changes colour or shape to indicate that the element is readily available for interaction by a single action. When it is selected, there will be an obvious resulting action;
- *activated:* in some cases it needs to be obvious which field or button is activated. For this reason visual feedback is given after the graphic element is selected, while the action is taking place or while its content is open. Such graphic treatment can happen for a fraction of a second before returning to the 'active' state (e.g. giving feedback that an action element such as a button was selected). Other kinds of elements can stay activated until it is pressed again (e.g. tabs of files that stay open, or 'toggling' buttons that stay pressed).

Highlight colours as applied to active and highlighted graphic elements should be brighter and sharper because they tend to be perceived as being larger and closer to the viewer. It is essential that highlights are also appropriate for use by the colour-blind (see earlier). Where a cursor is used this can also change to a different graphic over a highlighted screen element, something described further below.

Lowlights or inactive elements should have a lower contrast and duller colours than highlights, being greyed out or much paler.

Highlights and lowlights are complementary and should work in harmony with each other.

Care should be taken to make sure that any highlights are always more prominent and eye-catching than any lowlights, through their colour, size and shape.

Highlights, lowlights and active fields should be consistent in colour and representation throughout the interface as otherwise confusion may be the result. It would be strange, for example, for highlights to be green and lowlights red in one part of the interface, and for them to be the other way round elsewhere.

Inactive or unavailable fields or buttons in a functional interface should either disappear or fade out to reduce clutter and distraction. The decision whether they can disappear depends on the technical capabilities of the application platform, as well as on human-factors concerns: if the duration for which they remain inactive is judged to be relatively long, they can disappear in order to reduce the cognitive load of users.

Cursor and its changes

A touch-screen interface with soft buttons is an example of a situation where a cursor is not appropriate as it would be hidden by the user's fingers, but there are many combinations of application platform and input device where a cursor is a useful part of the interface.

The appearance of the cursor may change to reflect the system status, indicate how a certain task is done or give feedback of when it is being done. Some typical cursor shapes are:

- *arrow:* a normal cursor that moves around the screen;
- *open hand, closed hand:* the former indicates that the object under the cursor can be potentially moved or manipulated. The latter indicates that the screen object is currently 'grabbed' and is being carried somewhere;
- *selection square:* indicates that several screen objects can be dragged or an area of the screen can be selected;
- *line of text input:* indicates that in the particular area or field there can be text input;
- *watch, hourglass:* indicates waiting time.

A standard and successful way of indicating waiting time is the clock or hourglass cursor, telling the user that the system is processing information and cannot respond to other action simultaneously. There are many more scenarios and graphical representations of cursors possible, depending on the application needs and the system's possibilities. A designer can either use existing conventions or decide on a new way of displaying a cursor.

Text

The way text appears depends a lot on the screen resolution. Any given size of text will look smaller the more the resolution increases. Therefore it is very important to test how the text looks on the selected 'target' screen.

Making the text font even one pixel wider or one pixel higher has an additive effect and impacts directly on the number of lines of text that can fit on a screen, or the maximum number of characters that can be shown in some fields. Since the number of lines or the maximum letters per field can be crucial for some applications, designers may therefore need to do several proposals using different fonts before they can select an optimal balance between readability and the amount of information per screen.

A common trick is to play with the letter spacing and condensing the font, if the platform allows that.

Applications that use text appearing on screens of different sizes and resolutions require special care in selecting the font size. For example, if the display device is a TV screen, the graphics should take into consideration that this screen size may vary from less than 17 inches up to 31 inches. In such cases you have to ensure that the text is at least readable (if not optimal) even in the smaller screen sizes by testing it.

Guideline
- *Text should be large enough for the expected target audience and appropriate for the device screen(s) that may be used.*

If you can, try to use a proportional font, rather than a non-proportional one, since it is far more legible and allows more information to be fitted horizontally onto the screen. There are special screen fonts which are designed to increase readability in these situations and you could select one of those. Be careful with serif fonts, though, since in some screens their thin lines may flicker.

It is generally not a good idea to include text within graphics as images are somewhat heavier to load and the system can become slower. It also makes it more difficult (in time and effort) for the designer or developer to change anything or translate the text into another language.

Ensuring readability is not just a matter of using the right font at the right size but also of making sure that there is enough relative contrast against the background colours in both the normal and highlighted mode. It is also important to ensure that the text is readable by colour-blind and elderly people. The most readable contrast for text in daylight conditions is black or dark-coloured text on a light-coloured (pale yellow, creamy or white) background.

Readability problems, related to environmental lighting are most often related to:

(a) reflections, making the text hard to read;
(b) poor contrast in night conditions; when designing for an application to be used in surrounding dark conditions make sure there is a backlight on the display.

B   INTERACTION DESIGN GUIDELINES

6.8   Make the goal clear and do not forget it

There are a number of factors that should be defined (before you start) and borne in mind (while you work):

- goal;
- target audience;
- possible conditions of use;
- platform of use;
- atmosphere;
- feelings.

Keep moodboards – collages of related images – and other appropriate material around you when you are working to remind you of these issues.

6.9   Plan evaluation pauses

Try to step back and give yourself time to leave a certain task for a while after one iteration of development finishes. When you get back to the project you will be able to evaluate it more easily and see the problems and their solutions more clearly. Being continuously busy with one thing does not allow you to step back and get the broader view. If you leave it alone for a while you may find that solutions, different possible directions or obvious things that you skipped or missed because of the rush may appear.

Sometimes you may find that there is a balance to be achieved between the time spent in a project and the quality of it. These two can be inversely proportional. Try to feel comfortable with the time you have and make it creative, meaningful and productive. Although it sounds strange, pauses may help in achieving that. Try to focus, but not so much so that you do not see the 'big picture'.

6.10   Adaptability of the interface to the users

Think of the adaptability of the interface to the different users and their needs in terms of things like:

- colours;
- the size of buttons and fonts;
- contrast;
- readability;
- sound volume;
- audio style.

Different people have different preferences, levels of expertise or familiarity with similar systems, strength of eyesight, age, culture and so on. The aim is to make a relevant meaningful interface that will serve the purpose for which it was commissioned, if possible, for all these people.

6.11   Upgradeability of the interface

Ask yourself:

- When and how often are upgrades likely to be necessary?
- How can additions be made to the interface (e.g. in adding new features for later versions)?

Try to accommodate changes and additions within a flexible interactive structure. This structure should provide a basis for upgrading the system, rather than constraining it so much that any extension needs it to be redesigned from scratch.

6.12   Achieve a balance

A balance should be made between the interface and:

- the user;
- the content;
- the control device;
- the environment of use (lighting conditions, sound level, viewing distance from screen, etc.);
- the technical possibilities and constraints of the platform;
- the character of the product family. If the product belongs to a family,

there is a need for a robust fundamental concept upon which ery-thing can be based. This should promote consistency between all derived user interfaces (UIs), an overall brand identity and the identity of the product family;

- other issues related to upgrades, ageing products and the branding principles behind them: in some cases design concepts will have to address issues such as reuse of material, expandability into different platforms and expandability to include more functions (e.g. more downloadable functionality).

6.13 Appropriate elements of interaction

The interaction within on-screen interfaces can be characterized by the following elements:

- the frequency of interaction (how often something is used);
- the significance of interaction (how much the interaction affects the whole result achieved);
- the number of available choices;
- the way of interacting (clicking, dragging, dropping, squeezing, moving, etc.).

These factors are interrelated and should be balanced in the design. If this balance is done well, the result can be a pleasurable user experience; if not, the application can be tiring or boring to use. The top-priority elements (the most frequent and significant interaction elements) should be positioned in the most easily seen and accessed screen location or are bigger (according to Fitts' law).

A typical example of a poorly designed interface might have:

- a large number of similar looking choices;
- a simple click-only interface;
- a large amount of clicking needed to get at any content;
- no satisfying or involving results after all this effort.

Ideally interaction should be consistent and facilitate natural user involvement and effortless use.

The majority of users should benefit from the selections and menus. If some interaction elements are only used by a minority (less than 20 per cent), then they may be included only if they do not interfere with use of the other elements by the majority of users.

On-screen interaction can borrow elements from interaction in the physical world (e.g. squeezing, gravity or magnetic attraction) which make the user experience more entertaining without necessarily making it optimal or faster.

6.14   Width and depth of the menu structure

Depth of structures

The deeper and more distributed a structure is, the more difficult it is for users to know where they are in the structure and how they can navigate to another part. An excessively intricate structure may end up seeming random and disorientating. A good rule of thumb is:

Guideline
- *An interactive on-screen structure should be no more than three levels deep for novice users and not more than five even for experienced users.*

Hyperlinks from screens

The recommended number of hyperlinks to other screens or pages depends greatly on the layout of the screen. Generally the more links (or buttons) there are out in the page the more difficult and tiring navigation becomes.

For interfaces with button-like display elements a rough guideline is to use no more than about seven buttons – this is the approximate number of items that can be easily read and remembered. To help first-time users (especially) know the meaning of buttons with abbreviated captions, help text can be made to appear when the button is highlighted.

To some extent the use of pull-down menus, pop-up choice lists and hypertext allows a greater number of hyperlinks to deeper layers of the application to be comfortably used.

Menus and choice lists can each conceal many options, categorizing information into logical and easy to remember units. However, having large numbers of links on a page results in a certain amount of information overload – the page tends to lose its meaning and readability is reduced (because the eye stops at each of the highlighted words).

Navigational tricks

Direct access is only an option for the most important functions because of the constraint of the number of physical buttons or amount of screen area that can be provided. Indirect access is therefore usually required for lower level functionality. For some products (e.g. mobile phones) indirect access accounts for as much as 95 per cent of the functionality hidden into contextual soft-menus shown on the small screen.

Some ways of providing extra indirect control options is to arrange them in deeper layers of menus, as pull-down menus, have them activated by holding down a button or by using sequences of key-presses.

In the following some examples from the domain of mobile phones are given.

The Philips FiZZ mobile phone has ca. eighty functions with direct access only to direct dialling, redial, hang up and switch off. The rest – approximately 95 per cent of the functions – are indirect. The Philips SparK and Genie mobile phones have even more functionality. All the additional functions are accessed indirectly via menus.

In the case of the Philips Genie mobile phone, navigation is mostly text based, with soft keys: there are a small number of hardware buttons which are allocated to different functions the current function being shown by a text label on the screen adjacent to the button.

Like the Philips Genie, many other products have text-based navigation.

Guideline
- *For a user interface with a high number of functions use: text-based navigation; soft keys; animated navigation techniques (e.g. more graphics, scrolling panes, three-dimensional rotating objects).*

Other tricks (apart from pull-down menus, soft-keys and text navigation) that can be used to increase the virtual space beyond the limits of the screen space available include:

- a scrolling pane (with the visible window being the screen area);
- a carousel;
- scrolling menus;
- three-dimensional objects that can rotate at different angles;
- enlarging of selected areas of interest with the rest collapsing.

These techniques involve animation and heavy graphic effects that have to be shown by the platform at a reasonable speed and are consequently more appropriate for the more powerful phones or larger screens.

6.15 Response time

Make objects and actions visible. Try to be task oriented – follow a logical and natural order of actions for the user rather than a sequence defined according to arrangement within the system. This will also reduce the user's mental load.

Plan ahead, know what response times to expect and when possible try to design to help reduce this. Avoid sounds or graphics that will make an obvious delay to the system and try to use interface elements that use less memory and so appear more quickly. Actions should always be followed by feedback, and when long pauses of more than 1 or 2 seconds are unavoidable try to mask this with some form of entertaining animation to make the user lose track of time and reassure them that the system has not crashed.

Particularly long pauses should give extra feedback about why they are taking so long – describe in steps what the system is doing, and possibly an estimation on how much longer it will take, or how much has been done.

Once the waiting is over, give a clear – but subtle – indication (audio or visual) that the system has finished.

Part III
Input devices

7 Controls

Konrad Baumann

For mechanical or traditional electronic products, the choice and design of controls is determined mainly by technical constraints. For this reason, the function of many traditional appliances could be worked out from their controls. Thus, every telephone had a dial, every audio-amplifier had a control knob operating a potentiometer for volume adjustment, etc. In this way, the functions of the appliances became cultural standards or stereotypes (i.e. part of the general education of the users). Hence the controls were self-explanatory to a high degree, and operating them required relatively little time and attention, even if the degree of automation was low.

7.1 Is the diversity of controls lost?

Since the introduction of processor-controlled user interfaces (UIs), the diversity of controls is often lost to be replaced by a uniform set of keys. The predominance of keys in such appliances primarily has cost reasons; but the wish for visible modernity or aesthetics also plays a role, besides the requirement of low production costs. That is why aesthetical and economical arguments (besides the technical ones) support the trend toward a uniform design of controls as keys.

But while the self-descriptiveness decreases, the number of functions of most appliances increases steadily. The advantages gained through more automation and new functions are frequently neutralized by complex UIs. As a consequence, most of the functions gained by the processor control remain unused. But they are an important incentive for buying the appliance.

However, as soon as the novelty of processors loses its appeal for the customers, the design engineers will have to remember the existing diversity of controls and their respective ranges of applicability. In combination with present-day high-resolution displays, a well-programmed processor and a good industrial design, appliances having a highly visible ergonomic quality can be built. The latter will become an increasingly important buying incentive.

Guideline
- *Use a wide variety of existing control forms to make the UI unmistakable.*

7.2   Control elements for electronic devices

In this chapter the control elements relevant for electronic devices are described and evaluated in detail (Figure 7.1). They will be classified according to the mode of function as recognizable by the user. For that reason devices based on different physical principles may be treated under the same heading. This chapter presents an extensive survey of usable controls. It may serve as a construction set and decision aid in the development of new UIs.

7.3   One-shot key (with key click)

A key is a control which is operated by pressing with a finger and closes a contact as long as the key is pressed. A key with downward displacement or travel may be pressed down and automatically returns to its resting position after release. When operated, the key must have a noticeable click point and produce an acoustic signal (tactile and acoustic feedback) (Figures 7.2–7.4).

Curves A and B in Figure 7.5 represent favourable, C and D unfavourable force functions. The maximum of A and B should be situated between 250 g and 500 g. The steep decrease behind the maximum is essential. The diameter or the side length of a key should be between 12 mm and 20 mm, the distance between the centres of adjacent keys between 18 mm and 20 mm. The operation distance should be between 1.3 mm and 6.4 mm, the optimal value depending on the operation force. The surface of the key should be concave. Closely spaced keys should be separated by bridges (Burandt 1986; Woodson 1987; Gilmore 1989; Grandjean 1993).

7.4   One-shot key (push-button)

Keys without downward displacement or travel do not obey all ergonomic requirements. They are frequently used because they are cheaper, flatter or more attractive and resist environmental influences, but they are poor from the ergonomic point of view.

Because of the missing or inadequate feedback they give rise to more misoperations than keys with a key-click. They demand more attention by the user, because he/she must pay attention to confirmation of the operation.

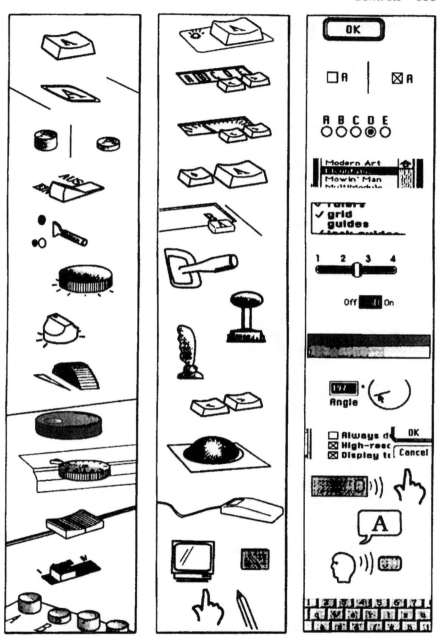

Figure 7.1 Overview of types of control: (left) standalone controls; (centre) controls to be used together with a display; (right) virtual and other controls.

Figure 7.2　Key with key-click.

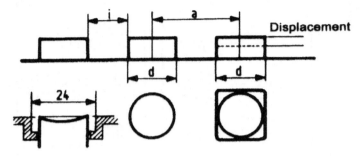

Figure 7.3　Key geometry (Burandt 1986).

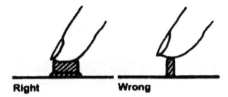

Figure 7.4　Recommended key size (left); and not recommended key size (right) (Grandjean 1991).

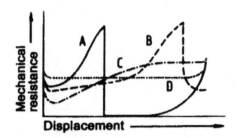

Figure 7.5　Force for pushing a key as a function of the downward displacement or travel (Burandt 1986).

As a compensation for the missing click point a beep is frequently used. In an office environment, this may be disturbing.

Foil keys (Figure 7.6) are operated with more power than keys with operation distance, and they may cause a numb feeling at the fingertips (Woodson 1987). Conducting contact sensors cannot be activated with

Figure 7.6 Foil key.

Figure 7.7 Bistable push-button.

gloves, hence they are inappropriate for public space (although they are often applied there, e.g. in lifts). They also have a minimal indication function and therefore must be made clearly visible.

7.5 Bistable push-button

A push-button is a control with two stable states which changes the state at every operation (switch function) (Figure 7.7). It must have a displacement except when there is a memory and a display for the state (compare with control signals). The positions of the button in the two states should be clearly different since they act as information about the actual state. The lower position belongs to state 'on' or '1'. For dimensions, operation force, click point and acoustic feedback the same recommendations are applicable as for the key with a key-click.

7.6 Bistable momentary toggle switch

The logical function of the toggle switch (Figure 7.8) corresponds to that of the bistable push-button. However, the toggle switch has some great advantages. The actual state, which corresponds to the angle between switch and casing surface, is visually and tactilely even more easily recognizable than with a push-button. The operation is less time consuming, and a very brief 'on' phase can easily be achieved. According to American convention, the 'on' position corresponds to the switch tilted to the right or upward or away from the user; according to European convention to the right or downward. Toggle switches are especially appropriate for

Figure 7.8 Bistable toggle switches (left part from Burandt 1986, middle part from Woodson 1986 and right part from Baumann and Lang 1998).

Figure 7.9 Trigger switches (A from Burandt 1986; B from Grandjean 1991; C from Baumann and Lanz 1998).

horizontal casing surfaces and for an arrangement in a row with common rotation axis. In some cases toggle switches with a third, intermediate state are used.

7.7   Trigger switch (bistable or with three switch positions)

The logical function of the trigger switch (Figure 7.9) also corresponds to that of the bistable push-button. When compared with the latter it has the same advantages as the toggle switch. However, it is also well suited for vertical casing surfaces. One disadvantage is that the label cannot be written on the switch itself but only on the casing. If additional trigger switches are closely spaced, confusion can occur. Therefore the labels should exactly obey the guidelines.

Special advantages of the trigger switch result from the larger lever-arm which makes greater operation resistance possible as well as a pronounced tactile and acoustical feedback (click). Hence trigger switches are especially suited for portable electronic appliances which are operated by the user while he/she is moving and not looking at the appliance. Examples are metal detectors and remote controls for model aeroplanes.

7.8   Control knob (continuous)

The control knob (Figure 7.10) is a control for the analogue (infinitely variable) adjustment of a one-dimensional variable. It permits rapidity of the adjustment process and precision over great ranges that is not simultaneously achievable with any other control. In electronic devices, the dimensions of the control knob are such that it can be operated with two or three fingers. Diameter and height should be between 10 mm and 30 mm. The handling surface has to be grooved, especially if the operating resistance (return force) is not small. A rotation angle up to 120° can be performed by hand in one turn. But larger or unrestricted rotation angles may also make sense.

Clockwise rotation should correspond to an increase in the variable. The zero point or the left edge of the adjustment range should be situated

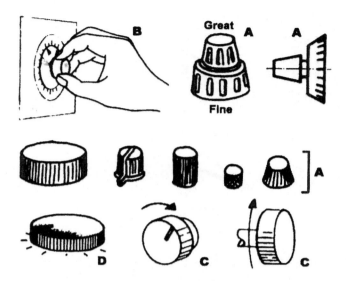

Figure 7.10 Continuous control knobs (A from Burandt 1986; B from Grandjean 1991; C from Woodson 1987; D from Baumann and Lanz 1998).

between the 6 and 9 o'clock position. At the left stop or at some other marked angle the control knob may markedly engage. In this case the device may have an integrated binary switch which switches the device off or brings the control into an undefined state.

The control knob must have a position mark which indicates the variable on a scale printed on the housing of the appliance. This is not necessary if the variable is represented on a separate display. Otherwise a scale label and a position mark on the control knob must be present. They must not be covered by the hand during operation. For that reason, arrow-shaped knobs or knobs with a disk below are especially suited. Control knobs and rotary switches with similar functions may be arranged concentrically above each other. This makes alternate operation without changing the grip possible, and the functional connection is clearly visible.

7.9 Control knob or rotary switch (discrete)

If the variable to be adjusted has three or more discrete states, a rotary switch suggests itself. The latter can be shaped like a control knob for continuous adjustment. However, the operating force should be higher. Correspondingly, strongly grooved shapes or arrow-shaped knobs are preferred. Between two switch positions the operating force as a function of the rotation angle should behave as in Figure 7.11. This will result in a marked tactile feedback.

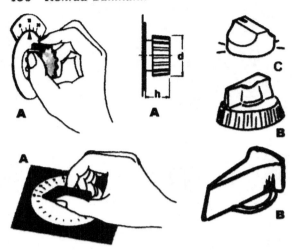

Figure 7.11 Control knobs with discrete positions – rotary switches (A from Grandjean 1991; B from Burandt 1986; C from Baumann and Lanz 1998).

The rotation angle traversed in one switching step should amount to at least 15° if the switch is read only visually, or at least 30° if the switch position is to be touched, without looking at the switch. The zero points or the minimal values of the variables of a switch should be situated between the 6 and 12 o'clock positions. Several rotary switches arranged in a row should have the same zero-position angle. If possible, the switch positions in the 'off' state should be directed along the line connecting the switches or orthogonal to it. In that case this state is recognizable at a glance even if there are many switches ('scanline').

7.10   Thumbwheel and shuttle (side-mounted and front-mounted rotary control)

A thumbwheel is a cylinder standing out from the casing with only part of its mantle showing (Figure 7.12). The mantle of the cylinder must have grooves. The wheel may be positioned at the side of the device and not in front of the user. Depending on the position, it is operated with the tip of the thumb or index finger. The shuttle (Figure 7.13), on the other hand, is a cylinder whose side plane is visible and almost flush with the surface of

Figure 7.12 Thumbwheel (left) and shuttle (right) on a horizontal device surface.

Figure 7.13 Shuttle in a combined control on a vertical device surface.

the device. The side plane of the shuttle has a smooth concave notch for rotating it with the index finger.

Thumbwheel and shuttle preferentially have a large or infinite range of rotation angle and require low operation force. As opposed to the control knob, these two controls are supplied neither with a scale nor with a position mark. The state of the variable must be recognizable for the user through separate feedback (e.g. display brightness, sound volume, a gauge).

The thumbwheel is especially suited for precise adjustment. The shuttle is appropriate for quick adjustment over wide ranges. At one end or in the centre of the range of possible orientations there may exist a marked orientation where the wheel noticeably engages. In this position a binary switch internally connected with the wheel may be operated (e.g. standby switch for the device or bypass switch for the input variable). An increase in the variable should correspond to a turn to the right, upward or away from the user for the thumbwheel, and to a clockwise rotation for the shuttle.

7.11 Thumbwheel (discrete)

For a stepwise adjustment of discrete states only, a thumbwheel must show a visible scale or position mark. Therefore it must be positioned at an edge of the casing in such a way that an inscribable surface is visible. A scale on the mantle is not good enough. In such a case only the actual value, rather than the whole range of values of the variable, would be visible.

7.12 Slider (continuous)

Sliders for analogue (infinitely variable) adjustment have the same function as control knobs (Figure 7.15). They are especially well suited for being mounted on horizontal or slightly inclined device surfaces, and with a

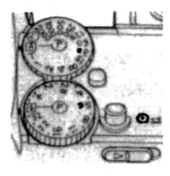 

Figure 7.14 (Left) Thumbwheels for setting of time and aperture on the electronic camera Nikon F-401. Both wheels are locked automatically when turned into their normal position and can be unlocked by pushing the small button between them. (Right) Thumbwheel with three positions (up/down/neutral) used for menu selection tasks in the mobile phone Sony CMD-Z1. At the same time, the wheel serves as a monostable key for confirmation.

Figure 7.15 Continuous slider.

sliding direction parallel to the user's line of sight. The reason is that pushing sideways or upwards is less powerful and less precise. The advantages of sliders take effect if at least two of them are parallel. In that case they can be moved individually, in pairs or synchronously in groups. With some practise, simultaneous opposite motions are also easily done. Therefore sliders are particularly used in audio mixers. In spite of their often large number they are readily surveyable. Rows of sliders may favourably be applied in combination with displays which represent block diagrams (creation of analogies).

Throttle levers in ships and aeroplanes are also frequently arranged parallel in pairs and can be moved simultaneously by hand. To be precise, these are rotatable levers with a similar appearance to sliders.

7.13 Sliding switch (discrete)

Sliding switches (Figure 7.16) for stepwise adjustment have the same function as rotary switches. However, adjustment is not possible as unerringly and precisely as that of rotatable controls. Moreover, positions cannot be judged quickly, neither visually nor tactilely. When adjusting discrete values, the user must look at the switch or count; however, a

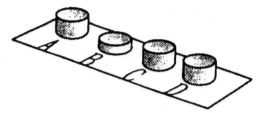

Figure 7.16 Discrete sliding switches (A from Baumann and Lanz 1998; B from Burandt 1986).

Figure 7.17 Radio buttons (key row with alternating function).

skilled user can operate a rotary switch with steps of 30° without looking at it.

7.14   Key row (alternating bistable keys, radio buttons)

A key row (Figure 7.17) consisting of a number of *n* keys with simple choice is a control with *n* discrete states. Thus the number of keys corresponds to the number of states. At any time only one key can be pressed (i.e. be in the 'on' state). By operating a key the state of the key operated immediately before automatically returns to the 'off' state. The key row does not correspond to *n* independent controls, each representing a two-valued variable, but to one single control with one *n*-valued variable. Key rows are frequently used in radio sets for the selection of stored frequencies. The advantage in comparison with the rotary switch is the possibility to switch from one state to any other without having to run through the intermediate states. Furthermore, each key may be arranged within a visible functional unit (modular structure) (e.g. together with a separate display).

7.15   One-shot key with control signals

There is a possibility to make controls with several stable states out of one-shot keys by (Figure 7.18) combining them with a memory and one or several display elements. The latter indicate the state of the control variable currently in memory. Light-emitting diodes (LEDs) are suited as a display as well as symbols on a liquid-crystal display (LCD) or keys illuminated from within. It is quite important that the connection between the key and the corresponding display element is clearly recognizable by the user.

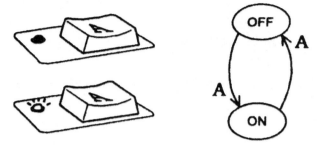

Figure 7.18 One-shot key with control signal in both states and corresponding state diagram.

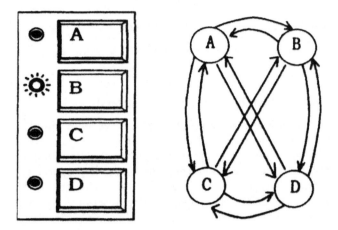

Figure 7.19 Key row (pseudo radio buttons) consisting of one-shot keys with alternating control signals and corresponding state diagram.

In the examples shown in Figures 7.18 and 7.19 controls with two and four stable states are constructed from LEDs, keys and a coloured border. They correspond in their function to a bistable key and to a key row with alternating function (simple selection), respectively. Hence the respective state diagrams are also the same.

The differently coloured LEDs must be positioned to the left of the keys or above the keys in order not to be covered by the right hand during operation. One advantage of these electronically controlled controls is that they can be switched by the device itself. The displays may also fulfil additional functions (e.g. a light diode may additionally serve as a warning signal by blinking).

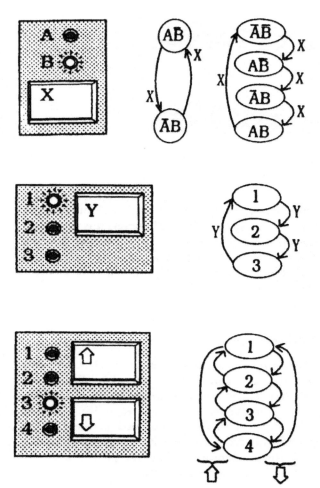

Figure 7.20 Possibilities to imitate rotary switches or toggle switches by the use of one-shot keys and control signals.

7.16 Arrow key(s) with alternating control signals

The function of a rotary switch (Figure 7.20) may be imitated by the use of one-shot keys, a memory and display elements. A rotary switch is operated by repeated rotation into a neighbouring state. Here, the two arrow keys correspond to the two rotation directions of the rotary switch; the states of the displays (the light diodes, in this case) correspond to the stable switch positions.

The control can have a cyclic state diagram, just like a rotary switch: a closed circle of states. In this case a single arrow key suffices. But if the switch is replaced by keys and if the number of keys is reduced, the

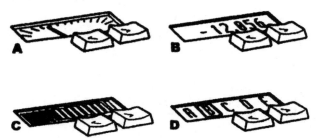

Figure 7.21 Arrow keys with (a) analogue, (b) digital, (c) pseudo-analogue, (d) alternating control signals.

operation of the control becomes lengthy. In order to pass through several states the user has to operate a key several times. However, it is possible to install a repetition function which is activated by prolonged pressing. Nevertheless, the operation of arrow keys generally requires more time and attention than the operation of rotary switches. A tactile feedback of the actual state is impossible; hence, during adjustment the user must permanently compare the actual state with the desired one.

7.17   Arrow keys with display element

Arrow keys (Figure 7.21) controlling a continuous or nearly continuous variable have the same function as a control knob or a slider. When they control a variable with discrete values they resemble a rotary or a sliding switch. The advantages and disadvantages of the respective electronic designs of the control are identical to those described in the previous paragraph. The electronic controls are more flexible, cheaper and controllable by the appliance, while the mechanical versions are quicker, less error prone and more comfortable. Thus, ergonomic considerations favour the conventional devices.

The advantages of both variants are combined in control knobs and sliders equipped with a motor for (remote) control. They are used, for example, in professional audio mixers for recording studios and as a volume control for audio amplifiers.

Arrow keys with display elements, in principle, have a similarity with the virtual controls to be described in Section 7.27. Arrow keys do not directly control the state of the variable but rather the speed of its change. The proper choice in speed of change will be considered later.

7.18   Multifunctional keys and shift keys

Monostable multifunctional keys are mainly used in keyboards (Figure 7.22). They act together with at least one shift key which switches the function of several keys at the same time. An example is the typewriter

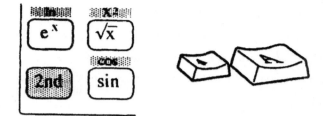

Figure 7.22 Multifunctional keys and corresponding shift keys.

keyboard, with one monostable and one bistable shift key for the capital letters and the computer keyboard having in addition 'control' and 'alternate' keys. Furthermore, some pocket calculators have keys with two or three functions generally labelled in colours.

More than three labelled functions per key are not advisable because in this case the keyboard would become difficult to survey. A multiple function without labelling contradicts the principle of self-explainability. But it may be assigned for the acceleration of frequently used processes ('hot key'). However, the same function must also be addressable explicitly (e.g. via a menu).

7.19   Soft key (software controlled key)

A soft key (Figure 7.23) is a semi-virtual control; that is, the labelling and the function of the one shot key do not exist physically. They are not part of the hardware but belong to the software of the appliance and may be changed according to its state. Soft keys are preferentially arranged so that the soft label and the hard keys can be associated.

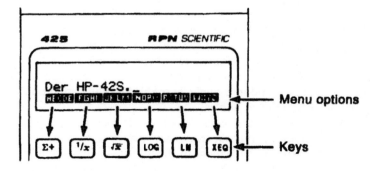

Figure 7.23 Soft keys in the upper line of a calculator keyboard (Hewlett-Packard).

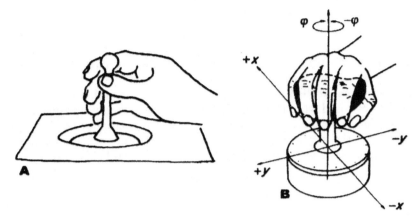

Figure 7.24 (Left) Two-dimensional isometric joystick or trackpoint built into a keyboard; (right) two-dimensional isotonic joystick in a model aeroplane remote control.

Figure 7.25 (Left) Two-dimensional and (right) three-dimensional isometric joystick (A from Baumann and Lanz 1998; B from Schmidtke 1993).

7.20 Isotonic joystick

An isotonic joystick 'is a control with the shape of a lever whose original position is vertical and which may be displaced in different directions' (Geiser 1990). It affords controlling one or two continuous variables and automatically returns to the vertical position when released.

The values of the variables are proportional to the angle of the original position. The name isotonic indicates that the return force (resistance) is approximately equally large for each displacement. The isotonic joystick is a quickly adjustable control but not a very precise one. Its advantages are small space consumption, easily and intuitively learnable handling and good tactile feedback about the state of variables.

It is appropriate for control and tracking tasks (e.g. steering aeroplanes, model planes, computer-simulated planes, cranes and lifting ramps). Its use is especially appropriate where the movement of the joystick is immediately analogous to the movement of an appliance part or to a visual display (Figures 7.24 and 7.25).

Figure 7.26 Six-dimensional control ('6D mouse') for steering an industrial robot. For every one of the three axes linear movement and rotation can be applied. The device is built into a remote control with screen and keys (Kuka robots). (The Kuka robot control shown has been redrawn by K. Baumann after Zühlke (1996b.)

7.21   Isometric joystick (force pick-up)

An isometric joystick is an analogue, self-centring control which measures the forces applied on it and yields up to three variables. Its properties correspond to those of the isotonic joystick. The name isometric indicates that the displacement is negligible. The third variable is gained preferably from the rotation around the lever axis. The force pick-up method is not used as frequently as the distance pick-up because it has no purely mechanically working counterpart.

A two-dimensional isometric joystick built into a keyboard is called a track-point. It has the advantage of small size and does not even require any space for being bent sideways when used. With an isometric joystick one can maximally control six variables – three translatory ones and three by rotation. With a control of this type the arm of an industrial robot may well be controlled (creation of analogies) (Figure 7.26).

7.22   Control column (discrete joystick)

For a digital control column, a three-valued variable with a zero state and two displaced states is related to each dimension. Without operation the variable is in the zero state. Operation causes a switch to be closed which yields an acoustical and tactile feedback (click). This type of control column may be considered as a discrete isometric joystick or as a discrete isotonic joystick (Figure 7.27).

Its two three-valued variables are much less precise than the continuous variables of joysticks. Nevertheless there are applications where the use of a digital control column is advantageous. For regulations with multiple integrating control circuits (e.g. large mass inertia in an acceleration process) a stable system is possible only if the user mentally integrates once or

Figure 7.27 Two-dimensional discrete joystick.

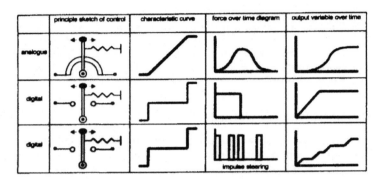

Figure 7.28 Principles of steering with analogue and digital controls. (Redrawn by K. Baumann after Schmidke 1993.)

repeatedly the time course of the control signal. This is much more easily done by counting control impulses than by multiplying duration and value of a continuous variable (Schmidtke 1993). Figures 7.28 and 7.29 compare analogue and digital controls in principle and by means of an example.

7.23   Arrow keys (one-dimensional or two-dimensional)

An arrow key is a one-shot key marked by an arrow (Figure 7.30). The use of one-, two- or four-arrow keys makes sense. The arrow keys should be arranged in such a way that the backward extended arrows meet in a point. Another optional arrangement is to have the arrow keys pointing left, downward and right in one row and the upward pointing arrow key above them in the middle. The control consisting of a combination of two- or four-arrow keys corresponds in its function to a one- or two-dimensional discrete control column.

Arrow keys have less favourable ergonomic properties regarding input speed and error security than the discrete control column but they are usable for the same tasks. For reasons of cost and space they are often applied instead of a trackball or a mouse. However, this is possible only

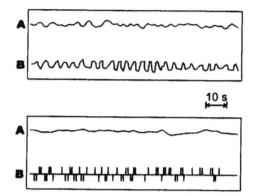

Figure 7.29 Control error over time A and control force over time B in an acceleration system using an analogue (upper diagram) and a digital (lower diagram) control. It can be seen in this case that better results are achieved with the digital control. (Redrawn by K. Baumann after Schmidke 1993.)

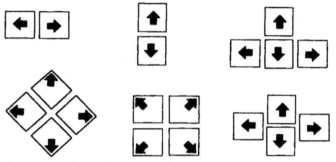

Figure 7.30 Arrow keys, possible arrangements.

at the price of a considerable loss of input speed or precision and requires more time and attention by the user. Only if the number of states of the controlled variable, or its speed of change is strongly restricted for other reasons, the use of arrow keys suffices for the control.

The processing of the input variables by the appliance is especially important when using arrow keys (repetition function, feedback). This will be described in detail in Chapter 8 on keyboards.

7.24 Trackball

A trackball is a two-dimensional, continuous control consisting of a freely revolving sphere and two mechanical or optical sensors (Figure 7.31). The sensors measure rotations of the sphere with respect to two orthogonal axes, thus describing a two-dimensional continuous variable. The diameter

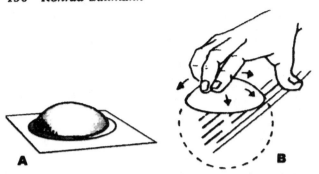

Figure 7.31 Trackball (A from Baumann and Lanz 1998; B from Woodson 1987).

of the sphere is between 2 cm and 10 cm. About one-third of the sphere stands out from a case surface. Mass and friction should be chosen in such a way that the sphere in normal motion immediately stops after releasing, but briefly (up to 1 s) rolls on when in quick motion.

The two-dimensional variable of a trackball can be adjusted very precisely, and in a restricted range also very quickly. As opposed to a joystick the trackball is not self-centring. It can stay in every chosen state. The two-dimensional range of adjustment is unlimited. The trackball may be integrated into an inclined case surface. Besides application of the trackball as a computer input device its traditional application is in flight control systems. As with the mouse the trackball needs to be combined with at least one monostable key.

7.25   Mouse

A mouse (Figure 7.32) is a two-dimensional, continuous control measuring the motion relative to a smooth surface (e.g. a table top), and in this way produces a two-dimensional continuous variable. To achieve this, mechanical or optical sensors may be used. The optical mouse is more precise and reliable than the mechanical one. But it only works on a special support and in the correct orientation.

The mouse is a quick and precise control that can be stationary in every state of the unlimited adjustment range. Disadvantages are the necessity of a table top and the possibility of the mouse detaching from the appliance. Usually the mouse is combined with at least one monostable key (one-

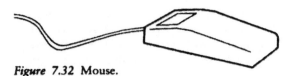

Figure 7.32 Mouse.

shot key) and so is well suited for pointing, selection and drag-and-drop operations on a screen-based graphical user interface.

With both trackball and mouse it is possible to process the variable in a velocity-dependent way. If the motion is quick then the variable changes more over a set distance than at slow motion.

Like the trackball, the mouse requires a display for representation of its variable. Ideally this is a graphical display in connection with a cursor or cross-hair. One problem with the mechanical mouse, as well as with the trackball, is the fact that the sphere may get dirty. For this reason their use in production plants is less advisable than in offices or laboratories.

7.26 Touch-sensitive devices

There is a variety of two-dimensional controls based on flat touch-sensitive devices. Feedback about the input process takes place via a display. The input surface is a pad, a tablet or a screen which produces a two-dimensional variable (i.e. a pair of coordinates) by touching with the position indicator. Depending on the design, the position indicator may be a special device (i.e. a pen) or the user's finger. There are many adaptations for such input tools, which use electric resistance, capacity, light or sound for recording the position.

The input surface is primarily suited for the input of coordinates (i.e. for drawing and for operating virtual controls), with appropriate programs also for input by handwriting. In the following subsections four basic types of touch-sensitive device are described and compared (Table 7.1).

7.26.1 Touch screen

A touch screen is a complete synthesis of control and display. It is especially suited for appliances with changing, novice users. It offers a maximum of flexibility in limited space.

Since the functional principle is practically unrecognizable (i.e. the device is not self-explanatory), the control must be made clear by the program belonging to the touch screen (the graphical user interface). Once the user has understood the principle he will enjoy the excellent visual feedback

Table 7.1 Four classes of touch-sensitive input device.

		Pointer	
		Finger	Pointing device
Input area	On the display	Touch screen	Touch screen with pointing stick
	Separated from display	Trackpad	Graphics tablet

Figure 7.33 Bang and Olufsen audio/video remote control. The glass panel is touch sensitive on the whole surface. An LCD is built into the upper and LEDs into the lower part of the panel.

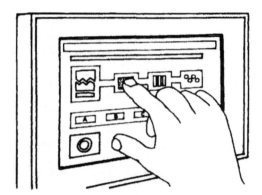

Figure 7.34 Control unit for industrial applications (Siemens Simatic operator panel) that has a 10.4 inch TFT (a type of LCD) touch-sensitive screen and an integrated Pentium processor without fan, and is resistant against moisture, dust, etc.

during operation. The touch screen is quick but imprecise. It is well suited for selection processes by the aid of virtual controls, menus and symbols.

If the screen is fixed vertically at face level the user's arms may become tired after prolonged operation. Furthermore, the user's hand can easily cover parts of the screen. With horizontally positioned screens in the workplace there is the possibility of operation by mistake. Since touch-screens have no movable parts they are frequently used in appliances in public spaces, like automatic teller machines (ATMs) or public information kiosks.

The *dimensions of hot spots* (virtual keys) shown on a finger-operated touch screen are generally the same as the pitch of hard keys, but 2 mm should be added to the diameter to account for parallax when aiming at the hot spot. Thus, the minimum dimension is 12 mm (vertical) and 14 mm (horizontal). Pheasant (1986) recommends that keys have a minimum diameter of 6 mm, and that there should be a minimum distance between keys of 6 mm. This gives a minimum pitch of 12 mm in all directions. However, this assumes either square or circular keys. In fact, the human finger is larger in the horizontal dimension than the vertical, and it is reasonable to reduce the vertical pitch by up to 2 mm (our own trial-and-error results) resulting in a minimum vertical pitch of 10 mm. For touch screens, we need to allow a larger pitch to allow for targeting errors (parallax, etc.). Adding 2 mm then gives 12 mm and 14 mm.

7.26.2 *Touch screen with pointer*

The first input technology used for this purpose were light pens in conjunction with a cathode-ray tube (CRT) display. Recognition of the position was taken over by the hardware part controlling the screen. Today, touch-sensitive devices are used in combination with a flat screen (LCD) and a cheap plastic pen (Figure 7.35).

This control is much more precise than its counterpart without a pointer; thus, its resolution is much better. The pen should have an easily accessible place for quick storage – otherwise it can be easily lost. Besides selection processes, the touch screen with pen is very well suited for drawing and for input by handwriting.

Figure 7.35 Apple Newton (trademarks) personal digital assistant.

Figure 7.36 Trackpad built into a laptop computer.

For more details about handwriting recognition see Chapter 9.

7.26.3 *Trackpad*

A trackpad (Figure 7.36) is an input area separated from the display, which is operated like a touch screen with a finger. It may even be separated from the appliance and be designed in the form of a tablet. The corresponding program may control the variable of the control element in different ways. Helander (ed.) (1988) refers to three modes: absolute, relative and joystick.

In the *absolute mode* of the control element, the coordinates of the screen correspond to those of the input surface or a section thereof. Touching the surface with the finger puts the variable onto the absolute coordinates of the touching point. Then, the program generally puts the cursor on the display onto the corresponding position. Thus, the absolute mode imitates the function of a touch screen.

In the *relative mode*, on the other hand, an initial touching of the input area does *not* change the variable. Only a movement of the finger along the surface moves the variable into the corresponding direction. Thus, the relative mode imitates the function of a mouse or a trackball.

Joystick mode: Finally, the input surface may also imitate the function of an isotonic joystick. Touching the surface corresponds to a displacement of the joystick to the respective position.

The centre of the quadratic or circular input surface corresponds to the resting position of the joystick. Without touching, the variable of the device returns to the central resting position.

The last two modes are well suited for small input surfaces whereas the absolute mode requires a larger surface. For selection processes the control element must be combined with a monostable key. This key is preferably positioned below the input surface, i.e. on that side of it facing the user. In this way the device may be operated by the index and the key by the thumb of the same hand or by the index of the other hand.

For a touch-sensitive input device there may be a need to have additional input information which replaces the separate monostable key for selection processes. This may be achieved by distinguishing between three discrete touch values 'no touch / pressure / strong pressure'. More commonly the monostable key-press is emulated by making the touch-sensitive device able to detect a click (short touch), double-click (two short touches in the same area within a short time interval) and grasp (short touch immediately followed by a long-lasting touch in the same area) operation. The grasp operation may be followed by a move and a release operation which results in a drag-and-drop operation commonly used in graphical user interfaces (GUIs).

The GUI developer must be aware of the fact that this software-aided functionality enhancement of a touch-sensitive device is not common knowledge of the average user. If no other way to perform the click, double-click, grasp and drag-and-drop operation is provided there is a need for special training or assistance for novice users.

The visual feedback about the actual state of the variable (e.g. about the position of the cursor) is much more important in the case of the described touch-sensitive devices than in the case of the touch screen. Thus, if a touch-sensitive device is used, it is advisable to make the cursor conspicuous during operation. For a touch screen, on the other hand, the cursor is unimportant during an operation because it disappears behind the position indicator (hand or pen). Therefore the selected symbols and menu options should be rendered prominent.

A small finger-operated touch-sensitive area is relatively imprecise because of the relative magnitude of the finger. In addition, lifting the finger can easily produce a small shift of the variable's position which must be corrected by the program if more accuracy is necessary.

7.26.4 Graphics tablet

A graphics tablet (Figure 7.37) is a pad of relatively large size on which the variable is entered by the aid of a mouse-like positioning device. The absolute position of the mouse is measured by the tablet's built-in hardware via inductive coupling. The mouse has at least one key and a transparent part provided with cross-hairs that makes exact positioning possible. The tablet has a grid with scales for the input of coordinates. The input of the coordinates is done by measuring the actual position when a mouse key is pressed and not by shifting the mouse on the surface.

The graphics tablet has higher precision than trackpads. For this reason it is mainly used as an input device for computer-aided design (CAD) software. Secondary (or multiple) input layers or input areas filled with virtual keys may be added to the tablet's primary grid of coordinates.

Figure 7.37 Positioning device used with a graphics tablet.

Specific functions or elements may be assigned to these keys (e.g. commands or predefined construction elements – shapes, mechanical parts, electronic devices, architectural construction elements) for CAD. They are printed on the tablet surface and can be selected by pressing keys on the mouse. Moreover, additional keys may be mounted on the mouse (e.g. a numeric keyboard).

Because the graphics tablet is one of the input devices with the most features, using it requires more time and training to get acquainted with than a trackpad, trackball or mouse.

7.27   Virtual controls

Virtual controls are generated by means of software and presented on a display screen. Their function, however, is usually analogous to real controls (like the ones presented earlier in this chapter). The transition from real to virtual controls is shown in Figure 7.38. In every row of Figure 7.38 there are three controls having the same functionality and the corresponding state diagram. In the leftmost column there is the 'real' control showing the selected state by the physical position of the moving part of the control. This gives immediate all-time visual and tactile feedback and is a real benefit for the user – not just the most expensive way of implementing the input device.

The middle column shows the version of the control storing the state of the variable electronically and giving visual – but no tactile – feedback via a control signal (LED or LCD). The next column shows the corresponding virtual representation of the control. It does not exist physically at all, but is implemented in software and represented on screen. It is operated by means of a cursor – the virtual finger of the user – and a separate input device (e.g. a trackball or a mouse with a key for confirmation of the selection).

Virtual controls have the highest flexibility – they can be hidden when not needed. However, to hide a control means that the user cannot have any feedback about the state of it. It is a challenge for the UI designer to overcome this flaw and make it a virtue.

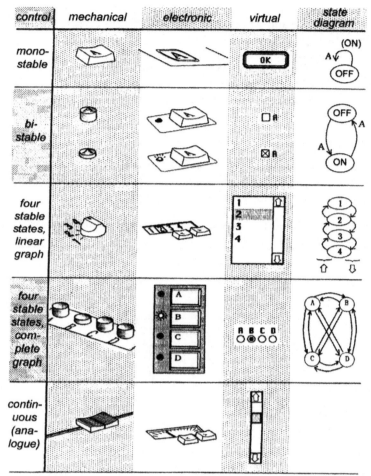

Figure 7.38 Transition from real to virtual controls giving five examples.

The rightmost column shows the state diagram corresponding to all three (real, electronic, virtual) representations of the control.

7.28 Safety and security

For security reasons it may be necessary to protect a control against accidental use (e.g. by a child). This can be done by making it difficult to operate the control, so that special knowledge or skills would be needed (e.g. some kitchen ovens having controls which the user has to push before use – only then do they come out of the surface of the appliance and can be turned on). Another example is the reset key of most laptop computers. As these appliances are designed for mobile use, it can happen that conventional keys are pressed by accident. To obviate this, the reset key is hidden under the housing and can only be operated by pushing a needle or a similar object through a hole.

In case of emergency switches protecting against injury, e.g. the emergency shut-off for laboratory equipment, quick and easy operation must be ensured. So such a control is protected against accidental use by a label and bright red colour, making it obvious that it is not a normal control.

7.29   Summary controls

Figure 7.39 summarizes the controls described in this chapter and Table 7.2 outlines some properties of input devices. More details about other input devices and methods will be given in Chapters 8–11.

Table 7.2 Summary of input devices to be used with displays and their properties (Greenstein and Arnaut in Helander 1988).

	Touch screen	Touch screen with pointing	Graphics tablet	Mouse	Trackball	Joystick
Eye–hand coordination	+	+	O	O	O	O
Free view to the display	−	−	+	+	+	+
Common housing	+	O	+	O	+	+
No parallax problems	−	−	+	+	+	+
Resolution (accuracy)	−	−	+	+	+	+
Flexible positioning	−	−	O	O	+	+
Small footprint	+	+	−	−	+	+
Short training time	+	O	O	O	O	O
Comfortable for constant use	−	−	O	O	+	+
Absolute coordinate input	+	+	+	−	−	O
Relative coordinate input	−	−	+	+	+	O
Can emulate other input devices	−	−	+	−	−	−
Suitable for:						
Pointing	+	+	+	+	+	−
Quick pointing	+	+	O	O	O	−
Pointing with confirmation	−	O	O	+	O	−
Drawing	−	−	+	O	−	−
Accurate drawing	−	−	+	−	−	−
Tracking of slowly moving targets	O	O	+	+	+	−
Tracking of quickly moving targets	−	−	O	O	O	+

control	principal sketch	max. dimension analogue data	number of stable states	preciseness tolerance error	speed
key (monostable, with path)		D 1	1	+	+
foil key (monostable, without path)		D 1	1	–	+
bistable push-button		D 1	2	+	O
bistable toggle switch		D 1	2	+	+
trigger switch		D 1	2..3	+	+
continuous control knob		K 1	∞	+	O
discrete control knob – rotary switch		D 1	2..10	+	O
continuous thumb wheel		K 1	∞	+	–
shuttle	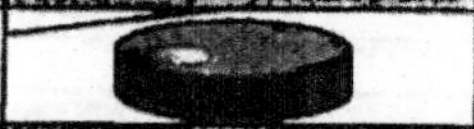	K 1	∞	–	+
discrete thumb wheel		D 1	2..20	O	–
slider		K 1	∞	O	+
discrete sliding switch		D 1	2..5	O	+
key row (radio buttons)		D 1	2..10	+	+

comments:

maximum dimension = maximum number of variables	'K' = continuous, analogue 'D' = discrete, digital	'+' = good 'O' = neutral '–' = bad

Figure 7.39 Summary of standalone controls and their properties.

control	principal sketch				
one-shot key with control system		D 1	2	+	+
arrow key(s) with alternating control signal		D 1	2.5	+	−
arrow keys with analogue display		K 1	30.∞	−	−
multifunctional key with shift key		D 3	1	O	−
soft key		D n	1	O	O
isotonic joystick		K 2	1	O	+
isometric joystick		K 3	1	O	+
discrete joystick		K 2	1	O	+
arrow keys		D 2	1	−	−
trackball		K 2	∞	+	O
mouse		K 2	∞	+	O
touch screen		K 2	0	−	+
touch screen with pointer		K 2	0	O	O
trackpad		K 2	0	−	O
graphics tablet		K 2	0	+	−

Figure 7.40 Summary of controls to be used together with a display.

control	principal sketch				
virtual monostable key	OK	D 1	1	+	+
virtual bistable push-button	☐ A ☒ A	D 1	2	+	+
virtual key row	A B C D E ○○○◉○	D 1	2.10	+	+
menu with alternating items	Modern Art / Mowin' Man Multidodule	D 1	2.30	+	O
menu with multiple choice	✓ rulers / grid guides ✓ lock guides	D n	je 2	+	O
virtual analogue slider	1 2 3 4	K 1	∞	O	+
virtual sliding switch	Off █ ▌ On	D 1	1	+	+
virtual input area		K 2	0	O	+
virtual shuttle	197 Angle	K/D 1	∞	+	+
dialogue box	☐ Always d OK ☒ High-res ☒ Display t Cancel	D n	je 1..2	+	+
movement sensor		K 3	0	−	+
voice input	A	K 2	0	−	−
EEG brainwave sensor		K	0	−	−
keyboard		see Chapter 8			

Figure 7.41 Summary of virtual and other controls and their properties.

8   Keyboards

Konrad Baumann

A keyboard is the combination of several keys located in a common housing. They are connected visually by one or more similarities (e.g. size, shape, close distance or arrangement). In general, there is also an intrinsic connection which means that the function of certain keys depends on the succession of their operation or from simultaneous operation of other keys (Figure 8.1).

8.1   Switch keys

In the most simple case the keyboard has n states attached to n keys and valid as long as the respective key is operated (key with simple function). If the set of symbols or signals to be produced by the keyboard is bigger than the number of keys, keys with multiple functions must be used. One-shot, bistable or temporary switch keys serve for switching (e.g. the shift, alternate (Alt) and control (Ctrl) key of a computer keyboard). A one-shot switch key, while pressed, changes the function of other keys. A bistable key changes the function of other keys until it is operated once more (e.g. the two switch keys for capitalization of the alphanumeric typewriter keyboard). Furthermore, there are temporary switch keys

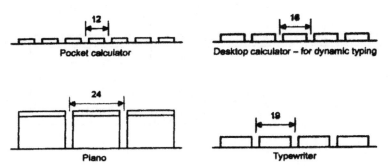

Figure 8.1 Keyboards for different applications as viewed from the front (measurements are in millimetres) (Burandt 1986).

which change the function of other keys for exactly one key-press. Bistable and temporary switch keys must be implemented either as one-shot keys with bistable control signal (as described earlier) or as bistable push-buttons. Bistable switch keys may be used everywhere while one-shot switch keys are suitable for keyboards for two-handed operation (typewriters) and temporary switch keys for one-finger operation (calculators).

The following sections outline standardized arrangements for frequently used types of keyboard.

8.2 Numeric keyboard

The arrangement of the keys in numeric keyboards is usually in the shape $3 \times 3 + 1$. This arrangement is suited for blind operation, especially if the keyboard is mounted horizontally and has a hand rest, keys with travel and roughly the dimensions shown in Figure 8.2. Of the two kinds of key assignment shown, the (1, 2, 3)-arrangement is generally favoured (Spinas 1983, Gilmore 1989; Burandt 1986; Helander 1988a). Experienced users achieve the same input performance with both, but the average user is quicker with the telephone-style keyboard and makes distinctly less mistakes. The position of the zero at the end is no longer considered a contradiction; as with the typewriter keyboard, it has become a convention.

8.3 Alphanumeric keyboard (typewriter keyboard)

QWERTY keyboard The arrangement of the keys in the typewriter keyboard is – for most letters and special symbols – an international standard. It is an American development that began around 1870. In

Figure 8.2 Two possible key arrangements for numeric keyboards: (left) telephone, (right) calculator.

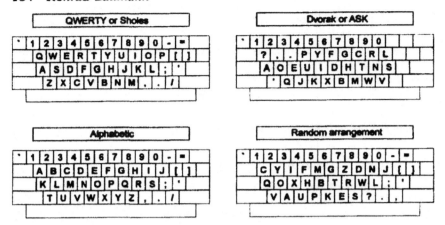

Figure 8.3 Key arrangements for alphanumeric keyboard: Sholes, QWERTY or American keyboard; Dvorak, ASK, or simplified American keyboard; linear alphabetic keyboard; random arrangement (all after Norman 1989).

mechanical typewriters it was necessary that succeeding letters lie on different sides of the keyboard as far as possible. This prevents collisions of the type levers in quick writing. Since this technically motivated arrangement is advantageous for the user too (a rare phenomenon!) the Sholes keyboard survived till the computer era (Figure 8.3).

Dvorak keyboard   The Dvorak keyboard is designed in agreement with purely ergonomic aspects. It is easier to learn and increases the writing speed by 10 per cent when compared with the usual keyboard, but apparently these advantages were not big enough for a change. Because there is such a large population of experienced QWERTY users in the world, QWERTY keyboards continue to be used. This is unlikely to change as long as keyboards are used for text input.

Alphabetically arranged keyboards   Alphabetically arranged keyboards have no ergonomic advantages. For qualified users Sholes and Dvorak keyboards are much better than alphabetic ones. Even beginners find the alphabetic keyboard to be only about as fast as the QWERTY keyboard (Norman 1989). So, despite all attempts at change, the QWERTY keyboard can be recommended for all types of appliances.

8.4   Alphabetic keyboard (upright matrix representation)

It is often not possible to install a normal typewriter keyboard, particularly in small and upright mobile appliances. Therefore designers use all kinds

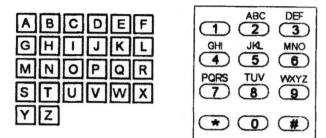

Figure 8.4 Different linear-alphabetic keyboard arrangements for small appliances.

of linear-alphabetic arrangements (see Figure 8.4). However, they are never a good solution, because the novice user does not know where the letters are located in each line. So he/she has to visually search every letter because this is still easier than calculating where the line break is (Norman 1989). So it is not even easier to become familiar with a linear-alphabetic keyboard than with a QWERTY keyboard. In this case the user would be better off if frequently used letters were placed where they are most likely found rather than by searching all over (Norman 1989).

Frequency centred keyboard As an alternative, a keyboard is proposed here which is designed for small appliances in upright format (when there is not enough space for a QWERTY keyboard) and for operation with one index finger – the other hand usually holds the device. The key arrangement is based on the letter frequency in a specific language. Frequent letters are placed around the centre where they are found most quickly (the centre is in the upper part of the keyboard because the display on top is taken into consideraton). Hence this arrangement is called 'frequency centred keyboard' (Figure 8.5). The key in the centre is the space key (the most frequently used key) which, at the same time, serves as optical centre.

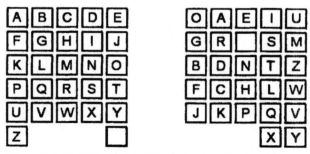

Figure 8.5 (Left) linear-alphabetic keyboard arrangement; (right) frequency centred keyboard arrangement optimized for the English language (Baumann and Lanz 1998).

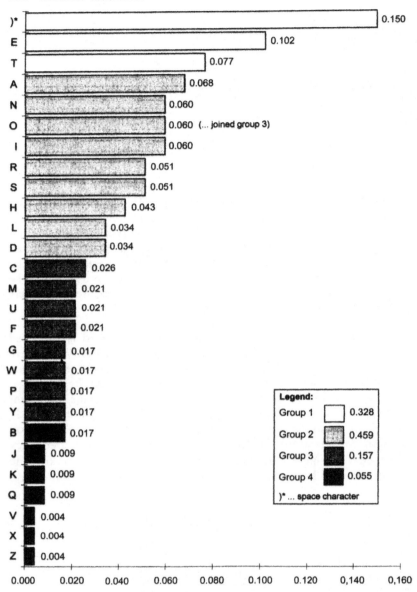

Figure 8.6 Character frequency for the English language (Baumann and Lanz 1998).

See Figure 8.6 for a bar chart of the character frequency for the English language. The other letters are placed around the centre according to their frequency in the specific language. Vowels are placed in the top line of the keyboard. The display is placed close to the keyboard because the user frequently looks from the display to the keyboard and vice versa. See Figure 8.5 for more details of this keyboard layout.

For novice users the writing speed on the frequency centred keyboard is about 15 per cent higher than with the linear-alphabetic arrangement. With practice the gain is up to 25 per cent (Baumann and Lanz 1998).

8.5 Chord keyboard

With a chord keyboard text is entered by means of pressing one or more keys at the same time, usually without looking at the keys. This enables a much greater number of letters or words to be entered than the number of keys. Chord keyboards, for example, are used by stenographers in courts or parliaments. Using a chord keyboard usually requires a great deal of skill and training and is therefore not recommended for the novice user.

8.6 Text input with word recognition

Text input via telephone keyboard The alphabetical keyboards shown so far have at least twenty-six keys. Text input with twelve keys only (telephone keyboard, Figure 8.4 [right]) is a very time-consuming method since the selection of many letters requires multiple operation of a key. Furthermore, a timeout function or a 'next' key has to be implemented to allow the consecutive input of two letters assigned to the same key. At least there is a standard key layout for this method. Acceptance of this method among users is surprisingly high. In northern European countries, having a high concentration of digital cellular phones, the exchange of 'short messages' on a daily basis can be seen. Because there is now a growing population of young users of mobile phones, the skill at using this kind of text entry is becoming remarkably high.

Word recognition Higher input speed can be realized by the aid of software-based word recognition methods (e.g. Texas Instruments' PDA Avigo in 1997 and 'T9' system for mobile phones). Text input takes place via a virtual keyboard with sixteen keys. Three or four letters are assigned to each of the eight keys, as shown in Figure 3.4 (right). The rest of the keys are used for special functions (space key, switch key for capitals, backspace, punctuation marks, number input, symbols and selection of words).

For each letter a key is operated only once. At the same time the appliance compares the key sequence with some thousands of words in memory and makes proposals for the word which the user is about to write. The right word is chosen by pressing the 'word selection' key, the full stop or the space key.

Of course, this method functions only in the language to which the appliance is adapted. Keys for language-specific special letters (é, ä, etc.) are not available. Those letters are nevertheless recognized by the appliance if they are present in a stored word. The user just has to input the corresponding 'standard' letter.

Guidelines
- *For the input of digits, both the telephone-type and the calculator-type keyboard are suited equally well.*
- *For the input of larger quantities of text, use the conventional QWERTY typewriter keyboard.*
- *The keyboard should not be arranged alphabetically. For the input of smaller quantities of text or in the case of lack of space, use a frequency centred keyboard as described in Section 8.4.*
- *Text input on a numeric telephone keyboard should be enhanced by a word-recognition method.*

8.7   Input of figures using arrow keys or analogue controls

Even in appliances without a numeric keyboard (e.g. a meter or a radio tuner) the input of numbers or the adjustment of numeric variables may be necessary. There are several possibilities for this, some of which are represented in Figure 8.7. Obviously the fewer keys used for the specific user interface (UI) the greater the effort needed for the operation. Clearly the effort also increases with the number of digits to be set.

Figure 8.7 shows two groups of solutions with fundamentally different input methods. In the first four pictures (*a–d*) the single digits are to be set separately, whereas for the other arrangements the variable may be increased or decreased as a whole number. In the first eight pictures keys are used as controls, whereas the last four pictures show (pseudo-) analogue controls. Roughly speaking, picture *a* shows the technically simplest UI, *l* the most costly. For the user, on the other hand, *a* needs the biggest effort or shows the poorest performance, whereas *l* is the most simple, convenient and easy to learn.

(a) Wristwatch-type UI   Only two keys are used – one for selecting the digit and one for setting the selected digit by incrementing it. Both operations cannot be reversed; if the user presses a key one time too many, the whole range of possible settings has to be run through again. As a consequence the values of the variable must be run through in a closed loop. The only comfort attribute is an auto-increment function which becomes

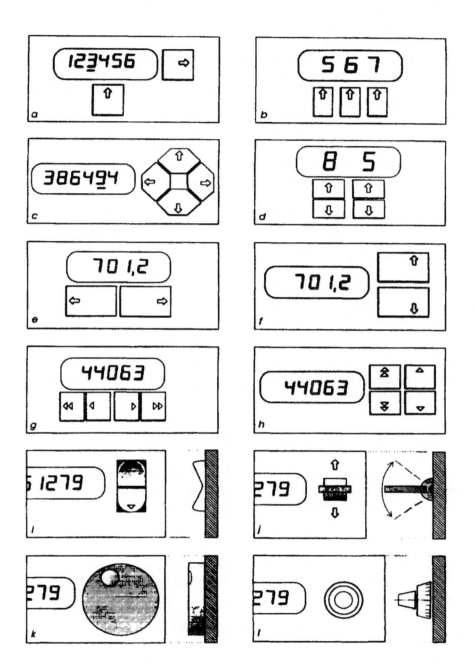

Figure 8.7 Input solutions for numbers.

active on a long key-press. In brief, low investment brings bad usability; only justified when there is an extreme shortage of space and keys.

(b) Sequence keyboard   The same interaction style as in (a) but using a dedicated key for every digit. If all digits are reset to zero before the user input, the digits can be typed in without looking at the display simply by counting the number of key-presses. This makes the sequence keyboard suitable for use in cars and plane cockpits. The average time for input is twice as long as with a telephone-type numerical keyboard. However, a driver spends much less time looking at the keyboard while using it. The amount of blind use is 60 per cent for the sequence keyboard, but only 15 per cent for the telephone keyboard (Geiser 1990).

(c) Four cursor keys   The same input style as in (a), but with keys for all directions. The average input time is smaller by 36 per cent as compared to (a). It is important to arrange the keys as shown and to make the surface of the keys suitable for blind use.

(d) Two-directional sequence keyboard   The same input style as in (b), but with keys for both directions. The average input time is smaller by 45 per cent as compared to (b).

(e) Frequency selector style   The keys change the least significant digit in both directions. The whole variable is changed in this way like a counter. This input method has an auto-increment function which becomes active on a long key-press.

(f) Frequency selector style   The same as in (e), but better usability by a different key layout following the 'up-down' analogy.

(g) Frequency selector style   The same as in (e), but includes dedicated keys for slow and fast scrolling. The fast scrolling increments the third digit, so it goes 100 times faster than the slow one.

(h) Frequency selector style   The same as in (g), but different key layout.

In the last four parts of Figure 8.7 analogue controls are used. Analogue controls are generally more expensive than simple one-shot keys. However, their usability is much better.

(i and j) Analogue controls   The analogue control in both parts of the figure is a one-dimensional isotonic joystick in two different physical representations. The variable on the display changes with a speed proportional to the angle of the current position to the middle position. When releasing the control it automatically returns to this middle position. It

makes sense to define a quadratic characteristic describing the relationship between the angle of the control and the change speed of the variable. The maximum changing speed is defined in a way that half of the range of the variable can be run through in 10 seconds (theoretical maximum time for setting a value). Near the middle position of the control there is only slow change in the variable making it possible to set the least significant digit.

(k) Rotary control This analogue control is very well suited for setting a value. For this reason it is of increasing importance. A suitable characteristic curve describing the relationship between speed of rotation and change speed of the variable has the following characteristics: the slowest change speed of the variable is 2 Hz (changes per econd) for the least significant digit and corresponds to a rotation speed of the rotary control lower than 0.2 Hz (cycles per second); the quickest change in the variable makes it possible to run through half of the range in 10 seconds and corresponds to a rotation speed of 3 Hz. Between these two values the characteristic is represented by a quadratic curve. A clockwise turn corresponds to an increase in the variable.

The rotary control has several strengths: it improves usability because of a quick and secure way of use; it is also self-explanatory and improves internal compatibility of the whole UI because it clearly indicates that there is a variable to be set in a quasi-analogue way.

(l) Two concentric rotary buttons This analogue control is used in aircraft cockpits for setting flight coordinates. It is also applied in UIs of electronic meters. Every one of the buttons changes a specific digit (e.g. the digits 10^0 and 10^2). If necessary a non-linear characteristic for the translation of rotating speed to speed of change of the variable can be added.

If certain values of the variable are needed repeatedly, it makes sense to add a memory feature. This feature enables the user to store a value for later reuse. Implementation of this feature can be done with one one-shot key (short press = retrieve value, long press = store actual value), or with two one-shot keys for these two operations. If more than one value is to be stored, a key on a numeric keyboard must be pressed within a short time interval (e.g. 3 seconds) after the memory key.

8.8 Key-press repeat function

For all arrangements of Figure 8.7 except for 'k' and 'l', the controls should have a repeat function. Thus, in each operation of a key the corresponding variable is changed by the smallest possible value. When the key remains pressed, the variable starts running after some time with a certain frequency. After further time intervals this frequency may increase.

For the input of figures and for the key-press repeat function the following guidelines are given.

Guidelines

- *Setting a given value by means of arrow keys should not take longer than 10 seconds.*
- *In the display, digits with a rate of change of more than 5 Hz should not be continually updated or set at value 0. Update frequencies of more than 5 Hz (changes per second) are no longer visible.*
- *The initial repeat frequency immediately after a key-press should be around 2 Hz.*
- *In the case of multilevel repeat functions, the next higher frequency should occur after approximately 4 s. The second frequency should be between 4 Hz and 10 Hz, with 5 Hz as the optimum.*
- *A multilevel repeat function should be similar to a quadratic graph.*

9   Alternative interaction techniques

Christopher Baber and Konrad Baumann

With an increase in the range of consumer products that incorporate some computer processing (or intelligence) comes a corresponding increase in the types of interaction that one can expect to use in order to engage with such products. On the one hand, this might imply greater use of the manual controls found on domestic products. However, the increasing complexity of domestic products might lead one to assume that such controls are already outmoded. Moving beyond manual control, one could conceive of various forms of keyboard for use on such products, or indeed the use of speech. These forms of control have already been dealt with in Chapter 7. In this chapter, some alternative interaction techniques will be considered. The intention is not to provide an exhaustive review so much as to overview the scope of such techniques.

Baber (1997) has argued that it is necessary to think of interaction devices as 'tools' (i.e. physical artefacts which can be manipulated and which can convey meaning in much the same way as handtools). He concludes that 'interaction devices can be developed as significant components of the computer systems, not only acting as transducers to convert user action to computer response, but communicating all manner of feedback to the user and support a greater variety of physical activity' (p. 276). The point of this statement is that one should be able to develop means of interacting with computers that extend beyond the conventional keyboard/mouse concept and which incorporate everyday objects. In the following section, some examples of how this might be achieved will be considered.

9.1   Capturing images

There are many occasions in which the user of a product might wish to capture some form of image. These range from the taking of photographs on vacation to recording information drawn on a whiteboard during a design meeting to capturing damage sustained to a component for maintenance advice. At one level, the digital camera, with its ability to download images to a host PC (or to save the images straight to disk) can be seen as a form of interaction device. The contemporary version of a digital camera provides the user with a means of transferring captured images onto a PC, which then allows manipulation and editing of these images. Furthermore, manipulation and editing is not confined to the realm of PCs (e.g. with the GameBoy Camera, users can take (admittedly low resolution) pictures and change them before printing). The recent launch of the Kodak PalmPix means that the Palm IIIc hand-held computer can be made to function as a digital camera and can be used to capture images when it is not being used as a 'personal digital assistant'. It is not unfeasible to assume that mobile telephones might follow a similar route, with some means of image capture becoming an integral part of the unit; after all, Nokia have demonstrated that their Communicator is capable of transferring images between telephone handsets.

From an interface design perspective, the notion of a camera as an inter-action device raises two significant questions. Leaving aside the matter of how one might control the capturing of images (a non-trivial matter), there is the question of how the capturing of images can (or should) be treated as a form of interacting with a computer. With a conventional instamatic camera, the interaction was largely confined to 'pointing and clicking' to take a photograph. It strikes us that 'pointing and clicking' with a camera could be seen as analogous to the primary form of human–computer inter-action (HCI) i.e. 'pointing and clicking' with a mouse). In other words, the very act of taking a photograph could be seen as a command to a computer (e.g. perhaps the command would be to 'tell me some more about this object I have captured'). In this respect, taking photographs could be linked to the field of augmented reality (see Chapter 11 on wearable computers).

In addition to using the act of taking a photograph to interact with a computer, there is the possibility that capturing images could form an essential component of shared work. For example, British Telecom reported a helmet-mounted camera for paramedics that would allow col-laboration between the paramedic and a specialist in a hospital (via radio communication). These ideas have been developed in the EU-funded HECTOR project. Collaboration using live video (from a person in the field) may well be beneficial in aiding a remote specialist to provide advice and guidance. This would suggest that image capture could play a key role in collaborative work and that the exchange of live images could become as useful as the exchange of text.

While the use of a camera as a means of 'capturing' images has been discussed in the previous section, one could anticipate that images could effectively be captured simply by looking at objects. This is one of the goals of developments of eye-tracking systems. The user will look at an object, typically an object represented on a visual display, and this will be sufficient to 'select' the object. *Eye tracking* technology is an advanced means for human–computer interaction. The position of the user's eyes is used to determine where he or she is looking. This information can help to automate several interaction steps in vehicle steering or camera use. A 'target' is selected automatically by looking at it for a defined time, which is certainly not only useful for military applications. The system can use this information for focusing the camera on the target, displaying or speaking background information about it, or storing the sequence of targets for analysis of the way of reading a website. Text input by eye tracking can also be the only way of communication for severely physically handicapped persons. Jacob (1990) has considered the problem of a 'Midas touch' for eye-tracking (i.e. every object over which the user's gaze passes could potentially be selected). One means of dealing with such a problem would be for the user to engage in some form of selection activity, say by pressing a button or by blinking.

A *virtual touch screen* is a device using projection for output and a camera for input. So the user interface (UI) hardware is mounted at a place not directly accessible to the user (e.g. in the ceiling). Virtually the display and the input devices are located within the reach of the user (e.g. on a desk with an inclined surface standing in public space). Such a system is used by Siemens for an information kiosk and by Sony for augmenting the computer desktop and merging it with the real desktop on which the computer is located. The Siemens virtual touch screen does not recognize static objects, but only the movement of the user's pointing finger. Research has shown that it cannot distinguish whether the user's finger touches the surface of the virtual screen – or points to a screen object somewhere above the surface – or whether he or she is just hesitating before the actual selection is performed. For this reason the pointing and clicking operation for the virtual touch screen is not performed as for 'real' touch screens, but the user has to move his finger horizontally over the screen object he wants to select. This implies a special layout of the keys – unlike in the principal sketch of Figure 9.1, the virtual keys are arranged in vertical columns.

For other types of advanced display see Section 12.5.

9.2 Capturing gestures

The idea of having a computer recognize and act upon gestures made by the user has been the subject of speculation and research in HCI for many years. One obvious range of developments includes gesture recognizers to interpret sign language. However, gesture recognition need not focus on so complete a vocabulary as would be needed for sign language and could concentrate on a very small number of gestures.

Gestures can be classified into four principal types (Table 9.1) (Koons *et al.* 1993).

In terms of *symbolic gestures*, there have been developments in datagloves (Zimmerman *et al.* 1987) that will allow the capture of hand gestures, and these gestures could take the form of sign language (Murakami and Taguchi 1991).

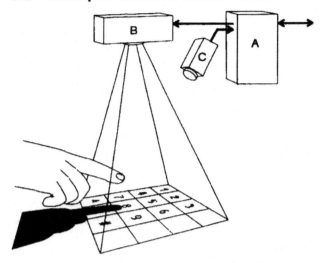

Figure 9.1 Virtual touch screen used for information kiosks in public spaces (Siemens, redrawn by K. Baumann).

Table 9.1 Types of gesture.

Type of gesture	Characteristics
Symbolic	To convey meaning
Deictic	To point to specific objects
Pantomimic	To represent the use of an object
Iconic	To describe specific properties of objects

Deictic gestures have been the focus of developments in HCI for several years. Such gestures could include simply pointing at an object, and having the object 'selected' (Bolt 1980).

The use of *pantomimic gestures* is still an area for future research; in virtual reality, one could allow the user to 'act' the use of a tool and for this action to be represented by a virtual object in the display (e.g. the user makes a hammering gesture and a hammer appears on the display and begins to hit another object).

In terms of *iconic gestures*, one could use hand gestures to describe object shape or trajectory. Alternatively, one could use a physical measuring device, such as a tape-measure, as an input device to a computer. Real-estate agents already use digital 'tape-measures' that can record a distance between two points by pointing at them, and it would be a relatively simple matter to enter this information directly to a computer (Lee *et al.* 2000). Of more interest is what uses one might put such technology to: for example, one could use a digital tape-measure to scale the design for a room plan to the dimensions of a particular apartment (and to use the same plan for different apartments by scaling them appropriately). In this way, a kitchen designer could, for instance, produce a scaled design directly from taking measurements (and possibly use the measurements to select automatically from a pre-stored set of layouts).

The notion of using physical objects for HCI has been investigated most notably in the graspable UI work of Fitzmaurice, Ishii and Buxton. This requires development of interaction devices that support a greater range of physical movement than conventional devices (Fitzmaurice *et al.* 1995; Fitzmaurice and Buxton 1997). The physical attributes of real-world objects can constitute a form of display interface for computer-based tasks. In developing a 'tangible geospace', Ishii and Ullmer (1997) tracked user interaction with real objects (such as plastic models of buildings on a campus), and map these tracked data into a computer world (e.g. to alter the perspective of a projected map display). The BUILD- IT project of Fjeld *et al.* (1998) also utilizes plastic models of buildings as a means of interacting with an architectural design package.

In studies of the manipulation of real-world objects to support interaction with objects in 'virtual environments' (VEs), performance using haptic augmentation is superior to that using a three-dimensional mouse, but remains inferior to straightforward interaction with the real objects themselves (Boud *et al.*, in press). One explanation for these differences in performance are the discrepancies between the visual feedback provided by the Virtual Reality (VR) system (e.g. in terms of frame-rate and user activity). This discrepancy led to marked differences in strategies for handling feedback. For example, when using real objects participants would switch their visual attention to the end point of the move and not track the movement of the objects in space; whereas, in the VR conditions, participants' head movements followed the path of the virtual object.

9.3 Capturing body movement

Well-known forms of movement sensors are light beams, infra-red sensors and pressure sensors (Table 9.2). In their simplest form a binary variable is generated as output. This signal may be used, for example, for triggering an automatic door opener or keeping a door from closing, switching a light on, switching a burglar alarm signal on, or switching an alarm clock off. More sophisticated types of movement sensors do not only recognize the presence or movement of a human body, but also the direction and speed of a specific movement. This is the case in a wristwatch that uses the significant turn of the left hand that indicates the intention of the user to watch the time, and automatically triggers the display light switch by this signal.

Table 9.2 **Types** of movement sensor.

Category, characteristics	Type and application of movement sensor
1 Detects a movement belonging to a user task – does not require any learning	Movement sensor for door opening
	Movement sensor for door closing prevention
	Movement sensor for burglar alarm
	Wristwatch with tilt sensor for light switch
	Tape-measure with three-dimensional orientation sensor
	Car immobilizer with chip card as a key
	Light sensor in a toy doll's eyes
	Eye tracking in vehicle steering
	Eye tracking in photo and film cameras
	Eye tracking in user testing
2 Creates a metaphor for moving money or information/performs a selection – dedicated movement is performed – requires some learning	Contactless fare collection via wristwatch or similar data carrier
	Contactless access control via chip card or similar data carrier
	Real or virtual data moving via memory stick or similar data carrier
	Movement sensor for switching alarm clock
	Pressure sensors in a toy doll's body
	Movement and direction sensor in toy (Frei *et al.* 2000)
	Eye tracking for text input

A similar approach is taken for the intelligent tape-measure by (Lee *et al.* 2000). This device detects its own orientation in space while the tape is pulled out for measuring the length, width and depth of a box. It transmits this information by wireless means to a computer system which uses it for calculating the volume of the box and fitting it optimally into a truck.

Contactless data carriers are used for automatic fare collection in public transport, for access control to hotel rooms or premises, or for electronic purse applications (Figure 9.2). As these data carriers may have the form of, or be built in a watch, or be stored into a purse or a pocket of the user, they lose the appearance of a tool and seamlessly become a part of the user's body. As a consequence, their operation is perceived by the user as an interaction of themselves with the system. However, some learning is required and the movement that enables the data exchange is a dedicated act by the user. This makes contactless technology similar to magic objects described in old fairy tales, as stated by Binsted (2000) from Sony at CHI 2000, because they also required some background knowledge and dedicated movements or words used at the right time and place.

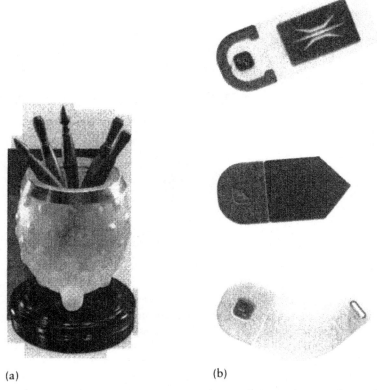

(a)   (b)

Figure 9.2 Contactless data carriers in form of cards, phicons (physical icons), and sticks for virtually moving software applications or data (Philips).

A contactless data carrier can help making data transmission between devices more transparent and give these hidden operations again some link to real objects and the real world. Bar codes may be used to tag physical objects and link them to URLs (Uniform Resource Locators), software documents or software applications. Contactless memory sticks may be used to virtually move applications between computer systems – in reality these applications move via a network. The same operation may be done without physical objects by using a projection screen that extends the desktop. Projects in this field have been shown at conferences by Philips, Sony and others.

Contactless data carriers are increasingly used for replacing the car key. An electronic car immobilizer is directly operated by means of radio-frequency identification (Figure 9.3). The user only has to carry a chip card on his or her body. When approaching the car this chip card is detected and gives access to the car. When taking the driver's seat the

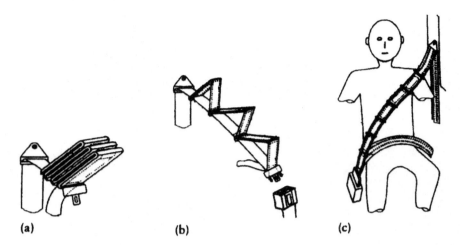

(a)   (b)   (c)

Figure 9.3 Antenna of a car immobilizer using contactless data carriers (Philips; Baumann 1996).

settings of car features (seat position, etc.) are automatically adapted to the specific user and the car is enabled to be driven (Baumann 1996).

In electronic toys different sensors may be used for interaction. Some of them detect changes in ambient light or orientation which occur simply by carrying the toy around or using it without the intent for intelligent interaction, others detect movement or pressure and require learning and dedicated operation by the user. Examples of these toys (Figure 9.4) are Microsoft's Barney, a talking dinosaur (Strommen 2000), Sony's Aibo, a robot dog and MIT's curlybot, a small device on wheels having one push-button and one light-omitting diode (LED), being able to record sequences of movements and play them back repeatedly (Frei *et al.* 2000).

9.4   Handwriting recognition

If people make gestures using some form of writing implement, then computers will be able to perform handwriting recognition. The text-input method using handwriting recognition was introduced first in personal digital assistants (PDAs) like the Apple Newton (see Figure 7.35 above). It became widely popular only in the PalmPilot (Figure 9.5) developed by US Robotics, the first commercially successful device of this type. The input device used in the PalmPilot is a touch screen with built-in track pad because the text-input area of the touch-sensitive device is not a part of the display area. Later, handwriting recognition was also introduced in PDAs by IBM, Hewlett-Packard and Handspring (Figure 9.6).

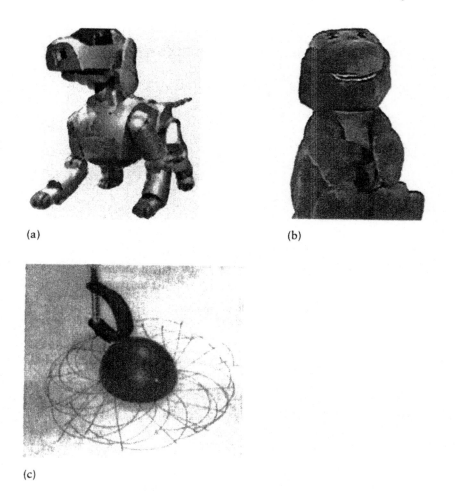

Figure 9.4 (a) Sony Aibo; (b) Microsoft ActiMates Interactive Barney (Strommen 2000); (c) Tangible Media Group, MIT Media Lab. Curlybot (Frei *et al.* 2000). (All names are trademarks.)

The main advantages of handwriting recognition are:

- the small size; and
- the low cost of a touch screen.

The size and weight of the PalmPilot were crucial for its success – besides the simplicity, short feedback time, right selection of features and relatively low cost. For details read the interview with Rob Haitani in Bergman (2000). Further advantages are possible blind operation (without looking at the device), and the possibility of making the device adaptive to the user.

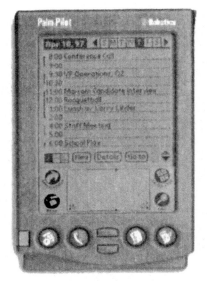 

Figure 9.5 (Left) An early PalmPilot PDA from 1996 featuring handwriting recognition; (right) a Palm (TM) Vx handheld from 2000. The fact that the UI did not change a lot over four years shows the quality on the original design.

Figure 9.6 A Handspring Visor PDA from 2000 (hat follows the same UI standard).

Challenges related to handwriting recognition are:

- required training of the user;
- special kind of writing that is necessary for some characters;
- limited text input speed; and
- processing power that is needed to ensure a short feedback time.

9.5 Tactile feedback

In addition to work on graspable UIs, there have been developments in tactile interaction devices, including modified mice (Gobel *et al.* 1994; Akematsu and Mackenzie 1996). The studies tend to demonstrate faster object-selection time, with a possible increase in target overshooting. This implies that users will be attempting to have tactile feedback complement their visual feedback (i.e. they will be watching the movement of a cursor, feel some change in the mouse and then change the course of movement). Virtual reality has seen the development of force-generating devices to provide program control over haptics (England 1995). One approach is to provide multipoint force input (e.g. using modified gloves that use miniature solenoids to provide multipoint vibration, piezoelectric coils, pneumatic air bladders, etc. to stimulate different areas of the hand). Another approach is to control force or position at an end effector, such as the fingertip. By constraining fingertip motion it is possible to simulate dynamic properties of the finger–object interface. Oakley *et al.* (2000) demonstrate that the provision of different forms of haptic feedback can significantly enhance user performance, particularly in terms of reducing error. These studies imply that whereas the tactile-feedback devices sought to complement visual feedback, the use of haptic feedback might allow modification of user action in advance of processing of visual information (Figure 9.7).

Of course, provision of tactile stimuli has been a subject of concern for system designers addressing the needs of visually impaired people. Braille displays can either comprise a pad of some 20–80 cells, or (in the case of a Optacon) can present individual letters serially. While one might anticipate that Braille (Figure 9.8) is an appropriate design for tactile displays, much research has demonstrated that visually impaired people experience problems reading this format on tactile displays (Thurlow 1986). Of particular interest would be the idea of using non-verbal symbols to present information (e.g. in the form of tactile icons). For more information about tactile displays see Chapter 14.

9.6 Speech-to-text conversion

While visually impaired people might be able to use tactile displays, the majority of computer systems for such users tend to rely on some form of

Figure 9.7 The SensAble PHANToM being used at the University of Birmingham to provide force feedback when visual and auditory cues are removed (source: http://www.vrweb.com/WEB/PRODUCTS/PHANTOM.HTM).

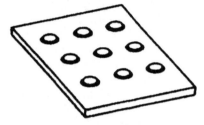

Figure 9.8 Braille tactile display element.

spoken output. *Newspapers for the Blind*, for instance, rely on pre-recording of speech (often by volunteers reading the newspaper) and the distribution of these recordings. One can imagine that benefits might be obtained by having a means of 'reading' the paper (e.g. to allow selection of articles or scanning of headlines). The conversion of spoken text to error-free written text is one of the biggest technological challenges still existing within the field of UIs. For speech-recognition technology, there are three performance goals – unlimited size of vocabulary, independence of the speaker, independence of background noise. With today's technology any two, but not all three of them, generally can be mastered today at the same time or with the same technology.

The ideal speech-to-text conversion program, something like the 'digital secretary', would need to master all three of them at the same time. For this reason it is not yet available for desktop computers. To have it included in PDAs we will have to wait even longer. Speech control and speech output are considered in detail in Chapters 10 and 14, respectively.

9.7 Other forms of audio

Speech is not the only means by which audio information can be presented to users. In the early work on 'earcons', sounds were associated with either objects of the display or with specific user actions, such as moving the cursor. In this manner, a host of ambient noises could be presented to users in order to provide support for the tasks in which they were engaged. Recent developments have focused on the provision of audio for people using mobile devices. One could use different types of sound to signal that a voice-mail or e-mail message had been received on a mobile device. This could allow subtle, non-intrusive sounds to be used to attract the user to the message without distracting the user from the tasks in which they were engaged. For instance, in the Nomadic Radio Project, Sawhney and Schmandt (1998) use the sound of a pebble being thrown into water to attract the user's attention to an incoming message. In an interesting extension of these ideas, the Nomadic Radio Project also presents audio cues to the user in a three-dimensional space. This allows the user to scan the day's messages in 360° to search for specific messages. As they scan the three-dimensional space, different messages come into 'audio focus' and the user can home in to the voice of interest.

An early example for auditory coding of information is shown in Figure 9.9. The device acoustically informs aeroplane pilots about deviations from the correct landing angle. This method has been used for landing under reduced sight conditions (night, bad weather) before the development of autopilot and electronic landing systems. If the angle is correct the pilot hears a continuous tone. If the flight direction has a deviation from the correct one to the right or to the left side, the tone becomes a series of short or long impulses, respectively. In addition and for error avoidance the information 'right' or 'left' is also displayed visually (Schmidtke 1993). For detailed information about sound in the user interface refer to Chapter 13 'Auditory displays'.

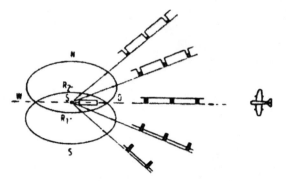

Figure 9.9 Auditory display for landing assistance using two superimposed sound signals with different length (Schmidtke 1993).

9.8   EEG input: control by thought

The electroencephalogram (EEG) can be used as an input device. To use this method it is necessary to fix several electrodes on the head of the person to be scanned. The signals from the grid of electrodes at any given moment form a distinctive pattern which is analysed by the computer using a pattern-matching method. Figure 9.10 shows some characteristic patterns appearing before and during some events. An event can be a movement by the person, planning of a movement, a thought or a sensorial impression.

The oscillations of the EEG signal consist of three superimposed parts of different frequency as shown in Figure 9.11. The measurable phenomena consist of changes in amplitude of these parts, as shown in Figure 9.12. As the signal processing has to be done in real time, a certain minimum processing power is required. To date, EEG input devices are only feasible in one dimension (i.e. for controlling a single variable).

In Figure 9.13 the device recognizes the brain activity related to thinking about the items 'to the right' and 'to the left' and is steering a cursor on the display with this signal. The processing software consisting of an artificial neural network needs some hours of training before optimum results are achieved. For severely disabled persons this input method can bring at least a minimum communication channel with others or with a computer.

There are several different concepts of brain–computer interfaces currently available. For example, it is possible to measure EEG signals of the brain and the neck and shoulder muscles at the same time. This can lead to better results than the brain signals alone. The EEG signals from muscles are easier to process and can be used with today's technology to steer electric wheelchairs, musical instruments and virtual-reality systems. In contrast to

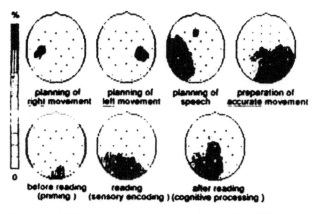

Figure 9.10 Brain activation patterns during different conscious brain activities (Pfurtscheller *et al.* 1993).

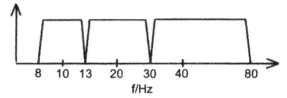

Figure 9.11 Frequency bands of the EEG.

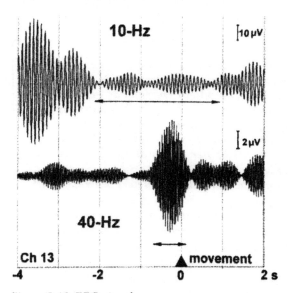

Figure 9.12 EEG signal.

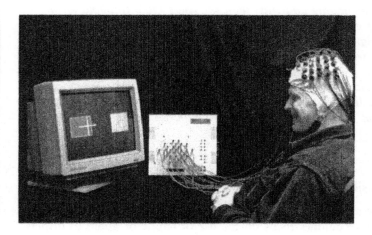

Figure 9.13 Early brain–computer interface used to move a cursor (cross-shaped) for a pointing task at two targets (rectangles) (Pfurtscheller *et al.* 1993).

the pure brain interface, however, these methods cannot be used by completely immobilized persons.

9.9   Personas: intelligent agents

Agents provided with an anthropomorphic body or appearing on the screen of an appliance are certainly the 'high end of user interfaces'. As a short list of guidelines would not do justice to this sophisticated topic, we give a summary quoted from Andre and Paiva (2000) instead:

> *Recent work on Intelligent Agents has shown a strong paradigm change in human–computer interaction: 'direct manipulation' was challenged by 'delegation'. Such paradigm change has given rise to several metaphors for human–computer interaction, in particular the personal assistant metaphor, the intelligent companion, the receptionist, etc. However, to perform tasks on behalf of the user, an agent has to be familiar with his or her habits and preferences and adapt its behavior accordingly.*
>
> *Furthermore, building embodied life-like characters animated with a set of rich behaviors is an area of research that has grown significantly for the last few years. The impact of these characters is strengthened by the fact that people in general tend to anthropormophize computers (Nass and Reeves) as well as by the richer communication styles that such characters can convey.*
>
> *The agent metaphor poses a major challenge to research on user modeling since it drastically changes the way humans perceive and interact with a computer. For instance, how to build up a relationship of understanding and trust with the user? How to avoid wrong expectations concerning the competence of human-like agents? And how to design agents that care about the user's experience with a system and are sensitive to his or her needs?*
>
> *To handle these problems, we have to continuously monitor the user's behavior and to use this knowledge to adjust. Thus, user modeling will be used as a means of achieving more effective human–agent interaction.*

9.10   Conclusions

In this chapter a variety of alternative interaction techniques have been considered. The techniques focus on the use of haptic or audio to display information, on the use of what might be termed 'everyday' objects and actions as a means of controlling (or otherwise interacting with) technology, and on electromagnetic brainwaves as an input. While this chapter is not intended to be an exhaustive review, it does present a view of the immediate future of HCI in which users move beyond the desktop and interacting with technology becomes merged with other activity. At one level this places HCI in the context of other forms of personal and domestic technologies. At another more interesting level this raises questions as to how people might interact with technologies of the future. Until now, HCI had often relied on people learning obscure command sets or learning to recognize words and objects on their computer screen. The most significant advance in HCI (the invention of the WIMP interface) is already some 40 years old. Thus, the future of HCI might be one in which people are encouraged (or at least allowed) to employ the skills that they have developed during their lives in order to interact with technology, rather than being forced to learn and perfect new skills.

10 Speech control

Christopher Baber and Jan Noyes

Speech control, as the term suggests, involves the control of a device through the recognition of human speech. Although human speech usually comprises words combined into sentences, it is possible to use single words or short phrases to issue commands to a speech recognizer. Further, the words do not need to be well formed in order to be successfully recognized by machines; this means that users who have difficulty in producing intelligible speech (e.g. through impairment, disability or environmental stressors) could also use speech control. However, it is necessary for the speaker to be reasonably consistent in the production of sounds. This is because recognition devices are programmed to process sounds, and commercial products lack the intelligence needed to differentiate between words and non-words. There has been a long debate in the speech community as to the importance of linguistic knowledge to support speech recognition. Intuitively, one might anticipate that such knowledge is essential and this was the underlying assumption of research funded by the US DARPA Program in the 1970s (see Erman *et al.* [1980], for further details). A speech recognizer (named *Hearsay*) employed a 'blackboard' architecture with several separate modules processing different aspects of the incoming speech and a central decision-making module to recognize speech; the system worked, but was a lot slower than real time and had a relatively poor recognition performance. In contrast, a speech recognizer developed as a university project (and named *Harpy*) simply treated speech recognition as a signal-processing problem, concentrating on the decomposition of the physical speech signal into discrete, time-varying parameters. This system produced near real-time performance, with a reasonable level of accuracy (e.g. *Harpy* recognized 1,011 words with a 5 per cent error rate) and can be seen as the forerunner of most commercial speech recognizers. For more details about *Harpy* see Klatt (1977).

Thus, it can be seen that in contrast to human to human communication, there are a number of benefits associated with ignoring linguistic knowledge in speech control. For example, individuals with speech that is unintelligible to fellow humans may successfully use speech control and, as noted above, this may be the case with some disabled users and particularly for people under high levels of environmental stress (e.g. pilots of high performance aircraft under g-force). Non-speech sounds (e.g. coughs, sneezes, grunts) or everyday sounds (i.e. doors slamming) can be recognized (using a 'babble model') and, if appropriate, not acted upon by the machine. Conversely, variation in the properties of the speech signal (e.g. changes in speech as a result of stressors [as discussed later in the chapter]) can significantly impair recognizer performance. Furthermore, while the lack of linguistic knowledge can be handled relatively easily in speech-control applications, recent developments in speech recognition for dictation might require additional work in natural language understanding in order to produce coherent text from the dictated speech. The word 'speech' is therefore not entirely accurate within the context of speech control as other sounds can be employed in a speech-controlled system. However, the user of speech control probably needs to speak in a manner that is different from talking to other people as the recognition process at the heart of speech control does not deal with 'speech' per se so much as a time-varying acoustic signal. Despite this inaccuracy, the word 'speech' is commonly used in all descriptions of devices of this nature.

There are many terms used to describe speech control. One of the more common is automatic speech recognition (ASR); others include electronic speech recognition, speech input, voice control and direct voice input (DVI). The same arguments apply to the use of the word 'voice' as those already detailed for 'speech' (although it might be slightly more accurate to use the means of producing the sound rather than the content of the sound). It could be argued that the term 'speech control' is one of the more useful since it encompasses the functions of these devices. Research into speech control has developed with one primary aim – to allow us to control a system or some part of a system through the medium of human speech. This would allow users to maintain control (or enter data) without the need to use their hands, which could be beneficial for people with movement impairment or for people in situations where their hands are already occupied.

10.1 Historical perspective

The concept of controlling electronic devices by human speech is not a new one and research on speech control began in earnest after the Second World War. In the late 1940s, companies such as Bell were actively investigating how to make devices that recognized and responded to human utterances. Baber (1991) discussed one of the first reported successes – a 'dog' named Rex, who reacted to his name being called, was developed around 1911 (strictly speaking this toy did not recognize speech but responded to any loud noise). A more serious application was a device developed by AT&T Bell Laboratories that recognized the digits 0 to 9 with 97–98 per cent accuracy (Davis *et al.* 1952). Thus, one of the first working examples of speech control was a device capable of recognizing ten words. The earliest devices were primarily research developments where activities focused on achieving accurate recognition. A couple of decades were to pass before these experimental devices became part of operational systems.

One of the first operational recognizers was reported at the manufacturing plant of the Owens-Illinois Corporation in the US (Martin 1976). In 1973, speech control was introduced into the quality control and inspection of components during television manufacture. Since the 1970s, speech control has been increasingly used in inspection tasks comprising part of the quality-control exercises necessary in industrial applications (Noyes *et al.* 1992). One of the main advantages of utilizing the speech channel for collecting inspection data is that it allows direct input of information to the computer system. Consider the inspection of the painted surfaces of new cars coming off the production line. An inspector will need to move around the car physically examining and checking off the various parts of the car's bodywork. Any slight defect will need to be noted for later attention. One way to collect this information is via pen and paper with inspectors working through a prepared checklist and marking their findings. This information is then collated onto a computer system both for quality-control purposes and later rectification of any defects. Speech control, whereby inspection information is input straight to the computer system via a radio-controlled microphone fitted to the inspector's headset, removes the need for the manual checklist (Usher 1993). Hence, a step in the data-processing sequence has been removed as inspection data is input directly into the computer system. Not only does this result in speeding up the inspection process, it also removes the opportunity for transcription errors when inputting data to the computer. Moreover, in this specific application, speech control is particularly useful since the environment is messy with paint. The speech link is close to being a transparent interface and is more suited to this environment than keyboards or paper that are likely to become soiled in the paint shop. Consequently, speech control can sometimes be more suited to adverse environments than other input devices. A further advantage as illustrated by this example is that speech control leaves the hands and eyes free for other tasks – the inspector does not have to carry a clipboard or similar, and is able to examine the car without encumbrance. A list of these and other benefits of speech control is given in Table 10.1.

10.2   Applications

Although the first examples of operational speech-control devices were in industry, there are other application areas that have benefitted from the introduction of speech recognition. These are summarized in Table 10.2. The disabled population has already been mentioned and for some individuals, the production of speech or vocal utterances may be unaffected by

Table 10.1 Summary of benefits of speech control.

- Removes a step in the data-processing cycle – this should lead to faster processing times, increased accuracy and greater overall efficiency as fewer personnel are needed.

- Suitable for adverse environments which are not appropriate for keyboards or other input devices.

- Mobility and portability – a radio link means the user does not have to be physically close to the centre of activity.

- Allows 'hands-free' and 'eyes-free' operation.

- Provides a further channel of communication in case of information overload.

- Can be used by individuals who do not have intelligible speech.

- Can be extended to recognize not only human speech, but other sounds and human utterances (non-words and babble) as well.

Table 10.2 Applications of speech control.

Application	Examples
• Industrial	Inspection tasks, quality-control checks and warehouse activities
• Disabled users	Environmental control units
• Avionics	Systems management, control and data input
• Office	Word processing and associated activities
• Telecommunications	The introduction of speech recognition in telephone/ Internet activities

their illness (e.g. quadriplegics, paraplegics, stroke patients and MS [multiple sclerosis] sufferers). Hence, the possibility exists for speech control of objects in their environment. These so-called environmental control units (ECUs) are operated by vocal utterances and may include opening and closing curtains and doors, turning on kettles, televisions and radios, etc. The ability to carry out these relatively simple physical tasks via the speech-controlled ECU and independent of a human carer has been shown to be an attractive proposition for the disabled population (Noyes *et al.* 1989).

Speech control has also been considered in the avionics and office environments. On board both civil and military aircraft, there are occasions when the crew are overloaded. They may be using both hands and feet to operate the plane, and speech provides a further channel for communication. Admittedly, crew already use speech in air traffic control (ATC) communications, but voice input has been considered *'for "hands and eyes free" systems management, control and data input'* (Cresswell Starr, 1993: 85). One of the problems associated with using speech control in the avionics environment is the criticality of the recognition performance. There may be certain operations where it is important that correct recognition is achieved as the lives of the crew and passengers may depend on it.

Many of the recognition difficulties associated with the avionics environment are also present in other application areas. The major difference between the cockpit and other applications is the safety-critical aspect, although it should be noted that some of the early experiments with speech-controlled wheelchairs also had safety implications (see Youdin *et al.* 1980). It is imperative when there is a safety element in the application that some form of reversion is present. In the case of the voice-controlled wheelchairs, this could be reversion to manual control of the chair. In contrast, the office environment does not have the safety element. Most speech-recognition research in the office has centred on speech control of word processing and related activities (see Noyes and Frankish 1989). Unlike the applications mentioned so far, word processing demands the use of natural language and this has a number of implications for speech control. In the first instance, an English-speaking adult will have a vocabulary of around 20,000 words. For some this may be closer to 30,000 words. Hence, the speech-recognition system has to be capable of handling a (very) large vocabulary. Not only are there implications for doing this in terms of speed, but the problem of maintaining accuracy tends to increase with the size of the vocabulary. One alternative is to introduce speech control into one aspect of the word-processing task (e.g. voice menus or voice editing of text). However, reaction-time research has indicated that voice responses are slower than manual responses when reacting to a visual stimulus (Teichner and Krebs 1974). Further, human speech has evolved 'to convey linguistic, symbolic, and categorical information (e.g. "acquire the red square"), rather than analogue information ("a little to the left; now up a bit")' (Wickens and Hollands 2000: 419). This has implications for speech control since voice editing would require very precise instructions and this is something that we do not do particularly well or naturally. Tucker and Jones (1993) considered the use of speech control in document annotation. They concluded that textual annotation was superior to voice since it was faster, more flexible, more appropriate for conveying spatial information and preferred by recipients.

A final comment concerns the use of speech control in telecommunications. To date, automatic speech recognition has had limited success in telephone-based activities. There are three primary reasons for this:

Table 10.3 Problems associated with speech control.

Problem	Reasons for problem, consequences, examples
• Achievement of correct recognition	May be an issue in applications with a safety-critical element
• Unacceptable recognition performance	May result in frustration for the user and reversion to another control medium
• Inability of technology to take account of other cues to help recognition performance	Context, non-verbal behaviours, existing knowledge about the speaker, linguistic information, redundancy in speech
• Innate characteristics of human speech	Tendency to be imprecise, which does not match the characteristics of how the technology is working
• Coping with natural language	Large vocabularies and unconstrained speech
• Use of speech control by the general public	People are untrained and not familiar with how the technology works
• Unrealistic expectations of users	Naturalness of talking might mean that users bring with them unrealistic expectations about human–machine speech interactions

- the unconstrained nature of the dialogue between the human and the system;
- the use of these devices by the general public;
- the nature of the communication channel that can add distortion to the speech signals and where voices can be faint or full of static.

All these features can impede the recognition process making it difficult to reach a satisfactory level of performance. Therefore, many telecommunications applications that involve a total speech-based system remain a development to be realized in the future. In contrast, use of speech recognition for part of the system (e.g. the voice-activated dialling services that became popular in the early 1990s) is now firmly established.

Although Table 10.3 lists a number of problems associated with speech control, it can be seen from the number of operational applications that there are many benefits. With judicious and careful selection of application, speech control can work successfully within the context of user interface (UI) design for electronic devices.

Guidelines
- A key feature of the successful, operational speech-control applications is that they are not safety critical.
- When applications that are safety critical use speech control, it is
  important that some means of backup (i.e. reversion) is designed
  into the system.

10.3   Technology

Very simply, current speech recognizers work by matching incoming
sounds to pre-stored models of auditory information and selecting the
best match. The technology may vary in the degree of sophistication of
the algorithms and mathematical models used, but the basis of recognition
is essentially a *pattern matching* exercise. It is not the place of this chapter
to expand on the technical issues associated with speech recognition, but a
brief discussion might be useful (the interested reader is advised to consult
Ainsworth [1988] for a comprehensive introduction to the topic).

In the example shown in Figure 10.1, the spectrogram has bursts of
energy (shown by dark lines) around the low- and mid-formant ranges.
This corresponds to the sound 'eh' (the initial part of the spoken 'eff' in

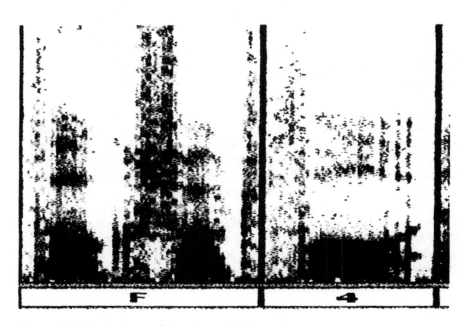

Figure 10.1 Spectrogram of the phrase 'F 4'.

this phrase). A period of silence follows in which pressure builds behind the teeth to be released in the 'F' (notice that the burst energy occurs at a higher frequency, but that there is a energy at low formants because the 'F' is voiced [i.e. the speaker continues the 'eh' sound]). Following a longer pause there is marked low- and mid-formant energy, which rises to the end. This corresponds to 'or'. The lack of a clear burst of higher frequency energy prior to this suggests that the speaker does not say 'four', but relies on the continued 'F' from the initial sound to lead into the 'or' (i.e. rather than saying 'ef', pause, 'four', the speaker says 'ef-or'). This example illustrates some of the problems that speech recognition faces in terms of determining the boundaries between speech sounds (there is continued debate as to how one should characterize speech sounds and not all speech scientists agree that a 'phoneme' is a unit of speech – thus we have adopted a somewhat idiosyncratic notation scheme to illustrate our example).

The earliest speech recognizers effectively worked by producing a template of individual words as follows. The incoming speech was segmented into short time slices. For each time slice, the intensity of sound for a set of frequencies, say sixteen, was measured. Thus, one had a model that took segments of the incoming message and determined how much energy was present for each of the frequencies analysed (the resulting model was not dissimilar to a speech spectrograph). The recognition process required this modelling stage to be repeated, and the resultant model compared with all stored models. The best match between models then resulted in the word being 'recognized'. Of course, the variability of human speech (both between and within individuals) meant that a straightforward comparison of models was very difficult. Often the models were modified to enhance the matching (e.g. through the use of major 'landmarks' in the models as anchor points). However, this still meant that any discrepancy between stored and incoming models would lead to problems; discrepancy could result from such simple matters as differences in regional accent, changes in the speaker's voice as a result of a nasal infection, or alcohol, or movement of the microphone, etc.

In the 1980s a significant step forward in speech recognition took place with the introduction of *statistical pattern matching* techniques. In essence, these techniques would break words down into discrete speech sounds, so-called phonemes, and then attempt to reconstruct the incoming word by comparing stored 'phonemes' with sounds in the speech. For a given vocabulary, there might be many possible combinations of phonemes and efforts are made to limit the combinations through the use of 'syntax' (i.e. rules of combination). This is the basis of almost all commercial speech recognizers. Given that the recognition process now relies on the reconstruction of speech, the recognizers tend to be better able to cope with variability. This has helped recognition performance increase to a point where it seems to have reached a plateau in the mid-1990s. Consideration of pattern matching helps explain why correct recognition is sometimes difficult to attain as matching of the incoming utterance depends on use of the model that most closely fits. Sometimes the system will be unable to find a suitable match and will select the best fit. Thus, it can be seen how misrecognitions occur relatively easily.

Similar to speech understanding in human to human interactions, perfect speech recognition is rare. This may not be a problem when we communicate with others since we are able to use other cues, such as context and non-verbal indicators (e.g. body language, facial expressions) as well as linguistic information (e.g. intonation of the voice) and implicit knowledge about the speaker. Hence, we can 'guess' parts of the conversation that we do not immediately comprehend. For example, the expressions 'ice cream' and 'I scream' could sound identical, but the context (i.e. the surrounding keywords) would provide the clues to the actual meaning. Words that share the same sound patterns but have very different meanings (homophones) are particularly problematic for machine recognition. As an example of this, Markowitz (1996: 6) gave the sentence '*I want to write about this rite to Ms. Wright, right away.*' Machines currently lack access to the rich source of information provided by other cues. Further, they cannot utilize the redundancy of human speech in the same way that we can. An example of this redundancy is the ability we have to make ourselves understood when in another country conversing in a language with which we are not very familiar.

A further problem with which the technology has to contend is that of *end-point detection* of vocal utterances. In human speech, we tend to run (see the dicussion of Figure 10.1) words together. This is usually not a problem for the listener, who uses the cues mentioned above, and is highly trained and skilled at interpreting and comprehending speech. In contrast, the technology does not have access to this 'extra' information with the result that end-point detection is problematic. An early solution to this problem (and one that is still popular in safety-critical applications) is to have isolated item recognition, where users input vocal utterances one at a time to the recognizer. Hence, the end points are marked by pauses. With practice, human speakers can become skilled at connected-item input as only the tiniest of pauses of a few milliseconds are needed. (Up until very recently, hesitations of at least 50 milliseconds were needed in connected-word recognition, but there are now systems where the pause has been almost eliminated.) The alternative is continuous item recognition where streams of vocal utterances are input to the recognizer as the speaker talks in a fairly natural manner. Because of the end-point problem, fully continuous speech (as used in everyday conversation) remains a challenge for technology.

Table 10.4 shows the main techniques used in the history of developing speech-control technologies. It should be noted that developments were not as staged as this since there was a renewed interest in acoustic-phoneme techniques in the 1980s. There was also a return to template matching around this time with the addition of more sophisticated techniques such

Table 10.4 Summary of developments of techniques in speech recognition.

Technique	Description	Current status
1950s to 1960s: Template matching	Pattern matching of speech data at the word level in the form of feature/parameter vectors	Use is declining although still found in many commercially available recognizers
1970s: Acoustic-phoneme techniques	Work on matching phonemes of which there are about forty in the English language	These techniques have great potential in terms of advantages, but rarely found in commercial recognizers
1980s to 1990s: Stochastic processing	Use of models for each of the items that will be recognized (e.g. hidden Markov models [HMMs])	Speed, efficiency and robustness mean that stochastic processing is used in most of today's commercial and research recognizers
2000–: Neural networks	Use of computer programs that emulate the activities of the human brain and are able to 'learn' new patterns (words) from experience	Many view neural nets as the recognition technology of the future

as *dynamic time warping*. Explained very simply, dynamic time warping stretches and squeezes the various parameters of speech along a time line to make the best match. The rapid advancements in computer technologies in the late 1970s contributed to speech control attracting a lot of research and development interest in the 1980s and many of the commercial recognizers developed around this time employed pattern matching as the primary technique. The pattern-matching process has a number of implications with regard to the vocabulary size since each utterance needed a template and larger vocabularies will entail a longer matching time. There is also greater scope for mismatches purely on the basis of the number of items to be matched, although the introduction of syntax can help manage these larger vocabularies. It should also be noted that a relatively small vocabulary can actually generate a large number of different word sequences. Hill (1980) gave the example of 13,000 distinct sequences being generated from a vocabulary of just 54 words in a fighter-aircraft application. Thus, a useful distinction can be drawn between the vocabulary of the speech recognizer (i.e. all of the words that can be recognized) and the active vocabulary (i.e. all of the words at a given point in the syntax). In conclusion, the size of the active vocabulary has implications for the successful use of the recognition technology. The vocabularies demanded by natural-language applications pose a number of problems in terms of processing time and accuracy. This is in contrast to small vocabulary sets of up to 100 items that are capable of achieving near-perfect recognition performance, and which are often more than sufficient for many speech-control applications.

Guidelines
- *Match the type of work the recognizer is intended to perform with the characteristics of the available technology.*
- *Get to know the hardware and software environment in which the speech-control application is intended to function – the recognition system needs to work in conjunction with other parts of the application.*
- *According to the application, decide whether the technology needs to be speaker dependent or independent, and whether it needs to handle isolated or connected (continuous) speech.*
- *Because of the problems of end-point recognition, isolated item recognition should be used in safety-critical applications.*
- *Recognition is more successful with smaller vocabularies (e.g. less than one hundred items). However, the introduction of syntax can increase the vocabulary set with no loss of recognition performance.*
- *Involve users in evaluating the usability of the system.*

10.4   Users

A further consideration in speech control is the user. Although most humans are skilled at talking, this is a skill that has been acquired and developed within the context of speaking to other humans. When one transfers this skill to an entirely novel domain (such as speaking to a machine) this can introduce problems. When users first encounter a speech recognizer, they bring with them certain expectations gathered from their knowledge and experience of talking to fellow humans. However, using a speech recognizer is very different from holding a conversation with a friend and these expectations may be a hindrance as the user perceptions will turn out to be very different (e.g. when we cannot be understood by a fellow human, we tend to talk slower, perhaps louder and attempt to make our speech clearer, perhaps by overpronunciation). When using a recognizer, these strategies will not work, because they do not mirror the way the recognition algorithms are working. At Bristol, when demonstrating speech-recognition equipment, a frequently heard comment from students was to exclaim that regional accents did not seem to matter when using the recognizer (but if one recalls that speech recognition might

Table 10.5 Characteristics of user communication.

Characteristic	Description and consequences
Skill	Skilled at talking to humans, but novices will not be skilled at talking to machines
Naturalness	Natural to talk to humans, but unnatural to talk to machines
Variability	Variability is a characteristic of human speech: we all exhibit variability in the way we speak, perhaps as a result of tiredness or illness or how we feel emotionally Intra-speaker variability coupled with inter-speaker variability make it difficult for machines to recognize human speech
Individual differences	Some individuals find it very difficult to use speech control – the so-called 'goats'
Fatigue	Continual talking results in voice fatigue Minute changes as a result of fatigue may not be noticed by a human listener, but will degrade recognition performance
Practice	With practice, most people can learn to constrain their speech; this might be necessary with some speech-control applications

be using phoneme models then accents might not create problems in the construction of models). In conclusion, speech is a very natural mode of human-to-human communication; however, this naturalness does not necessarily extend to human–machine interactions and indeed may work to the detriment of achieving acceptable recognition performance as the novice user brings with them unrealistic expectations about how the technology works. Table 10.5 summarizes some of the user characteristics contrasted with the demands of the technology.

Despite the mismatch between expectations and perceptions, many individuals do not have difficulties using speech control. However, it is thought that there is a group of people for whom the successful use of automatic speech recognition eludes them. Doddington and Schalk (1981) at Texas Instruments referred to these individuals as '*goats*' as opposed to the '*sheep*' who can successfully use speech control. The reasons why some individuals cannot use speech control must relate to the fact that they generate speech patterns that are difficult for the system to recognize. This may be because of unusual vocal characteristics or a lack of cooperation or consistency when using the pattern-matching technologies of the 1980s. Given that consistency is the key to the successful matching of templates, perhaps some people find it difficult to talk in a consistent way. Pisoni *et al.* (1986) suggested that the 'fault' did not lie with the user alone, but arose as a result of the interaction between the vocabulary set, the recognizer and the user. In terms of speech control, the answer may lie in improving the training of the users (i.e. giving them feedback when they achieve successful recognition) or improving the design of the system (e.g. adaptive templates).

It has already been stated that there are a number of industrial groups in the developed world where people have been using speech recognition as part of their everyday tasks for several decades. One group that we visited who had been using speech control since the early 1980s did explain that there was a need for *morning and afternoon templates* since the voice changed over the course of the day. The phenomenon of voice fatigue has been investigated by Frankish *et al.* (1992) who monitored voice drift over the course of using a speech recognizer. They found that after just 30 minutes of using a speech recognizer an increase in error rates occurs. This suggests that the deterioration in recognition accuracy is not simply a matter of 'fatigue' resulting from the strain of speaking, but that speech production is more variable than we might anticipate. Voice fatigue (arising from strain) may be a particular problem for disabled users who have to put considerable effort into speaking. Certainly, this was a finding from our work with disabled people. One solution is to have adaptive templates that 'track' the changes in the speech being input. Software techniques for doing this have been suggested by Frankish and Noyes (1993).

A final point concerns *speaker-dependent and -independent recognition*. For the most part, commercial products are sold as speaker independent (i.e. they are designed for any user to switch on and use). In practice, there is often a need for the user to spend some time speaking to the computer in order to help the stored models adjust their parameters to that person's style of speaking, accent and manner of inflection. Some recognition products are speaker dependent in that the user has to train the system to recognize his or her voice by providing one or more tokens of each utterance to be used in the application. This training is known as '*enrolment*' and needs to take place before the user can operate the recognizer. Although with speaker-dependent recognizers, individuals cannot immediately use the system (i.e. there is a cost in terms of training time), the benefit is increased recognition performance. This is in contrast to speaker-independent systems that are designed to be used immediately by any member of the population. Hence, it can be seen that the latter need more sophisticated technology since they have more speaker variables to accommodate. Two other approaches are to have multi-speaker recognition where the speech attributes of a known group are used. The process of enrichment requires prospective users of the speech recognizer supplying samples of their speech (e.g. asking the users to say all the words the recognizer will use, or reading a passage of text). The recognizer creates models of the sound patterns associated with these words (or else modifies existing patterns).

Guidelines

- *Speaker-dependent recognizers have a greater likelihood of achieving successful recognition since the users are cooperative in that they have to train to use the technology. However, the choice of speaker-dependent versus speaker-independent systems will depend on the application.*
- *Ensure cooperation of the users (i.e. dedicated users who are motivated to use the technology).*
- *Take into account the characteristics of the user group: individual differences in terms of physical, cognitive and linguistic capabilities, response to stressors and level of experience.*
- *Develop training regimes that provide a robust set of information for the recognition system, if speaker-dependent technology is to be used.*
- *Ensure that enrolment takes place in the same environment as the working environment of the recognizer.*
- *Consider ways to accommodate voice fatigue.*

10.5 Tasks

It can be seen from the successful applications of speech control that design of the task is of paramount importance. Tasks that require small vocabulary sets (e.g. less than 100 items) allow isolated word recognition (thus avoiding the problems of end-point detection) and are speaker dependent (i.e. have a dedicated set of users). These tasks are the most successful in terms of achieving good recognition performance. It has already been stated that although 100 per cent recognition might be desirable, it is difficult to achieve and perhaps unreasonable to expect given that perfect recognition does not occur in human conversations. This is probably one of the myths associated with speech control (see Table 10.6). For most applications, a high level of recognition (90+ per cent) would be needed to make the use of speech control an acceptable

Table 10.6 Some myths about speech control.

- Speech control is a natural interface for human–machine interactions
- With speech control, perfect recognition is achievable
- It is easy to talk for long periods of time
- Its sci-fi image implies that speech control is more appropriate for advanced, complex control tasks

alternative to manual or other means of achieving task completion. Possibly the only exception to this is use of speech control by the disabled population. In our own research, despite having low recognition rates of around 50 per cent, the users were still keen to use speech control. We found they were prepared to tolerate poor recognition performance because of the benefits that it gave them in terms of increased freedom and quality of life. Hence, it must be concluded that in most, but not all, applications, the level of recognition achieved is a major determinant of the task's suitability for speech control.

In speech control, a key component of the task is the *design of the dialogue* between the user and the machine. Given that there is a difference between speaking to a computer and speaking to other people, the question of how to design the dialogue between a user and machine is a non-trivial problem. To a certain extent, dialogue design is dependent on such factors as vocabulary size (e.g. the smaller the vocabulary, the more likely it is that users will have to use single words or very short phrases). This means that vocabulary size is related to the nature of speaking (i.e. isolated or connected items). A more important question is determining what the user can say. Given that 'syntax' can constrain the choice of words that can be spoken it is important to determine how the user could be informed as to what items can be used at a given time. One could simply state the desired response in a prompt (e.g. 'say "yes" to select this option'). However, a number of studies have demonstrated that users will often ignore this constraint and use words outside of the response set (i.e. out-task vocabulary). Thus, the issue of what influences the user's choice of words is important in speech control (Baber and Hone 1993; Baber *et al.* 1997; Hone and Baber 1999).

Given that one could ensure that users employed the desired vocabulary items, the next issue is how best to inform users that the recognition process has been successful. This represents an interesting HCI (human–computer interaction) issue that is not apparent in 'conventional' interaction techniques. One way of considering this is to think of feedback as comprising three levels: reactive feedback is the immediate response of a device to an action on it (e.g. when a button on a mouse is pressed, it moves); instrumental feedback is the immediate consequence of a control action (e.g. pressing the mouse button will 'select' an item on a visual display); operational feedback is the relationship between the action and the task at hand (e.g. selecting an item on the visual display will support dragging that item). For speech control, it is not possible to provide reactive feedback (the only reactive feedback that the user has is the production of speech). Typically, instrumental and operational feedbacks are combined (i.e. a word is recognized and the command performed). Attempts to separate instrumental from operational feedback (e.g. asking users to speak a confirmation word between the recognition of a command word and the performance of an action) often leads to frustration on the part of the user. Furthermore, instrumental feedback would require the computer to display the word that has been recognized, which will lead to the user's attention being divided between the task at hand and the utterance of speech. Evidence that this might be a problem comes from the finding that users will often ignore or misinterpret feedback showing the recognized word (Baber *et al.* 1992; Frankish and Noyes 1990). Thus, it is important that the feedback provided is able to maintain focus on the task at hand and also to indicate the success of the recognition process.

If the recognition process has not been successful, then it is necessary to effect some means of correcting errors. Typically, this involves either the speech recognizer following simple heuristics to cue corrections (e.g. if the match between incoming words and the stored model is below a certain level, then prompt the user to select from a set of possible alternatives). Alternatively, the user will need to engage in some form of error-correction routine (e.g. either to repeat the word or to issue a command to signal that the last recognized word had an error associated with it). In either case, error correction will interrupt the flow of user control and this can be a serious problem when using speech-driven dictation packages (see Baber and Hone 1993).

Guidelines
- *Select tasks carefully taking into account the criticality of achieving successful recognition.*
- *Decisions relating to the acceptable level of recognition performance will have to be made on an individual basis with respect to the nature of the task (e.g. regarding issues of safety).*
- *Carefully select vocabulary size and type.*
- *Carefully design dialogue, feedback, error-recovery procedures and reversion techniques.*
- *Take into account the nature of the task and the criticality of achieving correct input. For example, in some applications, it might be possible to ignore occasional errors as the efforts involved in correction may not be justifiable.*
- *Take into account task-related characteristics such as wearing protective clothing that may muffle speech, the embarrassment of users being seen talking to a machine in a public place, etc.*

10.6 Environment

Speech control has typically been developed to be used in benign office-type environments. This is because the possibility of environmental noise affecting speech recognition is very limited (although speaker-generated noise and other problems arising from signal to noise ratio, such as the

Table 10.7 Sources of noise.

Source of noise	Examples
User	Unintentional sounds such as lip smacks, coughs, etc.
Recognizer	Noise produced by the input device (the microphone or the telephone)
Machine itself	Noise produced by the machine where the speech recognizer is built in (e.g. by the loudspeakers of a TV)
Environment	Background noise that can be either intermittent such as a door slamming shut or continuous as in the case of the hum of machinery or aircraft engines

microphone being too far from the speaker's mouth, may well persist). Thus, commercial speech-control products tend to focus on office tasks, such as dictation or controlling desktop applications. There has been a long history of research into developing speech control for aviation and military systems as already discussed, and these environments represent a significant challenge to the use of speech. From a technical perspective, the environments can disrupt the production and receipt of speech to such an extent that the recognition process can be impaired. There is, however, a great deal of work into robust techniques for handling noise and other environmental stressors. Furthermore, it can often be effective to record samples of speech produced under operational conditions for the stored samples; in this way, matching between similar sorts of speech can be achieved and performance enhanced. Of course, this will work best in environments where the background noise is relatively stable and easily predictable. Table 10.7 lists the three main sources of noise that affect speech-control applications.

Noise is a significant problem for speech, and it is important to note that it affects both the production and receipt of speech (Baber and Noyes 1996). For the latter factor, it should be apparent that background noise could severely impair the capability of a microphone to receive incoming speech, or the ability of the recognition process to discriminate speech from other noises. For the former factor, it is clear that people modify their manner of speaking in the presence of noise; typically, speakers will raise their voice in order to make themselves heard over the background noise. This is known as the *Lombard effect* and is considerably more complex than people shouting when it is noisy. Rather people modify their speech in the light of feedback received from their conversation partners (i.e. they will raise their voice until their partners indicate that they can be heard). Thus, it might be possible for people in a very noisy environment to modify their speech only a little, if their partner indicates comprehension. One might expect speakers to raise their voice so that they can hear themselves speak; again, this would suggest shouting in noise. However, the need to raise one's voice will depend on the feedback received from the production of speech. Thus, one might not be able to hear what one is saying but can rely on physical cues from the movement of the speech apparatus to indicate that some noise is being made, but if the environment has a high level of vibration (or if the noise is of sufficient intensity to cause vibration) then the speaker might not have a physical indication that speech is being produced (i.e. they might not feel that they are speaking). Consequently, speakers cannot always rely on self-generated feedback. This means that the design of speech control requires feedback to be carefully considered. Typically, some form of side-channel is employed, i.e. to allow the speaker to hear what they are saying. While it might be possible to provide the speaker with feedback to indicate that they are speaking sufficiently clearly to be heard, it is a non-trivial problem to decide how a speaker ought to modify speech if it becomes ineffective. For example, what should a speaker do if shouting as loud as possible does not provoke a response in the conversational partner? Alternatively, how can a speaker exercise control over speech production in extreme vibration?

In addition to noise being generated within the environment, there is the very real possibility that noise could be produced by the object under control. For example, using speech to control a television set could run into problems with the 'noise' being produced by the working television. This 'noise' consists of the speech, music or other sounds from the TV. However, in order to function effectively, a speech-recognition system will need to compensate for this noise. Such compensation could be performed by minimizing the intrusiveness of the noise on the production of speech (e.g. through the use of a limited-range microphone). However, this might require the user to carry a separate control device. Alternatively, a signal-processing solution would track the 'noise' produced by the TV and subtract it from any sounds from the environment. The subtraction could be combined with keyword spotting such that the TV would only respond to words that it was programmed to recognize.

Murray *et al.* (1996) proposed that *stressors* can be characterized in terms of three orders: *zero-order stressors* have a direct impact on the speech-production apparatus and are very difficult to accommodate (e.g. vibration might lead to displacement of the tongue or perturbation of the vocal chords). However, it might be possible to modify one's posture, for instance, to reduce their effect. *First-order stressors* arise from unconscious physiological changes. For example, tightening of the vocal chords or increase in breathing rate. It might be possible to modify one's physiological state to reduce the effects of these changes (e.g. through breathing exercises). *Second-order stressors* arise from conscious physical effort (e.g. the Lombard effect). Finally, third-order stressors arise from an emotional response to contextual demands (e.g. panic). Third-order stressors, in effect, represent a combination of first- and second-order stressors in that they involve both conscious and unconscious responses.

Of particular interest in our work is the question of *workload* as a second-order stressor (see Baber *et al.* 1996). Clearly, as the level of information to which a person needs to attend or the time in which an action needs to be performed changes, there will be subjectively perceptible changes in demand on the person. We have shown that changes in demand can lead to definable changes in the production of speech; either in terms of not speaking at all, or in terms of changing the quality of speech (e.g. time pressure can lead to the speaker either failing to speak or to reducing the length of words and so reduce recognition accuracy). Furthermore, dual-task demands can also reduce word length and impair recognition accuracy. It is proposed that the dual-task results could arise from competing demands (i.e. as workload increases) so the task of producing consistent speech becomes, in itself, demanding.

Finally, a *cost–benefit analysis* weighing up the advantages and disadvantages of speech control compared with other input modes needs to be carried out. Speech control currently works very well for applications that have dedicated users, relatively small vocabularies and isolated item recognition. The days of communicating with a machine using natural language in an unconstrained way and as if talking to another human are still a long way off. In summary, the value of speech control is totally dependent on the application.

Guidelines
- *Take into account the amount and type of noise in the environment.*
- *Take into account other stressors in the environment.*
- *Take into account the amount of noise produced by the machine itself.*
- *Make a cost–benefit analysis regarding the use of speech control for the specific application.*

11 Wearable computers

Christopher Baber

11.1 Introduction

The idea of wearing technology can be traced to the development of spectacles (unless one counts clothing as a form of technology) and can be considered as a means of 'correcting' some problem that the wearer experiences. Thus, from one perspective, one can think of technologies that have been developed to correct visual problems (spectacles), auditory problems (hearing aids), physical problems (callipers and prostheses). More recently, technologies have been developed that can be worn inside the body (e.g. pacemakers for correcting heart rhythm). There is often sufficient intelligence in hearing aids and pacemakers for these devices to be termed computational (if not computers), and so it is not surprising that researchers have been making efforts to produce computers that can be worn. Please note that in this chapter the term 'computer' is used for all computational appliances.

From this perspective, wearable computers can be considered as means of correcting cognitive ability (e.g. by assisting the wearer to remember important information) (Rhodes 1997; Lamming and Flynn 1994). From a different perspective, technologies exist which either act directly upon the wearer (e.g. pacemakers, devices to inject controlled dosages of a drug to the wearer, etc.) or which are acted upon by the wearer (e.g. heart monitors, exercise watches, etc.). Thus, wearable computers could function as devices that are able to intelligently interact with wearers on a physiological level (Picard 1997). From yet another perspective, with the advent of the Walkman, technology has been developed to give the wearer access to entertainment and information while on the move. Indeed, the notion of providing access to information for mobile users can be traced back at least to the wristwatch. From this perspective, wearable computers could be used to augment the user's world (e.g. by providing information to the wearer in the form of circuit diagrams and instructions [Bass 1996; Bass *et al.* 1995, 1997] or by providing access to entertainment [developments in DVD have led to the production of a wearable video player from Sony]).

Wearable computers can be positioned within a space of technologies that are attached to or otherwise part of humans. These technologies tend to be semi-permanent fixtures (i.e. they are worn for hours or years) rather than being transitory attachments. The main purposes of the technologies appear to be to 'correct' human capability or to act directly on the wearer or to augment the world. From this point of view, wearable computers become an inevitable development of existing technologies. However, the notion of wearing a computer is still sufficiently novel for it to represent a radical departure from received ideas about computers and to raise the spectre of science fiction. Our perceptions of wearable computers have been shaped by the presentation of such technology in the movies (in much the same way that people will often think about speech technology in terms of the speech-driven computers in science-fiction films). Thus, people might visualize 'Robocop' or might imagine the 'Borg' in 'Star Trek'. If we turn to sci-fi films, we can identify two main trends in the presentation of wearable computers: they can be worn and used to augment the world (e.g. by over-laying text, graphics, etc.) and they can be assimilated into the wearer. In terms of the development of wearable computers, the first trend can be clearly identified and will be discussed more fully in this chapter. In terms of the second trend, it might be some time before we see fully developed 'Borgs'. However, the well-publicised exploits of Kevin Warwick[5] suggest that it is possible to implant some interactive technology into the wearer, and of course, the developments of medical devices, such as prostheses and orthoses, mean that wearers can be equipped with ever more sophisticated technology.

11.2   Technologies

Any review of technology in a field that is changing rapidly will inevitably be outdated on publication. However, I want to give a flavour of the sort of developments that are influencing the wearable-computer world, and some of the problems that are still to be tackled. Of particular importance is the question of whether wearable computers should be developed as miniaturized PCs or whether one can use other forms of technology (e.g. Baber *et al.* [1999a] distinguish wearable computers that use microprocessors from those that use microcontrollers). This means that a discussion of technology should not only focus on the available hardware, but should pay a great deal of attention to the power, processing and storage capability that an application will require.

It is possible (through PC-104 standards) to have the processing power of

5 Kevin Warwick, Professor of Cybernetics at Reading University, had a computer chip surgically implanted in his arm. The chip could be used to effect simple control of appliances in his office.

Figure 11.1 Birmingham University's wearable computer (PC104 processing with MicroOptical head-mounted display).

a laptop in a device not much bigger than a Walkman. Of course, one still requires some means of permanent storage, but with several megabytes of storage available on $1\frac{1}{4}$ in hard drives this is no longer a problem. Figure 11.1 shows our current wearable computer.

It is probable that further developments (e.g. by IBM) will lead to smaller versions of these devices. It is equally probable that hand-held appliances, such as mobile telephones, will be given the processing and storage capability of contemporary laptop computers within a few years. Thus, the question of how one might reduce a PC to a device that can be worn has largely been solved. Wearable computers are now commercially available (e.g. the Xybernaut [http://www.xybernaut.com] offers a 233-MHz Pentium processor in a case 18.7 cm × 6.3 cm × 11.7 cm that can be worn around the waist.

An alternative to providing PCs on the body is the move towards 'information appliances' (Norman 1998). These are devices that are designed to perform a specific function (as opposed to the general-purpose ethos of the PC). Stein *et al.* (1998) report a commercially successful product that combines a bar-code scanner with a 486-kHz processor into a product that can be worn on the wrist and hand. This product is designed for warehouse management and allows recording of stock details. This device can only be used for a specific set of applications, but it performs the work well and has proved very popular. Thus, a wearable computer need not be a PC that runs Windows and provides a spreadsheet or word-processing package. This, of course, raises the question of what metaphor a wearable

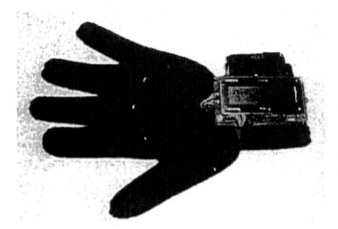

Figure 11.2 Glove interface (from Scott 2000).

computer should be designed to support (Rhodes 1997); clearly the 'desktop metaphor' of windows, icons, menus and pointing devices is less relevant to a computer that is being worn than to one on top of a desk. However, there is, as yet, no defined agreement as to what the metaphor for such computers should be. One might anticipate that 'information appliances' ought to generate their own metaphors (i.e. each application-specific product will be designed around the metaphor for the application). I feel that, just as the 'desktop metaphor' drove several generations of desktop PCs, so a new metaphor for wearable computers will be instrumental in driving this new era of computing. It remains a moot point as to what this metaphor might be.

One interesting set of developments concerns the use of specially treated fabric for developing electronic devices that can be worn. In this work, one sees the development of computers that can be worn in the same way as clothing (Figure 11.2). This is a departure from the notion of 'wearable computers' as products that are attached to the person, either by belts, straps or other harnesses. One could take contemporary fabric, such as silk organza, and attach surface-mount components, using the metal in the fabric to act as the circuit board (Post and Orth 1997). Alternatively, one could knit metal threads into the fabric (Farringdon *et al.* 1999) or attach overlapping piezoelectric film (Lind *et al.* 1997) to produce a fabric circuit board. Surface-mount components could then be attached to this material. Given the range of functions offered by surface-mount components (and the current capability of microcontrollers that could be used as the 'brain' of such a device), it is unlikely that these developments will rival the PC. However, it is clear that one can produce devices that respond to the physiology of the wearer (Healey and Picard 1998) or to the wearer's movement (Farringdon *et al.* 1999) or to changes in the environment (Schmidt *et al.* 1999).

There are, of course, problems still to be tackled. The most pressing is the need to provide reliable, long-lasting power supplies for such devices. There continues to be interest in using the movement of the wearer to generate power (Starner 1996). Furthermore, it has been accepted that running programs in Windows 95, 98 or 2000 tend to very power-hungry applications and that it is necessary to have some means of managing power demands. Linux has been used in many applications, often because it allows the developer to better control the running of different processes. However, recent developments in Windows ACPI could allow applications developed in Windows to be less power hungry. This is not to deny the flexibility of developing in Linux (nor to ignore the 'political' motivation for turning from Microsoft to Linux) but to suggest that one could use PC development tools to produce working prototypes for wearable computers. Of course, if one turns to microcontrollers, then the power-consumption problem is significantly reduced (e.g. one will run an application using a smaller battery and for a longer period of time), but with this comes an obvious reduction in processing and storage capability (although, given their different architectures, microcontrollers can often run faster than microprocessors).

The main point of this brief discussion of technology has been to highlight the question of whether a wearable computer needs to be a PC that one wears. I hope that I have suggested alternative perspectives (e.g. microcontroller- and fabric-based products) that will stimulate the reader into thinking outside the possibilities of PC and into new realms of development.

11.3   User interface

The user interface (UI) to a PC is, by definition, remote from the user; one sits facing a screen and uses a keyboard and mouse to create changes on the screen. With the notion of wearable computers comes the possibility of a more intimate relationship between computer and user. When the computer becomes attached to the user's body, then the distance between computer and user begins to reduce. Mann (1997) has considered a consequence of this reduced distance as the merging of computer with self into a cyborg. On a more prosaic level, the reduction of distance raises the question of how one could display information to the user and how the user could interact with the computer. Contemporary wearable computers tend to rely heavily on visual displays, usually mounted on the wearer's head and usually presenting information to just one eye. These displays can either be transparent (i.e. the wearer can see the world through the display and experiences the displayed information as being superimposed on the world) or can be opaque (i.e. the wearer's view of the world will be occluded by the display). The relationship between the world and a displayed image is at the heart of augmented reality. For example, one could track a visitor to a university campus (using global positioning system [GPS]) and present textual descriptions of buildings that overlay the physical buildings (Feiner *et al.* 1997). The problem of how well the description overlays the building relates to the registration between image and reality.

There has been an attempt to consider the potential role of audio for wearable computers (Sawhney and Schmandt 1998). One could anticipate the merging of mobile telephones with wearable computers to drive the audio display of information further. Finally, if the wearable computer is worn on the person then it might be possible to incorporate different forms of tactile display (Tan and Pentland 1997). There has been some work into tactile arrays (often worn on the back) for visually impaired users. My favourite example of a tactile display for wearable technology is from a James Bond film in which our hero has a wristwatch with a small paddle on its side; when the alarm goes off the paddle vibrates against Bond's wrist rather than making a noise. One might anticipate a growing interest in tactile displays in future products; after all, we already have pagers and mobile phones that will vibrate rather than ring. For more details see Chapter 14 on tactile displays and speech output.

In terms of interaction devices there appears to be a three-way split: advocates of hands-free wearable computers insist that speech recognition will be the main form of interaction; advocates of one-handed interaction have developed one-handed keyboards or selection devices; and advocates of 'minimal dialogue' suggest that there is no need for the user to explicitly interact with the computer. With reference to the latter example, the user could wear a badge to provide identification (Want *et al.* 1992) so that they could, for instance, have telephone calls and e-mails sent to their current location, or the wearer could have their physiological status monitored so that modification could be made to the environment (e.g. by changing the setting of a thermostat, or by playing soothing music), or to the individual (e.g. by adjusting the flow of an intravenous drug supply). The distinction between different approaches to interaction raises further questions about the relationship between wearer and computer.

11.4   Discussion

For researchers in human–computer interaction, the developments in wearable computers represent an interesting set of problems. On the one hand, it would appear that many of the assumptions, theories and methods that influence the design of desktop PCs may not be appropriate, and we need to develop new ways of seeing computers in order to effectively contribute to the design and development of these new products (Baber 1997). On the other hand, the integration of the computer with the person

Figure 11.3 Paramedic using first prototype.

raises many questions about the relationship between people and computers. Given that we can all become cyborgs, with wearable computers becoming forms of cognitive prostheses, there arise the perennial human-factors problems of how one should allocate functions and activity between the person and the technology. Furthermore, there are deeper existential questions about how much of one's self can be given over the on-body computer that can monitor, inform and (potentially) organize one's activity. Finally, there is the question of the role that on-body computers could play in social activity. One common concern relating to our work in developing wearable computers for paramedics has been the perceptions of members of the public should a paramedic (Figure 11.3) stand over them wearing a large head-mounted display and speak to a computer (rather than the patient). Our police work has been met with the response (in local newspapers) that we are developing the real 'Robocop', and one can appreciate the concern that police officers might be kitted out like the hero of Verhoeven's film.

The aim of this brief chapter has been to highlight some of the current concerns over wearable computers. In particular I have tried to develop two themes:

- Do wearable computers need to be PCs that can be worn?
- What is the relationship between user and wearable computer?

I hope that the discussion has suggested that the diverse approaches to wearable computers that we are currently witnessing can be used to herald the birth of a new generation of computational products: these might be limited-function information appliances (Norman 1998b), or they might be the result of the continued merging of mobile telephones, personal digital assistants and other technologies, or they might be a new species of products integrated into our clothing. As an early pioneer of ubiquitous computing pointed out, 'The most profound technologies are those that disappear. They weave themselves into the fabric of everyday life until they are indistinguishable from it' (Weiser 1991: 933).

Part IV
Output devices

12 Visual displays

Konrad Baumann

12.1 Data versus information

The selection and design of displays for an electronic appliance is much easier than that of controls. The reason is that there is only a limited number of different displays. In most cases, for a given variable, the only decision to be made is between digital and analogue representation. The main task is optimization of the representation since there are large quality differences between different displays.

Variables Displays serve as transmitters of information from the machine to a human user. As in the case of controls we say that the information is given as the value of a variable. A variable can be discrete (there is a number of possible states) or continuous (analogue, infinitely variable).

Data In the context of data, it must not be overlooked how complex and multiform these values may be. The term 'data', used frequently in this context, can give the impression that information is understood best when represented numerically (i.e. in the form of numbers). Thus, one talks about data banks, data processing and data transmission. But information is equal to numerical data *only for machines*. This is generally *not* the case for the user. Hence, when collecting data in the technical domain the first step is to digitize them (i.e. changing them to numbers). Information in the form of digital data is best suited for all purely technical processes like storage, processing or transmission (e.g. numerical data may be coded in such a way that they can be transmitted error free and securely).

Representation of data Errors about obtaining and digitizing data become a problem when humans become involved. This happens when data are represented by means of a user interface (UI). Here data must be represented in such a way that the person recognizes any restrictions properly and visualizes the given information as correctly as possible.

What is meant by the *information* given in nature or in our environment? Figure 12.1 gives a brief survey. The problem is that the UI is supplied with

Figure 12.1 Information versus data.

data but, at the same time, represents information. The developer of the UI must always be aware that the function of the appliance is *not* to represent data. The representation best suited to the user of the proper, original information may cover much more or less than the pure representation of the available data.

Environment and systems expert Frederic M. Vester states that most databases 'have become data cemeteries' (Vester 1984); the storage of data, which works very well, is considered their main function. But the information content is used only minimally. Vester therefore postulates more intelligent and user-centred programs as well as a dynamic, interdisciplinary and task-oriented handling of the data.

Not only should access to the stored information be made easier, the faculties of the computer for gaining new information on system connections should be used too. Hence the design of a good UI must not be restricted to shaping the user interface. It starts with processing, combining and selecting data in order to prepare the representation of the information in the best possible way.

Other appliances What displays should an electronic appliance show when its main function is *not* to present information? If the number of display elements is reduced, control is shifted to the human memory which is a bottleneck of the UI (Seeger 1992). On the other hand, an appliance with many display elements is difficult to survey. So, just the information which is important at the moment should be presented to the user. Control tasks should be automated as much as possible.

Just as for controls, different forms of representation should be used for qualitative variables. In the following, criteria for the selection of the proper display element for each variable will be described. A person can perceive information in a visual, acoustic and tactile way. Correspondingly, there are displays for each of these senses. According to the kind of representation, we distinguish:

- visual displays;
- auditory displays; and
- tactile displays.

12.2 Analogue versus digital: A fundamental question?

Since the advent of digital displays, the question has arisen whether they are better than conventional ones. But the analogue versus digital coding of variables was not the actual point at issue. It was not a discussion about usability, but about the old, conservative, traditional versus the new, progressive, innovative. With matters of taste being at the base of this controversy, no final agreement could be expected. In the long run, the market and hence the customer make the final decision.

In the case of some consumer appliances (e.g. the wristwatch or the speedometer in cars), we have seen the traditional analogue forms replaced by digital ones. However, the analogue display has now become an internal part of digital devices. First, the analogue watch, which had nearly been superseded by the digital one, reconquered the market. In computers an analogue control – the mouse – prevailed. In meters and in consumer electronics, control knobs were originally used for the adjustment of analogue variables, later mainly (digital) arrow keys and finally again pseudo-analogue rotary controls. Digital displays prevailed in meters and radio receivers but not in speedometers and revmeters of vehicles.

In fact, analogue displays are better suited for supervision and controlling tasks, especially when there is more than one variable, and when the supervision of the variable is not the user's main or only task (e.g. a speedometer when driving a car). On the other hand, digital displays are better suited for exact reading of a value, because the numeric form of a value does not allow for reading errors.

The same applies to input devices. Analogue or pseudo-analogue input is better for setting analogue values (e.g. the sound volume of an audio amplifier). Digital input is better for setting exact values (e.g. series of numbers typed in by a cash-desk employee in a shop). In complex steering tasks (e.g. aeroplanes) digital or pseudo-digital control may be more efficient than analogue. The reason is that the variable to be controlled (flight direction) changes slowly and fairly long after the correspondent steering activity. This is because the variable has an integrating characteristic. So when making a correction to flight direction it is better to make short but strong (pseudo-digital) steering impulses rather than long and small ones.

There is little competition between analogue and digital coding. For every application one concept or the other is superior. In the following two sections (Sections 12.3 and 12.4) we discuss many important applications with respect to the most suitable concept.

Guideline
- *Use analogue displays for supervision and controlling of one or more values.*
- *Use digital displays for exact reading of a value.*

12.3 Analogue display elements

In analogue displays the value of the variable (e.g. a physical quantity) is mapped onto a distance or an angle. Every possible value is related to a point or angle, the relationship being fixed by some function (linear, quadratic, logarithmic).

Analogue displays have either a mobile pointer or a mobile scale.

Display elements with a mobile pointer

In display elements with a mobile pointer an increase in the variable is indicated by a movement of the pointer to the right or upward or clockwise. The most common types are shown in Figure 12.2.

The pointer The pointer should be slender and simply shaped. The tip must be clearly distinguished from the rest of the pointer. Its width equals that of a scale mark (Burandt 1986). The base of the pointer may project beyond the axis. However, this part should amount at most to one-third of the length of the pointing part. The greater total length has the effect of making movements of the pointer easier to observe (Woodson 1987).

The scale The threshold at which an angular deviation is definitely recognized is about 10 degrees. Thus on a full circle 36 distinguishable classes may be defined, or preferably just 30. For quick and absolutely reliable

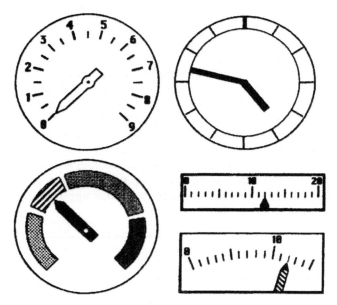

Figure 12.2 Analogue displays with a moving pointer (recommended designs).

reading the familiar twelve-part division in 30° steps should be used (Burandt 1986).

It is especially important that neither scale marks nor numbers are covered by the pointer tip. Thus in a circular instrument the numbers are situated outside the scale marks (Woodson 1987).

The scale marks There are very precise and almost identical guidelines for the design of a number scale in the literature. The numbered main marks should be situated at integer multiples of 1 or of 10.

Advisable are: 0, 1, 2, 3, ...
                      0, 10, 20, 30, ...
Possible also: 0, 100, 200, 300, ...
Scales like this are
    not advisable: 0, 2, 4, 6, ...

Scales without integers are also rejected. They require more time for reading and cause more mistakes. The only permissible exceptions are scales for time and for angles, which are numbered in multiples of 3 and of 15, 30 or 45.

Between two numbered main marks there is either a median size mark or a median and eight small marks. Other divisions (e.g. four marks in intervals of 0.2), are not advisable. If there is only the median size mark in the middle of the interval the user is nevertheless able to read it with an accuracy of one-tenth of the interval by estimating the proportions.

The proportionality factor The height of the main marks H is calculated from the reading distance D with the aid of a proportionality factor P:

$$H = D/P$$

The recommended values are $P = 90$ (Grandjean), $P = 130$ (Woodson) and $P = 150$ (Burandt). All other dimensions of the scale can then be derived from the value of H. They can be seen in Figure 12.3.

The choice of P must take into account whether the display is used under unfavourable conditions: in a car, under time pressure, with bad illumination, or by a weak-sighted user. In these cases a small P value should be chosen.

The contrast But in any case the scale should have as much contrast as possible. Black marks on white ground are more legible than an inversely designed scale. If a dark background is necessary, as in a car by night, the width of marks, pointers and numbers must be doubled.

The beginning of the scale The smallest value ('beginning') of the scale of circular displays should be situated between the 6 and 9 o'clock position.

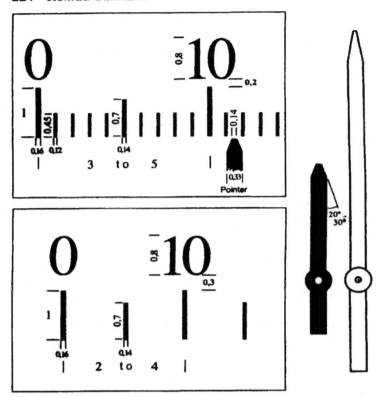

Figure 12.3 Scales, pointers and their recommended dimensions.

This convention is especially important if the display is observed from a larger distance so that the user recognizes the pointer position but not the scale.

Colour as redundant coding   In addition to or instead of a scale, coloured sector markings, symbols or an inscription with text may be used. In order to avoid excessive display complexity, scales and numbers should be used only if values really have to be read. For simply monitoring tasks a display with coloured sectors (e.g. labelled 'normal', 'warm', 'hot') may be better suited. In this way reading time and error rate are reduced. For the colours of the sector markings there are standards (for details see Chapter 15). Of course, the symbolism of colours may be used at other times too, like colouring of controls and control lights.

The issue of colour-blindness has to be kept in mind (see also Chapter 6). So, colours must never be the only way to determine the meaning of a display element or to distinguish between them. Compare this with traffic lights where the information is coded by colour *and* by position of the

Table 12.1 Symbolism of some colours for analogue displays (Burandt 1986; DIN VDE 0199 standard).

Colour	Meaning	RAL colour code
Red	Actual danger, alert	3,000
Yellow	Possible danger, be careful	1,004
Green	OK, normal operation, release	6,001
Black	Information, hint	9,005

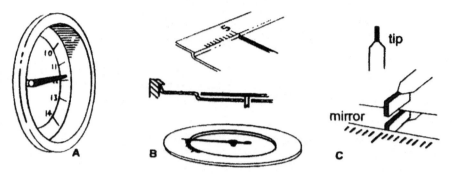

Figure 12.4 A display housing that makes it impossible to read the display from an inclined viewing angle (Woodson); B scale and pointer in one plane (Woodson 1987); C scale with mirror (Burandt 1986).

display elements. As the position of display elements is normally not standardized in a UI, a scale or labels should be used in addition to colours.

Sight angle Displays should be legible also in cases when the user's line of sight is not orthogonal to the surface. To avoid parallax errors the pointer and the scale must be in the same plane. An edge and a mirror scale may also be used. The display should not be hidden by its own casing if looked at from an inclined angle (Figure 12.4). This fault is frequently observed in digital displays as well.

Symmetry The arrangement and numbering of scales need not create any symmetry. As shown in Figure 12.2 (bottom) a scale may well represent the domain 0 ... 15 without the number 15 being written explicitly.

The displays in Figure 12.2 are drawn according to the rules mentioned in Figure 12.3. The reduction in reading time as a result of lucid displays is considerable and can be shown easily by means of testing.

Shape Round displays are read more quickly and are less error prone (Figure 12.5). For them the pointer is completely visible, and its position

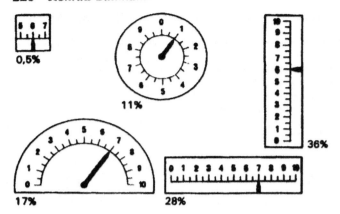

Figure 12.5 Error rate when reading the displays for a period of 0.12 seconds. The display element with a mobile scale has the lowest error rate of 0.5 per cent (Grandjean 1991, after Sleight).

or motion is easily recognized even from a larger distance. Hence ergonomically good scales (also virtual ones) should be round – especially when the UI has only one or a few displays, as in a car.

The application of linear displays can be justified for multiple displays. They can form an analogy to sliders (e.g. for an audio-mixer UI) and create a similar effect as described later for the column chart and bar chart (Figure 12.22).

Display elements with a mobile scale

In display elements with a mobile scale an increase in the variable is represented by a motion of the scale to the left, downward, or for round scales in all cases counter-clockwise.

The window   For the shape of the scale and pointer the examples given in Figure 12.3 are recommended. The scale may be attached to a disc or to a cylinder. The casing has a window through which the scale is visible having a length at least twice the distance between two numbered marks. Should only one numbered mark be visible the user would be unable to see whether the displayed value is bigger or smaller than the numbered value.

The pointer   The pointer is mounted firmly in the middle of the window. Beside the small space consumption, this is the big advantage of these displays. If a rapid reading is necessary, no time is lost in locating the pointer. For this reason the display with a mobile scale has the smallest error rate in experiments of short duration (Figure 12.5).

Figure 12.6 Displays with a mobile scale.

The scale For cylinder scales running horizontally the pointer is at the lower edge of the window; for vertical ones, at the right edge. For disc scales whose marks diverge, the pointer is always fixed at the outer edge of the window, hence the numbering is always inside. In most cases the window shows a partition of the disc scale at the 12 o'clock position (as in Figure 12.6, middle). But the 3, 6 or 9 o'clock position is also a possibility.

The cylinder counter (Figure 12.6, right) has an intermediate position between analogue and digital displays. The numbers should not move continuously, but stepwise, such that most of the time they are all in a row. Groups of three numbers should be separated by spaces. To the left of the highest non-vanishing digit there should be empty places rather than zeros (Burandt 1986).

Comparison of analogue display elements

All displays with a mobile scale are less legible from a distance than displays in which the direction of the pointer is observed. Since the advent of digital displays, they have superseded mobile scales. The situation is similar for linear scales with mobile pointers. We can summarize: If analogue displays are used for ergonomic reasons, only the three circular shapes, shown in Figure 12.2 with three-quarter or full circular scales, should be considered.

Inscription The inscription of displays should be clear and easy to survey. Every display must be provided with the unit of measurement (kg, °C, A, V, etc.) of the displayed quantity. This measurement is represented by the respective standardized abbreviation, and in the same colour and character as the scale.

In addition, a short, clear denotation of the displayed quantity should be attached to the casing above or below the display. It should not be abbreviated, as opposed to the dimension. An abbreviation used in the computer program or internally by the producer is also unwise. The denotation should help the inexperienced user to form a picture of the system, therefore it should be as self-explanatory as possible.

Denotations of all elements of the UI must be consistent. They should show analogies between displays and controls.

Guidelines
- *Between two numbered main marks there is either a median or a median and eight small marks (other scales are not recommended).*
- *The height of the main marks H is calculated from the reading distance D and the proportionality factor P, where $90 < P < 150 (H = D/P)$.*
- *The pointer must not cover marks or numbers.*
- *For circular displays, the scale should begin between the 6 and 9 o'clock position.*
- *Displays should be legible even if when the user does not look orthogonally at their surface.*
- *Colour can be used for redundant coding of display sectors.*

12.4   Digital display elements

In a digital display element (Figure 12.7) the variable runs over a finite number of possible values. The most simple display element is the binary display (signal) with two states. They may be represented by two states ('on', 'off') of a bulb, an LED (light emitting diode), or a liquid-crystal display (LCD).

Binary display elements

Only the 'on' state is absolutely free from misinterpretation. The appliance must be constructed in such a way that a defect or a breakdown never leads to an uncontrolled flash-up of a signal. The 'off' state is not so easily interpreted. In principle, it may have three causes:

- the appliance is switched off;
- the display does not work;
- the display really displays 'off'.

Actually, a binary signal is suited only for displaying a single state or event.

There are binary variables whose two states are about equally important and should be equally noticed by the user. In such cases it is better to take a binary display element consisting of two alternating lights (e.g. LEDs). One example of a redundant (carrying minimum information) and therefore

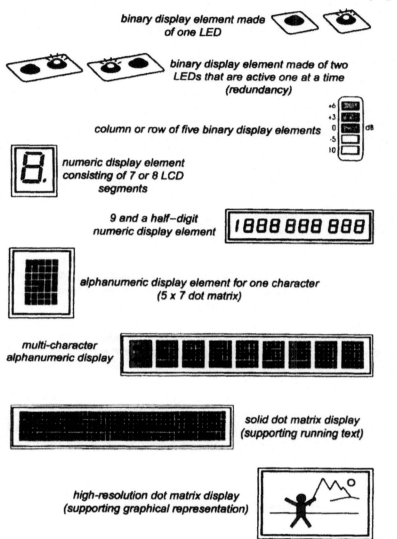

binary display element made of one LED

binary display element made of two LEDs that are active one at a time (redundancy)

column or row of five binary display elements

numeric display element consisting of 7 or 8 LCD segments

9 and a half–digit numeric display element

alphanumeric display element for one character (5 x 7 dot matrix)

multi-character alphanumeric display

solid dot matrix display (supporting running text)

high-resolution dot matrix display (supporting graphical representation)

Figure 12.7 Digital display elements.

unmistakable representation are traffic lights. This display, consisting of three binary signals, is used to represent three active states ('go', 'get ready to stop or leave', 'stop'). If there is a broken light bulb or the system is shut off, this can be recognized and cannot be confused with one of the three possible states in active mode.

Redundancy By means of a display consisting of N binary signals $2N$ different states can be represented. The representation is less error prone and unburdens the user if not all of these states are used for expressing information. This is already shown by the example of the binary switch. Just as in telecommunication engineering, the basic principle is: the more redundancy (i.e. the less information is transmitted) the less error prone is the transmission.

In our case this means: the state of the appliance which the user can only read from the combination of the states of a number of binary signals is easily overlooked by the user. Frequently things are involved which appear obvious to the developer of the appliance ('if *A* is active while *B* is not, there can only be *C*'). But such relationships are often far from obvious to the user because he/she has not yet made a mental model of the appliance, or has made a completely wrong mental model (see Chapter 6).

If, on the other hand, redundant (i.e. superfluous) signals are added to the necessary ones in order to indicate such an event, even an inexperienced and distracted user may not overlook it.

Features of binary display elements If LEDs and bulbs are used, a number of other features must be fixed: magnitude, shape, colour, possibly the blinking frequency, brightness, labelling and arrangement. It is desirable that binary display elements displaying completely different variables differ clearly in magnitude, shape and colour, possibly also in the kind of blinking. A row of uniform LEDs, differing in colour only, may indeed look well ordered and aesthetic, but a rarely active error signal is then easily overlooked among the other, routinely active signals.

Size The size of binary display elements depends on the distance from which they are read by the observer. A standard value is given by doubling or trebling the size of the capital characters of a text which can be read comfortably from the same distance. Signals aimed at attracting the user's attention must be four to eight times the size of uppercase text.

Shape Either symbols or clear basic forms should be used. Equally shaped display elements should represent comparable variables. Analogy

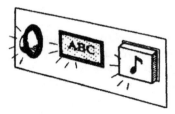

Figure 12.8 Three UI elements with backlight. The left element is definitely a binary display. The right element is easy to recognize as a push-button. But what is the middle element?

Figure 12.9 Consideration of appropriate viewing angle for displays. The viewing angle should be made easy by the form of the display (left and middle) or it should be possible to adapt it by rotating the whole display (right).

to controls may also determine the form of a display. However, it must be obvious at a glance whether an element is a display or a control (Figure 12.8).

The shape of the display together with the shape of the surrounding casing should allow a sight angular range of nearly 180 degrees, both vertically and horizontally (Figure 12.9). Glass coverings and deepenings can greatly reduce this range. Spherical segments projecting from the casing, as well as revolving and swivelling parts, are clearly visible. With stacked devices it may happen that the user standing in front of a stack looks at the lowest one, or sitting looks at the highest one. With flat table appliances, on the other hand, the visual direction may have a big horizontal variation. That is why the visibility of a display from a wide range of directions is so important. For the choice of colours, the guidelines in Table 12.1 may be used.

Blinking frequency The blinking frequency should be not less than 3 Hz (cycles per second, c.p.s.) and not more than 10 Hz, with an optimum of 4 Hz (Burandt 1986). Frequencies of 5 to 10 Hz are perceived as excited and alarming. Blinking should be restricted to signals which are active only rarely and for a brief period, and which require an immediate reaction of the user.

Brightness The brightness of displays is an important ergonomic factor. For the choice of brightness, the colour of the surrounding casing and even more importantly the brightness of the environment is decisive. Since eyes adapt to the latter, the same LED may be nearly invisible by day but much too bright by night. A control knob for adjusting the brightness is a necessity for appliances used in different light conditions. Automatic adaptation to the surroundings with the aid of a light-sensitive device is a better solution.

Labelling (inscription) The labelling (see also the subsection 'Inscription' on p. 227) should be clear. It should be designed together with the general concept for the arrangement of all controls and display elements of the UI. Inscriptions (labels) of controls generally should be positioned above the control so that the hands of the user will not obscure them during operation. For display elements the inscriptions may be below. In any case, all inscriptions should be in the same orientation (i.e. either above or below the display elements). An inscription must never be equally distant from two display elements to avoid ambiguity.

Arrangement   The arrangement of the elements should not be monotonous but characteristic and easy to remember; it should contribute to giving the user a picture of the system. Related elements should be arranged in groups or on a common coloured area or within a borderline. Symmetry is, in general, unhelpful to the human memory. It renders recognition of elements more difficult and hence is to be avoided.

Digital display elements for more than 1 bit of information

Digital display elements for representation of more than 1 bit of information consist of a row or a matrix of elements – bulbs, LEDs, LCDs, etc.

Row of display elements   A row or column of 1-bit elements (Figure 12.10) is the most simple imitation of an analogue display element. Even if it may be constructed from analogue electronic devices, the kind of representation is discrete, or digital, to the user. The row must consist of four or five elements at least. In this way the value of a continuous variable is approximately represented by a column. Only changes in the variable with a signal amplitude bigger than some threshold value are mapped.

The elements are not aimed at representing an arbitrary pattern but rather a column, beginning at one end of the row and ending at some definite element. Hence the information content of a row of N elements is not N bit but only the dual logarithm of $N + 1$ bit, $ld(N + 1)$ bit. In other words, only the uppermost element of a column represents information while the elements below are active 'automatically'.

Compared with an analogue display element the row has advantages and

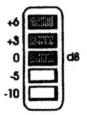

Figure 12.10 Column of 1-bit display elements (LEDs).

disadvantages. The disadvantage is limited resolution. The advantage is the much greater maximal speed of change. Hence the name 'peak value display' because signal peaks are displayed in real time (e.g. in audio amplifiers or tape recorders). An analogue pointer instrument, on the other hand, has an integrating behaviour.

However, with a row of elements it is possible to display an approximation to the intensity envelope simultaneously with the momentary value. This is achieved by the uppermost element staying active after a peak value, even if the column has already dropped and, only after some time (2–4 seconds for audio signals), returning to the column. With several rows a dynamical bar chart may be formed. Even if a single row is not well suited for precise reading, a bar chart enables the reader quickly to picture the relative magnitudes of several variables for himself (e.g. the intensity distribution of an audio signal over different frequency ranges).

Numeric display elements

A numeric display element, by mainly using seven segments (binary display elements), forms an angular image of the numeral '8'. All other numerals are represented by activating only some of these seven segments. Originally the angular numerals were considered strange and futuristic, but now users have accepted these widespread display elements. The shape of the segments (see Figure 12.9, right), the slight inclination to the right and the natural way of creating the numerals have become standards.

For bigger numbers, rows of these display elements are used (Figure 12.11). As for cylinder counters, a marked distance between groups of three numerals is desirable. Unfortunately, prefabricated numeric elements are generally used which do not fulfil this ergonomic requirement. To the left of the biggest non-vanishing digit no zeros should be displayed but rather empty places.

Thirteen out of the twenty-six letters of the alphabet can be expressed readably by numerical display elements (A C E F H I J L O P S U Y). This may suffice for writing some hints or menu items; however, legibility is not very good. Whenever text must be displayed frequently a dot matrix should be preferred.

With a 14-segment numeric display element ('starburst') it is possible to represent all roman characters (but not special characters [e.g. é, ä]), however, legibility is limited.

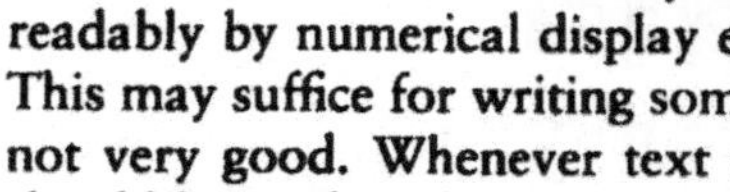

Figure 12.11 Representation of figures on a 7-segment numeric display.

Alphanumeric display elements

An alphanumeric display element for one character consists of a matrix with at least 5×7 points. Usually alphanumeric displays consist of at least 16 or 20 matrices.

A 5×7 matrix allows all letters, numerals and special symbols to be displayed in a legible manner. Figure 12.12 presents a proposal for a symbol set of a 5×7 matrix. Design and selection of symbols for low-resolution displays should be according to the following guidelines:

- the symbols should be legible from some distance;
- any two symbols should be clearly distinct;
- the symbols should be aesthetic and neither right-heavy nor left-heavy.

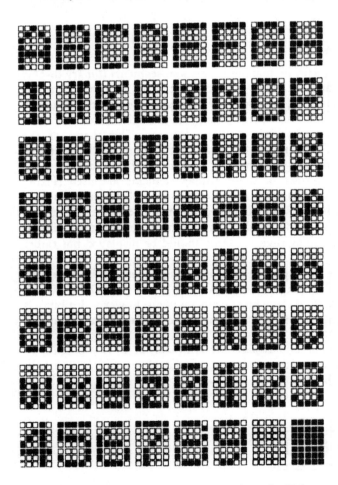

Figure 12.12 Alphanumeric character set for a 5×7 dot-matrix display.

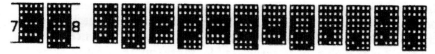

Figure 12.13 Characters of a 5 × 8 dot-matrix display that differ from those of a 5 × 7 dot-matrix display.

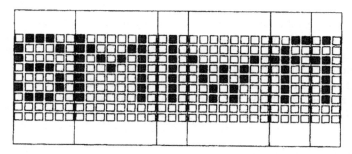

Figure 12.14 Some characters in a proportional font.

A distinctly improved legibility, compared with the 5 × 7 matrix, is achieved by the use of a 5 × 8 matrix. The display is enlarged by one row of points at the bottom. This permits the representation of letters with descenders. This way, not only individual letters but also whole words are read more easily. Further enlargement to 5 × 9 points would be just as good. A 5 × 8 matrix uses nearly the same set of symbols as for a 5 × 7 matrix. Figure 12.13 shows the few symbols with descenders which change by extending the matrix. For all other symbols the lowermost line remains empty.

Coherent matrix If the individual 5 × 7 matrices in an alphanumeric display are not isolated from each other but are designed coherently as an N × M matrix (N > 25, better N > 50, M = 7 to 9) the applications of the display are considerably extended. It then becomes suited for representation of proportional fonts, running text and to some extent graphical symbols (Figure 12.14). While the visible part of the display remains nearly unchanged the circuitry and programming effort increases. The N × M matrix is an advantage to the user: it allows the use of the much more legible proportional font. This is a set of characters in which the area per letter depends on the form of the character. Such a font is more legible and at the same time requires less space. In a proportional font it is also reasonable to apply different sets of characters (e.g. bold and cursive). The minimum height where this is possible amounts to 8 or 9 dots. (Since the width of the characters varies, only the height need be given.) At the same time, 9 dots is the smallest height used in present-day screen fonts.

Unit for font magnitude   The unit 'point', used in typesetting, corresponds to $\frac{1}{72}$ inch or 0.3528 mm. The pixels of liquid-crystal computer screens are frequently arranged in this raster. From a distance of about 30 cm a 3-mm high font is easily legible. Furthermore, a fairly agreeable font requires a matrix with at least 8 or 9 lines. These requirements can be met since a height of 9 points corresponds to 3.175 mm.

Scrolling text   The use of a coherent matrix as a display has many advantages. Most important are better legibility of the proportional font and the possibility of using different fonts. Additional advantages are that it is suitable for scrolling, fading out and graphical symbols.

If the text to be represented is longer than the display, scrolling text may be used. Usual scrolling directions are horizontally from right to left and vertically from below to above. Text scrolling horizontally makes sense only if it does not scroll too quickly for reading. This is the case if the step width is no larger than one-fifth of the breadth of the symbols and the scrolling speed does not exceed half the maximum reading speed of an attentively reading user. It may also be limited by the drag effect resulting from the afterglow of the display pixels. The horizontal motion may take place without interruption.

Text scrolling vertically should be stopped in the middle of the display for a time allowing quick reading of the line. The motion itself can be arbitrarily quick, independent of reading speed and drag effect. Between the text line scrolling off vertically and the next one there should be a space of one-quarter to one-half the height of a character.

Change of text lines   The change of text lines may also take place in a way that the new text moves over the previous one like a curtain or a roller blind. Another possibility consists in filling the lines or columns of the matrix one by one with the new text. To the user this appears like unrolling a new poster over an old one. Finally the change may happen pointwise, with the points selected randomly, which has the appearance of fading out in a movie.

Graphical display elements

A matrix may be used to represent simple graphs, like bar charts or function curves. If we increase the number of lines and columns we finally obtain the best and most versatile display (i.e. a high-resolution graphical dot matrix [Figure 12.8, bottom]).

Like the one-line displays considered so far it is generally implemented as a liquid crystal matrix preferably illuminated by backlight. The possibility of representing sixteen or more greyscales or colours is useful for graphics representation; if just text or numerals are to be represented one may dispense with both because the only usability requirement is strong contrast. When lit, text is best represented black on white; when dark, white on black (see Figures 6.4 and 6.5).

For text, best readability is achieved with a proportional font. This is a font where the width of the characters is unequal (i.e. the character 'i' requires less space than the 'm'). A display for a reading distance of about 30 cm is called a high-resolution display if the dots are no larger than $\frac{1}{72}$ inch or approximately 0.04 mm. This is equal to the measurement '1 point' used in typesetting. Easily legible text on such a display should have a height of at least 12 to 14 point (4.2 to 4.9 mm). With a resolution of 14 dots and more, nearly all well-known fonts can be represented clearly, affording great flexibility in design.

The representation of proportional fonts is becoming a technical standard for displays and display-controlling software.

Maximum character density The maximum density of characters in text representation is chosen depending on the kind of information, the occupation of the user and the arrangement of characters. (In Figure 12.15, the density is the ratio between the number of 'x' characters and the number of 'space' characters visible on the screen.) In general, the density should not exceed 30 per cent and certainly never 50 per cent (Gilmore 1989) (Figure 12.15). If the display has to show not just text and numerals but also virtual displays and controls, it may be very difficult to keep the character density below the recommended value. This problem may be solved in different ways: by good grouping, by the use of windows, by selective representation of the information and by the use of symbols instead of text.

Grouping Grouping represented elements strongly influences how the information is perceived. In Section A of Chapter 6 we saw how much grouping affects the appearance of on-screen representation. Gestalt psychology (Figure 12.16) has shown the aspects to be considered in grouping graphics and alphanumeric information:

> The gestalt psychological principles state that elements appear related and are perceived as a unity or as a figure if at least one of the following relations exists between them: Proximity, good continuation, similarity, and closeness. Elements whose main axis is vertical are perceived most easily as a figure, those with a horizontal main axis rather good, and those with an inclined one rather bad.
>
> (Wandmacher 1993; Lauter 1987)

In grouping the following principles must be observed: simplicity, regularity, symmetry and equilibrium. The application of these principles reduces search, reaction and decision times by up to one-third. The better the principles are observed, the more impressive the figures appear. Since it

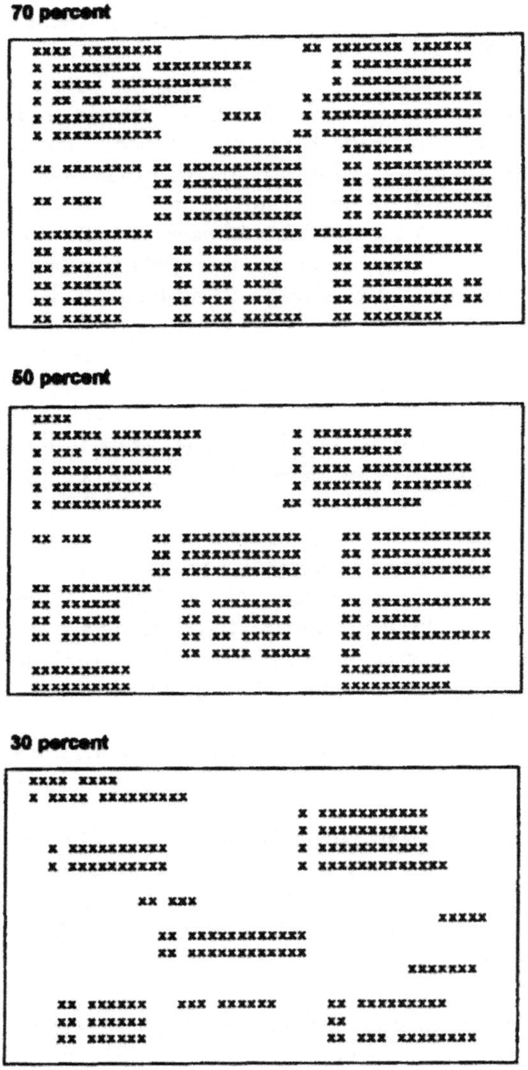

70 percent

```
XXXX XXXXXXXX              XX XXXXXXX XXXXXX
X XXXXXXXXX XXXXXXXXXX        X XXXXXXXXXXXX
X XXXXX XXXXXXXXXXXXX          X XXXXXXXXXX
X XX XXXXXXXXXXXXX            X XXXXXXXXXXXXXXXX
X XXXXXXXXXXX         XXXX    X XXXXXXXXXXXXXXXX
X XXXXXXXXXXX                 XX XXXXXXXXXXXXXXXX
                    XXXXXXXXX    XXXXXXX
XX XXXXXXXX XX XXXXXXXXXXXX     XX XXXXXXXXXXXX
            XX XXXXXXXXXXXX     XX XXXXXXXXXXXX
XX XXXX     XX XXXXXXXXXXXX     XX XXXXXXXXXXXX
            XX XXXXXXXXXXXX     XX XXXXXXXXXXXX
XXXXXXXXXXXX       XXXXXXXXX XXXXXXX
XX XXXXXX     XX XXXXXXXX     XX XXXXXXXXXXX
XX XXXXXX     XX XXX XXXX     XX XXXXXX
XX XXXXXX     XX XXX XXXX     XX XXXXXXXXX XX
XX XXXXXX     XX XXX XXXX     XX XXXXXXXXX XX
XX XXXXXX     XX XXX XXXXXX   XX XXXXXXXXX
```

50 percent

```
XXXX
X XXXXX XXXXXXXXX        X XXXXXXXXXX
X XXX XXXXXXXXX          X XXXXXXXXX
X XXXXXXXXXXXX           X XXXX XXXXXXXXXX
X XXXXXXXXXX            X XXXXXXX XXXXXXX
X XXXXXXXXXXX          XX XXXXXXXXXXX

XX XXX        XX XXXXXXXXXXXX     XX XXXXXXXXXXXX
              XX XXXXXXXXXXXX     XX XXXXXXXXXXX
              XX XXXXXXXXXXXX     XX XXXXXXXXXXXX
XX XXXXXXXXX
XX XXXXXX        XX XXXXXXXX     XX XXXXXXXXXXX
XX XXXXXX        XX XX XXXXX     XX XXXXX
XX XXXXXX        XX XX XXXXX     XX XXXXXXXXXXX
                XX XXXX XXXXX    XX
XXXXXXXXXXX                     XXXXXXXXXXX
XXXXXXXXXXX                     XXXXXXXXXXX
```

30 percent

```
XXXX XXXX
X XXXX XXXXXXXXX
                        X XXXXXXXXXX
                        X XXXXXXXXXX
X XXXXXXXXXX            X XXXXXXXXXX
X XXXXXXXXXX            X XXXXXXXXXXXX

        XX XXX
                                XXXXX
        XX XXXXXXXXXXXX
        XX XXXXXXXXXXXX
                                XXXXXX

XX XXXXXX     XXX XXXXX     XX XXXXXXXX
XX XXXXXX                   XX
XX XXXXXX                   XX XXX XXXXXXXX
```

Figure 12.15 Examples of character density at 70, 50 and 30 per cent (Gilmore 1989).

is difficult to formulate such shape-related aspects as guidelines, a graphic designer should be consulted in each case, or the draft itself should be presented to such a person.

The desktop metaphor Most of the ways of interacting with the desktop were shown to a wide audience for the first time by Doug Engelbart in

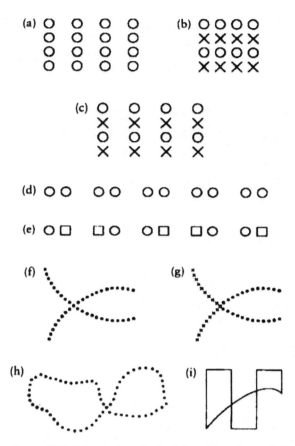

Figure 12.16 Examples of principles of gestalt psychology: (a) and (d) proximity; (b) similarity; (c) and (e) proximity is dominant over similarity; (f) good continuation; (g) good continuation dominates similarity; (h) closeness (i) good continuation dominates closeness (Wandmacher 1993).

1969. They were commercially applied for the first time in the Xerox Star (1983) and Apple Lisa (1983) operating systems and finally gained worldwide acceptance in the Apple Macintosh (1985) (Figure 12.17) and Microsoft Windows (1988) operating systems. The central element of a virtual desktop is a two-dimensional background area (the desktop itself) on which the user may interact with objects by means of a pointer operated by input devices. Other prominent elements, icons, menus and pointing which led to the expression 'WIMP' interfaces.

Elements of windows Besides the proper display area a window generally features a title bar which is always above, carries the window title and affords dragging and dropping of the window by means of a pointer.

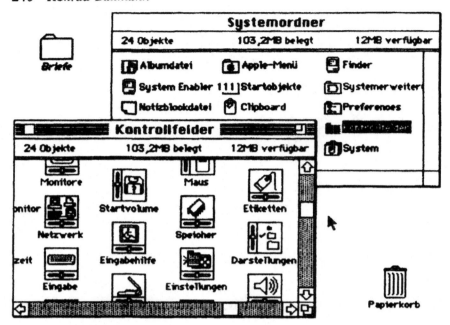

Figure 12.17 Desktop and windows of the Apple Macintosh operating system
(© Apple Computer, Inc. Apple, the Apple logo, and Macintosh are
registered trademarks of Apple Computer, Inc.)

Furthermore, the window may have one or two scrollbars (see Section
7.27 on virtual controls) located at the right and bottom edge that
afford moving the document area displayed by the window, a virtual
handle for changing the window size and virtual keys for minimizing,
maximizing (or returning to the previous size) and closing of the window.

Symbols and icons   Icons or pictograms are representational, whereas
symbols are abstract. All these graphical elements contribute to reducing
the complexity of a screen interface. They should be employed whenever
an object appears repeatedly. The user recognizes a symbol or an icon
much more quickly than he/she identifies the corresponding expression
(item) in written form. Symbols are suited for the interaction method of
direct manipulation (e.g. for dragging and dropping with the aid of the
pointer). Symbols may be considered as virtual-display elements or as
controls. In this case their information content amounts to 1 bit indicating
that something is at a certain position or in a certain logical state.
Nevertheless it is generally preferable to use a symbol or icon in conjunction with text so that novice users can learn the association.
  The totality of symbols used in a UI should derive from a single graphical

concept. It is not always easy to find the right balance between an abstract and a detailed graphic representation. In general, the form of a consistent set of symbols should be upright, rectangular and of equal size. The graphic design should be simple and based mainly on black and white or greyscales. Colours should be reserved for the most important symbols.

Virtual-display elements The prominent position of the graphical dot matrix is not only a result of its greater information content but even more of its suitability for the representation of virtual displays and controls. By using virtual elements the UI of appliances with many features can be made easy to deal with.

The different kinds of virtual controls and their properties are described in detail in Chapter 8. Every real display element can be represented virtually; moreover, the possibilities for graphical representation of information are nearly unlimited.

Freedom of creativity is much greater for a virtual-display element than for any other: it can emphasize some aspect of the information displayed; a number of variables can become interrelated; a variable can be represented at the same time in different ways to meet different users' requirements; the information content of the variable can be filtered (e.g. limited) in an appropriate way.

Icons Symbols and icons are the simplest virtual-display elements. Every icon may be a display element and a command at the same time. This means that, while primarily indicating the presence of an object or a certain state of the interface, it may also allow certain operations by the user.

An icon with a single way of representation shows its information content simply by its presence on (or absence from) the virtual desktop. The next higher degree of sophistication is an icon with two ways of representation (e.g. Apple's wastepaper basket is permanently visible on the desktop in either empty or full state).

Virtual analogue-display elements Virtual analogue-display elements may be designed following the same rules described for real analogue-display elements. Again, a circular shape with mobile pointer is to be preferred; see Section 12.3 for details about the shape of scale and pointer. Take care, however, that low resolution with a dot-matrix display may distort the shape of the scale or pointer and reduce its readability.

Virtual digital-display elements Virtual digital-display elements are not very conspicuous on a big screen; so, they should be provided with a conspicuous border, a shadow or a bright-coloured surface. It makes sense for the user to be able to adjust the size of the characters or the whole element.

Figure 12.18 Icons of the Apple Macintosh UI (© Apple).

Icons (Figure 12.18) and digital displays are the only virtual displays where a shadow or a three-dimensional effect can be applied. All other kinds of virtual displays are so complex in themselves that utmost simplicity and clarity of the design are advisable.

Pie chart   The pie chart (Figure 12.19) is familiar from business reports. It is used whenever parts of a whole are to be shown. In a pie chart it is readily recognizable whether the value of a variable is larger or smaller than one-half or one-quarter of the totality. However, a comparison of different parts is difficult if there are more than six parts or if the parts do not differ much (Wandmacher 1993).

The pie chart should not be displayed in a three-dimensional way, but only in a two-dimensional way. The segments should differ clearly in colour or pattern. In addition, they should be provided with thin black borders. Their brightness should not contrast too much. A black segment or a white one would be too conspicuous among hatched or patterned segments. The contrasts should not be too small either, considering the possibility of colour-blindness. It is not advisable to use the same hatching in different directions for all segments – from a distance, they will all look equally grey.

The pie chart has no scales. It is used when precise reading is unnecessary or when the values are indicated additionally (e.g. in numerical form).

Spider-web chart   Although the spider-web chart is little known it is a very expedient way of display. It is suitable for the simultaneous visual representation of several (from three to about twelve) variables. The values of the variables are represented by the length of rays originating from the centre of the chart. In addition, the envelope of all rays and the enclosed area are made visible. The scales can be linear, quadratic or logarithmic. The units of the scales should be adjustable in such a way that normally all rays are of the same length and thus form a regular polygon. Deviations of some variables from the norm lead to characteristic polygon shapes and are easily recognized.

The polygon should clearly contrast against the coordinate circles. The area included by the polygon must not hide the circles, however. The names

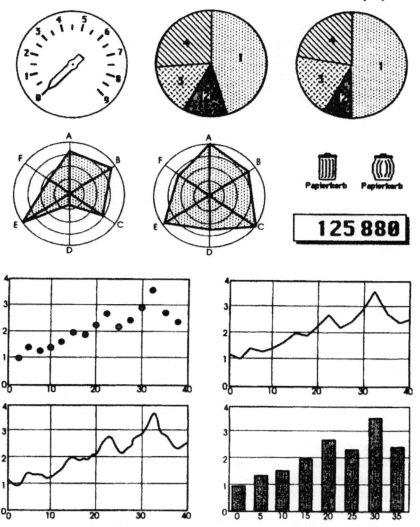

Figure 12.19 Virtual-display elements: virtual analogue display, pie chart (twice), spider-web chart (twice), icon (twice), virtual digital display, dot chart, line chart, curve chart, column chart (wastebasket icon is the copyright of Apple Computer Inc.).

of the variables indicated outside the circles must never be hidden or rendered illegible by the polygon.

Cartesian coordinates The following charts are based on Cartesian coordinate systems. For easy reading, a coordinate grid is recommended. The distance between two lines should be five times the required reading accuracy, because one can easily divide a distance visually into five equal parts. In the design of the chart the scale marks should not be chosen narrower than necessary for a rough estimate by the user.

The border of the chart should be drawn twice as thick as the coordinate lines, function curves and borders of bars should be at least three times as thick. The labels of the coordinate lines should be at the left rim and at the lower rim of the window. The coordinate lines located at the edge of the chart should be labelled, too. The name and dimension of the variable should be indicated between the last and the last-but-one labelled graduation mark.

Dot chart   The dot chart is the simplest representation of Cartesian co-ordinates. The diameter of the dots should be equal to the intended reading accuracy, the distance between the coordinate lines should be five times as large.

Dot charts are also suited for displaying pairs of values which do not describe a function of the abscissa ('x' axis) but a 'dot cloud'. Superposition of several sequences of measurements by use of differently coloured or shaped dots is possible. No more than four colours or shapes are advisable. A problem is posed by coinciding dots.

Line chart and curve chart   A line chart (Figure 12.20) is obtained by connecting the dots of a dot chart by straight lines. If the density of dots is large enough the line chart becomes a curve chart. These are the two most effective methods of representing quantities in a graphic way (Wandmacher 1993, after Cleveland 1985 and after Schutz 1961).

Reading accuracy is greater than with dot charts, and the features of the function curve are recognizable more quickly.

Several curves may be superimposed on the same chart, but they should clearly differ in colour or otherwise. The curves must be identified by labels in the diagram or by a separate legend. To achieve this, dots and lines may be used together.

Compared with bar and column charts, the advantage of the dot, line and curve charts consists in the fact that the ordinate ('y' axis) does not need to start with zero. Thus any section in the range of values of both axes may be selected in order to show a part of the curve in detail.

Column chart   A column chart consists of vertical columns with equal width and equal distance whose height is proportional to the value of some variable. Although the same information is expressed in a dot chart the column chart gives a different impression. The attention of the observer concentrates less on the shape of the function but rather on the integral (i.e. the area under the function curve). The same difference exists between a curve chart and a profile chart in which the area under the curve is emphasized by colouring.

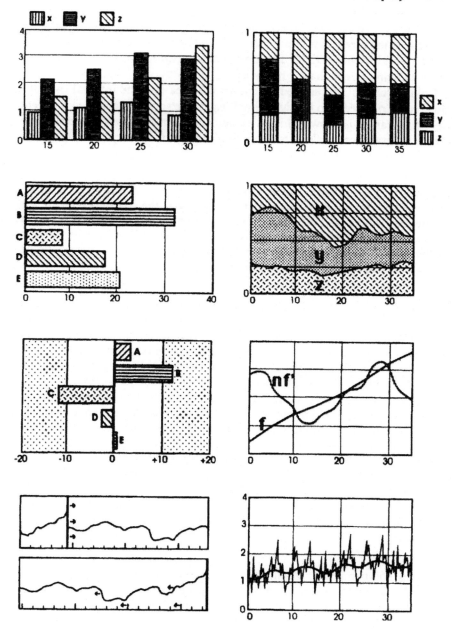

Figure 12.20 Virtual-display elements: grouped column chart, stacked column chart, bar chart, stacked profile chart, deviation bar chart, line chart with amplified differential curve, two refreshing methods for curve chart, filtering of a curve chart.

An exceptionally high column is much more conspicuous than a single dot which might even be overlooked. In a column chart comparison of values is sometimes impaired as a consequence of the systematic error in estimating areas (Wandmacher 1993). The observer sometimes perceives the highest value as being larger than it really is.

Column charts are justified whenever the area (i.e. the integral over the given function) needs to be highlighted (e.g. if the sound intensity in different frequency ranges is shown). In this example, columns are best suited because they are a graphical analogy to averaging over the frequency band. Columns should never be displayed in a three-dimensional way.

Combination of column charts with dot charts   In a frequency display, columns and dots may be combined in a way that the columns indicate the momentary values and the dots above the columns indicate the peak values over a certain period of time. Quickly varying columns give the impression of the area and emphasize the momentary values while slowly changing dots show the envelope and can be read individually.

Details of the column chart   For column charts the (y ordinate axis) must always start at zero; whenever this is not the case the whole kind of representation becomes senseless, even misleading, because the values are no longer proportional to the area of the columns. (Unfortunately this error is often made in business graphics, or, even deliberately, in advertisements comparing two products.).

Each column has its own label below the abscissa. The distance between the columns should be not less than one-third and no more than four-thirds of their width. They should not lie flush at the edge of the chart, so that the horizontal coordinate lines remain visible between the columns. The columns should be filled with coloured or patterned areas which differ clearly in brightness from the background. The areas may be opaque such that the coordinate lines are partially covered. If the columns represent the values of the same variable at different times all areas should be equally patterned; however, if the columns represent different variables the colours and patterns may be different.

Grouped column chart   Analogous to charts containing several curves, grouped column charts may be used. A group of two to four columns with different colour or pattern belongs to each value on the abscissa. The columns of the same group touch each other. The legend for identification of the columns should be outside, preferably above the chart. For the choice of colours or patterns the same guidelines hold as for the pie chart. Grouped column charts draw the attention of the viewer to the higher columns and to the comparison of values within a group. In contrast, charts with several curves show better the individual shape of the curves.

Stacked column chart Another variation is the stacked column chart. The values of each group are not presented side by side but one upon another. This representation should be chosen if the sum of the values of each group (each stack) is constant in reality or if it is not relevant in this context. In this chart the legend should be positioned outside the right border of the chart.

Bar chart A column chart presented horizontally is called a bar chart. Here the roles of abscissa and ordinate are exchanged. This chart is used mainly for representing the values of different variables. In order to avoid misreading, each bar should be differently patterned. The horizontal axis of bar charts must start with the value zero at the left end. If this is not possible a horizontal dot chart should be chosen instead.

Profile chart If in a line or curve chart the area below the curve is coloured or patterned, it is called a profile chart. It resembles a cross-section through a tract of land. The profile chart has much in common with a column chart. The viewer directs his/her attention to the area, and he/she tends to overestimate the curve peaks. This chart should be applied only if the integral over the curve is a meaningful quantity. Like the column chart, the profile chart is suited for stacking. The labels of the areas may then be integrated in the chart.

Care must be taken that the viewer does not take the stacked profile chart to be a three-dimensional chart. Therefore the areas should not become gradually darker from top to bottom or vice versa. Otherwise the impression of a landscape with near and remote hills may arise which may lead to misinterpretation of the values.

In the example shown in Figure 12.20 the sum of the variables x, y, z always amounts to 1. But, of course, it also makes sense to represent variables whose sum is not constant. However, take care that the co-ordinate lines remain visible under the patterned areas.

Deviation bar chart A bar chart may be made such that deviation of the variables from certain normal values is highlighted. In our example (Figure 12.20) the chart was transformed in such a way that the deviation of the variables from the value 20, situated in the middle, becomes visible. The deviation bar chart is presented immediately below. Designations and patterns of the bars remained the same, but the vertical axis was shifted to the centre. In addition, the deviation range of more than 10 units is supplemented by a slightly patterned area. This example shows that pre-processed data is more easily grasped by the user than in the original chart.

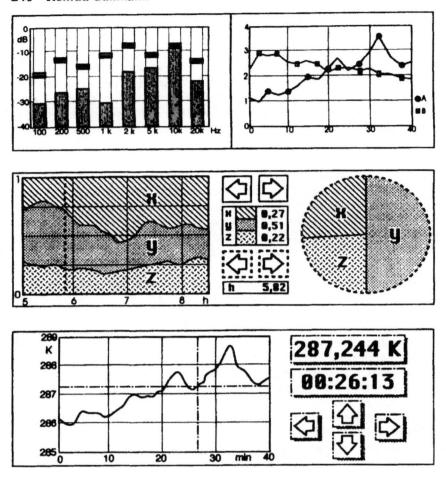

Figure 12.21 Application-screen examples for measurement devices using different types of virtual display.

Refreshing of virtual displays   Refreshing of virtual-display elements may take place in different ways. If the technical conditions permit, refreshing can be done in real time (i.e. continuously and in invisibly quick and small steps). However, it may be reasonable to keep the presentation of data static and to refresh only in greater time intervals. Curves, profiles or dots may be drawn slowly from left to right above the old data. If a variable is presented as a function of time, the whole display together with the abscissa may be shifted to the left, and at the right edge be supplemented by the new data. In this case the displayed domain embraces a constant span of time reaching into the past.

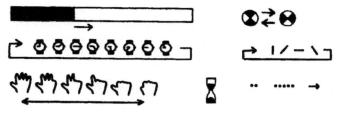

Figure 12.22 Examples of action-feedback signs.

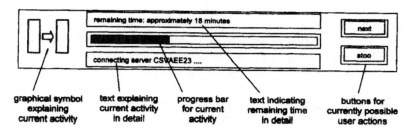

Figure 12.23 Elements of the feedback window of the Lotus Notes communication software.

Action-feedback signs The function of action-feedback signs (Figure 12.22) is to indicate the activity of the appliance to the user in case of a prolonged technical process like measurement, calculation, data transmission or waiting for an event. In an earlier section on feedback (p. 99 above) time we stated that action signs or information on the current system state must be used in all processes lasting longer than 3 seconds.

In Figure 12.22 action-feedback signs are ranked according to quality. The first position is held by the progress bar which is the only action sign indicating the ratio between time passed and total duration of the process. The same information can be displayed numerically ('*34% of document printed*'); however, this is not as well suited for this purpose as the analogue progress bar (Figure 12.23).

All the following action signs are displayed at the position of the pointer. In this way it is guaranteed that they are immediately noticed by the user, in spite of their small size. Thus the pointer, while being a virtual control, has a second function as a display element. Actually, the wristwatch and the counting hand are small animations; their amusing appearance possibly makes the waiting time appear shorter.

The rotating ball is the simplest moving graphical action sign. The rotating line and the dot row are suited for displays without graphical function and are used in text-based UIs.

In some computer programs the hourglass (a static symbol) is still applied instead of an action sign. It does not show a process but rather the beginning of a process as a single event. Static symbols are inappropriate as action signs for processes lasting longer than 3 seconds because they do not even show whether the device is still working. Instead they indicate that the UI is temporarily switched off by the device.

Screen savers   A special kind of action sign are the so-called screen savers indicating that the appliance is switched on but idle because the human–machine dialogue has been interrupted by the user for a certain time (usually a few minutes) and may be continued at any time. They were created for saving cathode-ray-tube (CRT) displays against burn-in of static patterns. However, they may survive in the era of liquid-crystal displays (LCDs) as a welcome possibility to introduce some pleasure, fun and personalization lacking in most UIs. Screen savers can be much more elaborate than simple action signs (e.g. in the form of realistic animations).

Visual clarity and simplicity   It is the task of the developer of appliances to reasonably use the diversity of virtual-display elements, but without giving the impression of treating the matter frivolously. In the visual design of parts of the UI it is especially important to maximize clarity, simplicity and self-restraint. The use of three-dimensional effects is not advisable. Symbols, icons and windows may (if necessary) be rendered prominent by the use of shadows simulating illumination from the upper left corner of the display. Three-dimensional pie, column or profile diagrams falsify the representation since a comparison of lengths and areas is no longer possible. For all designs, especially when going beyond the concepts presented in this chapter, experts should be involved (human-factors engineers, psychologists, product designers, graphic designers). For more details on screen design see Chapter 6.

12.5   Situation-analogue displays

In situation-analogue-display elements either the scale or the information or both are represented as similar to or taken from real objects (Figure 12.24). The range of application of situation-analogue displays is where the information presented is related to the environment or is a part of it, as in vehicle and aircraft steering and in systems serving the representation of virtual reality.

Virtual reality   In virtual reality (VR) systems visual, and maybe acoustic and tactile, information on the surrounding environment of a person or a vehicle is displayed in a way as close as possible to reality. The application of this style of human–machine interaction ranges from car driving and flight simulators to full-immersive VR systems. Both the displayed information (the position of the car, plane or body) and the 'scale' (the

Figure 12.24 Situation-analogue display.

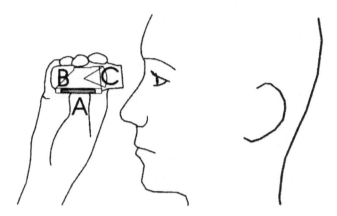

Figure 12.25 Virtual display. By means of optical lenses this display gives the user the impression of viewing a much bigger display at a much greater distance. A display, B mirror, C optical lens.

surrounding environment) are displayed in a virtual way (i.e. on screen). For more information see Chapter 9.

Head-up display A 'head-up display' is a device (Figure 12.25) which mirrors information via the windshield of a car, of an aeroplane, of a helmet or a pair of glasses into the field of vision of a person (e.g. a driver or a pilot). It is a situation-analogue display whenever the visible environment is used as the scale for the displayed information (e.g. a situation-analogue representation of the braking distance of a car might display it as a perspective bar virtually lying on the real visible road in front of the car). A situation-analogue warning of reductions or obstacles might be done by virtually highlighting them in bad visual conditions. Route information generated by a geographic information system may also be displayed in situation-analogue form.

Augmented Reality   Recent developments in head-mounted displays and wearable computers (see Chapter 11) have resulted in the possibility of fusing computer generated images with objects in the real world. This merging is known as augmented reality (AR) because the computer-generated information is used to augment or supplement the real world. In a standard application, a visitor to a city might wear an AR kit. As the visitor approaches a building, the AR system presents information pertaining to that building on the head-mounted display. The visitor sees the presented information superimposed on top of the building.

Object display   In a situation-analogue display not only the scale, but also the information, may be part of reality. In this way, physical phenomena or objects outside the proper electronic device become part of the UI; for example, the light conditions in a room, water surface waves projected on a wall, traffic noise, an extractor fan or other motor-driven physical objects may be used as a display (Ishii and Ullmer 1997; Lee *et al.* 2000). In a similar way, physical objects like pots with a cover and bottles may be used as commands. So, the UI breaks the limits between the machine and the real world.

The advantage of objects as part of a UI is that the possibly disturbing presence of control lights, displays and keyboards in a private or office environment is unnecessary. Information provided by electronic devices is made available in a less obtrusive way. The user is able to watch the status of an electronic device in the same way as he/she watches over a pet – without interruption of a primary task and perhaps even in an unconscious way.

Anthropomorphic display   Computerized toys were among the first technically advanced exponents of anthropomorphic UIs. First elaborated and commercially successful examples were: the Microsoft Barney, a toy dinosaur capable of talking and understanding a certain range of words, sensing light intensity, movement and pressure; the Sony Aibo, an automated pet dog capable of movement; and MIT's Curlybot. For more details about head-up displays and anthropomorphic UIs see Chapter 9.

13 Auditory displays

Othmar Schimmel

13.1 What is a sound?

The basic function of all our combined senses is to receive and process information about our surrounding environment and provide feedback on the actions we perform. Our sense of hearing is complementary to our other senses. Whereas our vision is limited to approximately 160° in front of us, and touch, smell and taste are limited to a close range, our ears are able to detect minor changes in air pressure 360° around, over large distances, without the need for orientating ourselves to hear. It constantly updates the status of the surrounding physical space, indicating possible threats to our existence, assisting in navigating through spaces, in our hunt for food and of course in communication. While this is no longer a real necessity for survival, we haven't lost this very refined ability for it is not rudimentary at all; we still use it every day in interacting with our environment.

The most information in sound is processed in time, other than vision, where a short glance at an image or even several images simultaneously can give all required information on the content. With the exception of the conclusion whether an acoustical signal is present or not, one has to listen to the time-based syntax of the sound to be able to determine the semantics of it. The sound emitted by a stimulus changes over time as a result of its physical characteristics, being a vibration of air. This is graphically represented as an amplitude-over-time diagram (see Figure 13.1). This characteristic of time relates also to the perception of the sound: as soundwaves travel through air at approximately $340 \, \mathrm{m \, s^{-1}}$, the minute time difference in receiving a sound in each ear gives a strong cue on the localization or movement of a sound source.

These wave patterns in air are transformed by the ear into nerve pulses to be processed by the brain. In our brain we are able to distinguish coherent patterns and derive meaning from them according to our sociocultural background. The sound in itself has no meaning, it is merely the representation as its graphical equivalent.

Figure 13.1 Example of an amplitude-over-time diagram of a piano tone (10 seconds).

The most basic waveform in vibration is the sine waveform. According to the mathematical theory of Jean-Baptiste Fourier all periodic waveforms can be described as a combination of sine waves. Complex waveforms are composed of a fundamental frequency combined with multiples of this frequency, the so-called harmonics or partials (see Figure 13.2).

The physical unit to describe frequencies is the number of cycles per second, and is called hertz (Hz). The higher the number of cycles, the faster the vibration, the higher the frequency. The exact ratio to the fundamental frequency, the amplitude, envelope and phase of each harmonic component determines the compound waveform, and therefore the acoustic and perceptual characteristics of the sound, the timbre. In nature perfect periodic waveforms do not exist, only constantly evolving and changing waveforms. Artificial sounds, mechanical or electronic like a buzz or a hum, are often perfectly periodic and therefore become uninteresting after a certain period of time: they do not provide new information about the subject and are no longer of interest to our active perception.

Humans are able to hear frequencies approximately between 16 Hz and

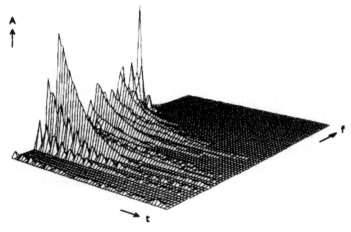

Figure 13.2 Example of a Fourier analysis of the piano tone from Figure 13.1.

20,000 Hz, diminishing at the higher frequencies when growing older. The subjective perception of this physical quantity is described as hearing the pitch of a sound. We are not capable of discriminating the exact frequency of a single sine tone or individual frequency components of a complex waveform, but are only aware of the general pitch and timbre of a waveform, often in comparison to previously heard sounds. This sensation is derived from the recognition of the nerve impulse patterns triggered by the frequency component(s) of the sound in our ears, translated in our brain.

The excitation of an oscillating object, transduced in air, has a certain amplitude. This is represented as the intensity of the sound on a logarithmic scale, and called decibel (dB) sound pressure level (SPL). The threshold of hearing for a 1,000-Hz sine tone is referred to as 0 dB and the level of 125 dB is about the threshold of pain. Sound radiated to all sides freely can be considered as uniformly distributed energy over the surface of an imaginary sphere that will decrease proportionally to the distance from the source. A distant sound is perceived less intense than the same sound closer by, because of the spreading of energy over this expanding imaginary sphere. The subjective perception of this intensity is called loudness, expressed in sones. One sone is defined to equal a 1,000-Hz sine tone at 40 dB SPL and, given the logarithmic nature of intensity, an increase of approximately 10 dB in intensity results in a doubling of loudness.

The perception of loudness is not only dependent on the distance from a sound source, but also on frequency and duration. Human ears are more sensitive to certain frequencies than to others, the most sensitive area ranging globally from 1,000 to 5,000 Hz (see Figure 13.3), where the hearing threshold (the dotted line) is even below the reference level of 0 dB. A low-pitched sound has to have a higher amplitude to be perceived equally loud as a higher pitched one, although the difference decreases with higher intensities. For sounds shorter than 1 second, as most user-interface feedback sounds are, loudness increases with duration. The shorter a sound, the more intense it should be in order to be perceived as loud as a nearly identical sound with a longer duration. Therefore it is very difficult to make absolute judgements on loudness. Relative judgements on loudness are in practice limited to the distinction of just a few different levels.

Natural non-periodic sounds are not adequately described by the mathematical technique of Fourier analysis in terms of frequency and intensity of their content. The dimensions of pitch and loudness seem hardly applicable in the perception of our everyday acoustic environment. For instance, what is the pitch of a fridge door closing or the loudness of inserting a key in a keyhole? These attributes are not important in everyday listening for the perception of the events taking place, but in artificial situations such as a user interface (UI) the physical properties and the structure of a sound become main assets for determining the represented functionality.

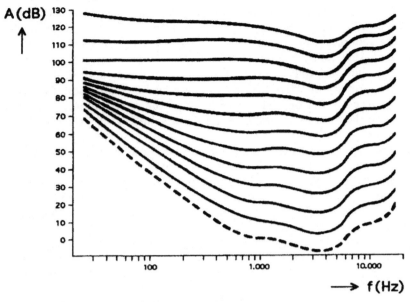

Figure 13.3 An equal-loudness contour diagram.

As a result of its complex physiological nature, briefly touched upon here, the application of sound in UIs is a complicated matter. As technology emerges, the potential for addressing this sense in a cost-effective way is increasing. In the following sections we take a closer look at the deployment of auditory feedback in UIs.

13.2　Why use sound?

As a carrier of information, sound has always played an important role in our lives. Because of its ubiquitous nature sound conveys information in several ways:

- it supports visual stimuli and strengthens the means of information transfer and guides perception, often on a subconscious level (e.g. the sound heard when touching an object gives information about the material and construction of the object, like knocking on a table or wall);
- it provides immediate feedback on performed action, also often perceived on a subconscious level (e.g. the sound of the door closing properly behind you upon entering your house);
- it attracts attention and can summon required action, especially when a visual cue is likely to be missed or trade-offs on visual display have to be made, which displays conditioned behaviour on a more

conscious level (e.g. the ringing of a doorbell or telephone urges you to open the door or pick up the phone).

In this respect the distinction is made between intentional and unintentional sounds. The sound made by two objects when brought into contact is mostly unintentional, it is the by-product of an action performed on those objects. The sound of a doorbell is intentional, it is the result of a specific coordinated action using a dedicated tool. This division is, however, not always very clear. The sound of an engine may not be intentional for it is a by-product of the mechanical process, but the way it actually sounds can be carefully designed to contribute to a qualitative positive experience and build a brand image, like the patented sound of the Harley Davidson motorcycle. In this example a car mechanic would also listen in a different way to the sound of the engine than a regular driver would. The information provided can be pluriform and the perception is highly context dependent. In the case of product design this should be taken into account.

Aside from carrying specific information sound can also deceive perception or create otherwise (un)pleasant effects, partly subconsciously, partly consciously (e.g. putting on an audio CD not only evokes emotional responses according to the nature of the auditory stimulus and set a mood, but can also suppress present environmental sounds and improve concentration on a specific task, like studying). By manipulating room acoustics the impression of a different physical space can be given; a small noisy apartment could sound like a large quiet estate. Noise cancellation, to reduce sounds from the outside world, is considered a serious principle for enhancing perceived environmental quality.

Electronic devices do not have a long history, and familiarity with operating newly developed technologies has suffered from the exponential increase over the last decades of applications and functionality, without actual relations or analogies to the real world and without much commonality in interfacing principles. In general, it is therefore necessary to assist the user in the man–machine interface by properly addressing all the senses in a natural manner to enhance the interaction with the electronic device. The application of artificial sound in human–machine interfaces is a logical element in addressing all senses, not in the least because it is getting technically more feasible and can be deployed very effectively at relatively small costs.

13.3 How to use sound

Once the need to use sound in an interface is established the next step is how to use sound. Not every aspect of the interface might be relevant for acoustic representation. So, it has to be determined what functions would clearly benefit from having a sound and what information an auditory interface is to provide. Sound is adequate in displaying status or background information of ongoing processes, unveiling complex multidimensional data structures and supporting computer interaction, but having explicit auditory feedback on every action could drive the user and the people around mad in a very short time. In applying auditory feedback one objective is to have some kind of acoustic metaphor as a representation of a real-world event. As stated before, the sound in itself doesn't have any meaning, it is the vehicle for information transfer. The more intuitive the link between the representation and the represented event is, the more powerful the information transfer will be (e.g. the crowing of a cock has a strong historical association in our waking-up experience and will do very well as an alarm clock sound, although a harsh repetitive beep will do the same trick).

Unfortunately there are not many interfaces that can rely on these stereotypical communication protocols which lean on a rich tradition in associative acoustic events. New technologies require new ways of communication between machines and human users, and the often very abstract representatives have to be transformed into clear auditory representations. The obvious solution is to make use of functional similarities to other real-world events in the representation, like the sound of an analogue light switch on activating a button on a computer screen, without any actual relation to switching on a light or the function of the button. But a too obvious link to real-world events might obstruct the clarity of the interface in suggesting a similarity in functionality, while being very dissimilar. In order not to lose strength in the translation, it is in these cases preferable to have simple neutral sounds with no clear representative, which can easily be understood in their meaning after a short learning period. Often the information that is given is so generic (a certain criterion is met) that sounds can easily be customized without really confusing the user. This requires some effort on the user's side to acquire the conditioned behaviour, but has the great advantage of being unambiguously clear once learned.

When using different sounds relating to different functions there has to be some consistency in defining these abstract sounds and mapping them to the interface. This will shorten the learning process because of the sounds being highly mnemonic and therefore beneficial to the user's experience with the interface. The consistency in this matter can be maintaining the same pitch with different temporal patterns for comparable feedback messages; a low-pitched sound for all positive feedback and a higher pitched sound for all negative feedback. Other qualities like urgency or importance can then be stressed by the speed or repetition of the temporal sequence. The use of too many different sounds should be avoided for the same reason of maintaining the ability to differentiate between them. In many cases 'less is more'.

Mapping of auditory display for processes differs from sounds for discrete events since, instead of providing feedback of a certain state being reached, the whole trajectory of a process has to be displayed. This can be seen as a succession of multiple discrete status sounds which should fit together as a coherent-sounding feedback. Also here intuitive links can be made, like an increase in an acoustic parameter for an increase in data value, but in most cases the mapping of abstract data to acoustical parameters is not that obvious. If, for instance, frequency was to represent temperature, the perception would be the higher the temperature, the higher the pitch. However, if frequency was to represent the size of an object, then the increase in pitch would intuitively be perceived as a decrease in size, since small objects sound higher pitched than large ones. Should a visually horizontal progress bar be expressed in sound via an increase or decrease in pitch, increase or decrease in volume, or a changing timbre? And at what frequency, volume or timbre should it start and end? Since there is no uniform answer to these questions of mapping acoustical parameters for status review, it is of great importance to study the objective of the auditory display to come up with the best representation, polarity and scaling factor for the specific application.

The running context is very important in the understanding of the information provided. For instance, timing is crucial: the auditory feedback has to follow directly upon performed action or when action is required. Hearing short beeps after having already pushed multiple keys that do not actually make sound themselves gives no indication whether the sounds represent the recognition by the machine of the keys being pressed or that something else is brought to the user's attention. The expectation of a sound to happen should be followed by a system as best as possible. Incongruent information is likely to distract the user too much from performing adequately and will not contribute to the interaction, the only exception being unexpected or alienated sounds from overruling alarm messages that should have the undivided attention of the user.

13.4 What sound to use

The auditory display that can be used in an interface is highly dependent on the chosen hardware of the product. Not only the technique for sound reproduction or generation determines what can sound and how it will sound, but also the physical construction of the product has its limitations on acoustic performance through application of specific materials and construction methods that induce damping or amplification of certain frequencies, the so-called frequency response (see Figure 13.4). This is something to beware of when designing an electronic device that has to produce acoustic output as well, because it could literally ruin the intended auditory display.

Though it is virtually impossible to define beforehand how a product's frequency response will be, there are several methods known to measure and enhance the acoustics of product hardware. Measuring the frequency response over the whole audible spectrum gives a clear indication of which frequencies to use to obtain the required sound level (e.g. a

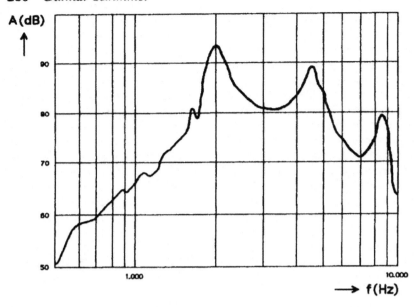

Figure 13.4 Example of a frequency-response curve of a small electronic device.

warning sound or a telephone ringer), and thereby defining the auditory feedback by constraining the sound design to the limitations of the hardware. Other solutions have more impact on the physical design, like the placement of a small speaker at the front of a device instead of at the back or side to result in a higher sound level that can be reached to address the user with the same amount of energy. Simply making holes in the casing can sometimes provide the solution to the problem of having too little energy to sound loud enough. Making a bridge between the front and back surfaces of a product to make it more stiff, or filling cavities with damping material, may get rid of unwanted resonances.

When the auditory display is not clearly defined (e.g. in systems with spoken output or with a large set of various sounds with different spectral content), it is best to get the frequency response as flat as possible (all frequencies sound equally loud) to have the highest qualitative result. Basically it is a method of trial and error to come up with the best way of emphasizing or de-emphasizing the required frequency ranges for a particular application.

There are two methods in practice for implementing auditory display in electronic devices, both with their own advantages and disadvantages for design and application. Playback of pre-recorded sounds, or sample playback, is a very powerful way of incorporating any kind of sound in an interface because in fact any sound that can be recorded can be used to represent required feedback. This enables intuitive links between representations and the representatives by deployment of real-life sounds. Provided that the physical and technical design of the product allows it, most of the audible frequency range can be reached, and there are hardly any limitations to the spectral content of the sound.

This method, however, consumes large amounts of memory for digitally storing the sounds recorded, since a waveform (see Figure 13.1) has to be converted in detail into a large list of numbers. CD-quality audio takes about $88 \, kb \, s^{-1}$ per second (for comparison: this chapter in text equals about 25 kb of electronic data), and downgrading saves space but heavily degrades the acoustic quality in terms of frequency range and signal-to-noise ratio. Current sophisticated compression techniques to save storage space and bandwidth for transmission like MPEG, have reduced the memory space needed to about 10 per cent of CD-quality audio for the same perceived acoustic quality, but real-time decompression in hardware is not yet commonly available in electronic devices because of the high constraints on processing power. Besides the problem of storage, sample playback puts heavy constraints on the rest of the physical design to sound qualitatively reasonable. It is like having an impressive hi-fi system in mind and ending up with a small AM radio.

An easier way of achieving auditory display is through sound synthesis. Calculation of a waveform on the basis of algorithms and input parameters is highly efficient with regard to hardware constraints and memory space needed. It takes a general-purpose microchip to perform the calculations, a small software program containing the algorithm and a table of input parameters to output (a series of) specified frequencies, durations and amplitudes. However, it does have quite some limitations on the sounds and frequency range that can be produced. These are the sounds that we know as the artificial beeps from earlier generations of computers, digital watches, mobile phones, electronic alarm clocks, cash-point machines and so on. Despite the negative association of sounding cheap and the fact that they are abused so often, they really can do the job of representing the generic information of abstract representatives in electronic devices, and prove a challenge to the sound designer to capture the true essentials for information transfer.

Because of the inability to use real-life sounds and the limited potential for timbral manipulation, the design principles, although not clearly defined even if you can speak of them as such, are often related to music composition. Since we are dealing here with electronic devices developed by Western or Western-oriented cultures, the application of music theory is also oriented towards the Western tonal system and tonality. In this way intuitive and traditional values are transcribed into the interface (e.g. like a fast upward-going melody in major key representing positive feedback, and a slower downward-going melody in minor key representing negative feedback).

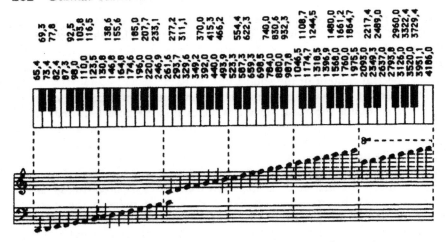

Figure 13.5 Table of notes from the musical scale and their frequencies.

Because of their lack of clarity, compared with real life sounds, these short tunes are open to personal interpretation of musical style and require some learning from the user to become unambiguously clear. Recommended practice is simply to design a sound or tune according your own concepts, and test its perception with several different people on the required characteristics, especially since it is a costly affair to infringe on preconceived and copyrighted melodies. Unless you are willing to pay extra for having a well-known popular tune in your product, this do-it-yourself method is not only more cost effective, but also easier to adjust and customize to fit the requirements after user testing.

Basically all it takes is to find out what frequencies which can be output by the device are usable (e.g. within a musical scale [see Figure 13.5]), and use them to generate several different successions of them. These can be as long as it takes to get the message across, though the shorter the better. Sometimes one note is sufficient, in most cases it takes several to be clearly distinguishable within a set of auditory feedback signals, and often an alarm sound has to be continuously present until being shut off. Using intuitive principles like upward melodies for positive and downward melodies for negative feedback messages (or as an analogy to 'switch on' and 'switch off'), loud and shrill timbres for alarm sounds and softer and comforting timbres for feedback and attention signals, fast rhythms for urgency, the design can simply be tailored to the application.

Prototyping of all kinds of auditory display can be done on a computer, using sound recording and editing software with tone-generation capabilities and a playback system, like the internal speaker if this resembles the intended application of a small electronic device, or a soundcard combined with external amplification for hi-fi quality. A hardware synthesizer or sound sampler with keyboard can also be used to fulfil the purpose of prototyping, but often lack the flexibility in manipulating and storing the sound for further processing.

In some cases the way the prototype is made resembles the structure in which it will be implemented, as with lists of frequencies, durations and amplitudes. These lists can then often be copied and pasted directly into the final software. In other cases some transcription from the prototype format to the implementation format has to be made. Since most electronic products have their own development trajectories, it is recommended to try beforehand to get the prototyping structure and the final implementation structure to resemble each other as much as possible, to have an efficient system for designing the auditory display without any loss of information in the transcription. Best is, of course, to prototype and fine-tune on the final device itself, but this will require a proven good design and more valuable time in the development phase.

13.5 Examples

To illustrate the theory above the example of a telephone ringer is given, one of the best known ringers today: the Nokia 'Grande Valse'. When rebuilt in MIDI sequencing software on a computer, which is used for recording the output and controlling the input of MIDI synthesizers, this ringer is best graphically represented best using the piano-roll notation as in Figure 13.6. For generating the ringer the designer probably used a similar software package to prototype, edit and define the ringer structure. This can be done with any existing or yet-to-be-developed ringer or melody, since it gives drag and drop as well as numerical control over all the musical parameters of the notes, such as pitch and duration. Because it is more about defining the structure, rather than the sound itself, the sound output of the MIDI synthesizer only has to globally

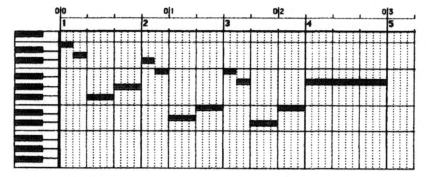

Figure 13.6 Graphical representation of the Nokia 'Grande Valse' ringer.

Table 13.1 Frequency-duration table for the Nokia 'Grande Valse' ringer.

Note	Frequency H_2	Duration ms
1	1,318.5	125
2	1,174.7	125
3	740.0	250
4	830.6	250
5	1,108.7	125
6	987.8	125
7	587.3	250
8	659.3	250
9	987.8	125
10	880.0	125
11	554.4	250
12	659.3	250
13	880.0	750
14	0.0	3,000

match the Pulse Width Modulation (PWM) blockwave of the telephone hardware.

The vertical axis in Figure 13.6 is the piano keyboard, representing the pitches of the ringer melody. The timeline on the horizontal axis shows that the total duration of the sound is 3 seconds (not shown here is that the interval before repetition is of the same length). From this timeline the durations per note can be derived and, combined with the transcription from pitch to frequency for each note, the result is Table 13.1.

If these ideal frequencies cannot be output by the chipset, they have to be rounded off to the nearest value, as is truly the case in this example. Also if the time grid cannot be as accurate as required, because of the duty cycle of the processor chip, concessions on the timing have to be made as well. This may result in a slightly different table, but the general structure for implementation will be clear. These tables often have to be reworked, as the frequencies available and the time units are described in a hexadecimal number (e.g. the lowest frequency possible could be 01 (00 would be silence) and the highest FF, and time a multiple of 10 ms). Assuming this way of encoding for all the notes in Figure 13.5 the final ringer definition could look as shown in Table 13.2.

Note that 125 ms is alternatingly rounded off to 13 and 12 (0D and 0C), and in order to have a silence of 3,000 ms two lines of code are required to fit 255 (FF) and 55 (37) to have a total sum of 300. Applying such a very efficient technique for ringer implementation, in which the sound is calculated by the chip on the basis of a simple algorithm driven by lines of code, a 6-s sound structure is reduced to 60 bytes instead of about 540,000 bytes for

Table 13.2 Possible software-code table for the Nokia 'Grande Valse' ringer.

35	0D
33	0C
2B	19
2D	19
32	0D
30	0C
27	19
29	19
30	0D
2E	0C
26	19
29	19
2E	4B
00	FF
00	37

CD quality or about 50,000 bytes for MP3 audio. Although there is a trade-off in sound quality, storing this ringer in MP3 audio format would require 833 times more memory, in CD-quality audio format 9,000 times more memory!

Another example shows a possible set of feedback sounds for a fax machine for domestic or small-office use. This product features an answering machine and a detachable hand-operated scanner. The system sounds shown in Table 13.3 give additional feedback in situations where the user may need it (e.g. when pressing an on/off key the user may not be aware of the fact that the function is already switched on or off). In addition the user does not always look at the display when doing a key-press. So two clearly distinguishable and easy-to-remember sounds for the 'switch on' and 'switch off' commands makes the UI more error prone.

Furthermore, this device has functions where the user is busy with paper handling and scanning. In these situations the visual information channel is overloaded and no attention can be given to visual displays, especially for novice users. As the hand-operated scanner is a new feature for users in many markets, special attention has been given to the development of good feedback sounds for the scanning operation. Find a short description of these sounds in Table 13.3.

13.6 Summary

Auditory display offers great potential for enhancing a UI in an efficient way, especially when carefully combined with other sensorial modalities. Sound can support or surpass other senses, such as the visual sense, for

Table 13.3 Example of a set of feedback sounds for a fax machine.

No.	Feedback sound	Use	Requirements	Recommendations
1	Key-press acknowledge	After every key-press (maximum delay 20 ms)	Short and unobtrusive	Single tone between 500 Hz and 2 kHz, shorter than 50 ms; may also be an imitated key-click
2	Key-press negative acknowledge	After a key-press which is not appropriate at the given moment	Clearly negative, but not too alerting	Single tone slightly longer than tone 1 above, frequency clearly different from tone 1
3	Alert	If an error condition occurs (e.g. paper out)	Asking for the user's attention, even if user is away from the machine	Imitation of a military-horn signal saying attention, quick succession of short sounds at one or multiple frequencies
4	Switch on	Switching something on (e.g. answering machine)	Sounds like a starting activity	Short raising-note sequence
5	Switch off	Switching something off (e.g. answering machine)	Sounds like a stopping activity	Short falling-note sequence
6	Task fulfilled	Successful transmission or reception of a fax message	Sounds like a successful positive experience	Melodic structure like at the end of a tune
7	Task not fulfilled	Transmission or reception error	Sounds like a failure or negative experience	Falling-note sequence in small intervals
8	Process OK	During normal operation of hand scanner	Sounds like the operation of a mechanical device (e.g. motor)	Repeated short beeps at low frequency with equal pauses between them
9	Process warning	When buffer of hand scanner is more than 80% full	Sounds like the quick operation of a mechanical device	Repeated short beeps at high frequency with short pauses between them
10	Process overflow	When buffer of hand scanner is full	Sounds like an alert if a problem has occurred	High and long single continuous tone

certain applications, and in many cases 'less is more'. There are no specific design basics or guidelines other than some psychoacoustic and musical principles, and the design is strongly bound to the technical constraints for sound (re)production of the product. Trial and error via user testing is suggested as the way to find the best possible auditory display for the intended application.

14 Tactile displays and speech output

Leo Poll

Until recently non-visual user interfaces (UIs) were traditionally associated with computer applications for the visually impaired. Advances in miniaturization and mobile technology in general have increased to importance of these interfaces in applications and devices for the able-bodied.

This chapter starts with a discussion on mobile devices highlighting why non-visual UIs are gaining in importance and how they can improve the overall usability of certain devices. This is then followed by a more general discussion on existing non-visual UIs for the visually impaired. Finally some guidelines are given which are derived from applications for the visually impaired but are applicable in non-visual UIs for the able-bodied.

14.1 The growing importance of non-visual user interfaces for mobile devices

Evolution of mobile media devices

The very first personal cassette player, called the Walkman, was introduced by Sony (Figure 14.1). The concept of being able to play your own music 'on the move' proved to be very successful. Other companies, such as Philips, quickly followed suit and the race was on to create the smallest, lightest, least power consuming, feature-laden device.

Basic use is based on the UI of a home cassette recorder. Buttons for play/stop/fast forward/fast backwards, volume control, etc. are provided. Over the years many other features have been added such as the option to record, reduce noise, equalizers, auto-reverse, etc. None of these features is unique, and can also be found on cassette recorders in the home.

Feedback on the operation of the cassette recorder is explicitly or implicitly given by:

- the presence or absence of sound/music while playing;
- a see-through window through which the user can see the action of the tape;

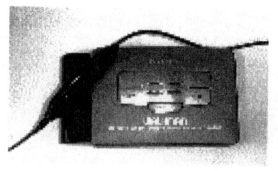

Figure 14.1 Sony 'Walkman' with electronic buttons on top.

- tactile feedback when using the controls (buttons) that has a direct link with the mechanical operations;
- sounds generated by the mechanical components of the player and cassette.

Advancements and price reductions in digital technology resulted in the introduction of the digital compact-disc (CD) player. The advent of CD players was inevitably followed by the introduction of mobile equivalents. The UI of the CD player is based on the UI of the cassette player. However, there is an important difference. Because of the digital nature of the device, there is no direct link between the way the device works in a mechanical/physical sense. Moreover, all buttons are electronic controls. Button-presses are interpreted by the digital processor and translated into internal system actions. This has important consequences:

- the UI paradigm of the personal cassette player/cassette recorder can be maintained even though the device works differently in a physical sense;
- controls can be placed in arbitrary positions on the device;
- multiple functions can be assigned to a single control (e.g. pressing the button labelled (>|) once moves to the next track whereas holding the same button instructs the device to fast-forward the current track);
- context-of-use specific controls can be developed that support the on-the-move specific use of the device (e.g. controls can be added to the headphone cord).

The growing digital nature of the device reduces the mechanical feedback in the form of tactile, visual and auditory information. As an alternative, displays (liquid-crystal displays [LCDs], light-emitting diodes [LEDs]) are used that provide the user with more abstract visual feedback. Moreover,

Figure 14.2 Diamond Rio solid-state player.

the presence of a digital processor that translates user actions into system actions allows for the addition of more complex features such as shuffle play, repetition of tracks or whole CDs and user programming of track order.

A further increase in digital compression rates allowed for the introduction of solid-state players (Figure 14.2). These devices no longer contain any moving parts but store the audio in solid-state memory. The lack of any mechanical parts means that the size of the device can be reduced even further. No moving parts also means that there is no tactile or auditory feedback from the player. Feedback is only given visually on a small display. This is considerably less than the feedback that was given by the first Walkman (see previous list).

Conclusion 1   Status feedback on current mobile media devices is primarily presented in a visual way.

Mobile information and communication devices

Over the last 10 years a whole new breed of personal devices such as Personal Digital Assistants (PDAs) and mobile phones were introduced. Originally these devices only offered a few dedicated functions. Increased computing power, memory capacity and market demand have caused an explosion of the number of functions on these devices. Some of the many functions offered on mobile phones are:

- a diary;
- a 'to do' list;
- a calculator;
- games (e.g. Tetris).

Most of these applications can be considered as scaled-down versions of applications found on personal computers (PCs) and as such only use a small display to present information to the user. The most recent PC application that made it onto the mobile phone is a web browser. For the latter, a special hypertext mark-up language was developed (WML) but all information is still presented in a visual manner.

Conclusion 2 *All current mobile information devices present infor-mation mainly in a visual manner.*

The mobile context of use

To optimize the on-the-move usage of mobile device, attempts have been made to:

- optimize the battery life;
- reduce the form factor (size);
- minimize weight.

Great progress has been made in the area of miniaturization and battery life and mobile devices have definitely become more portable. Unfortunately, miniaturization often affects the usability of a device in a negative way especially because, at the same time, the number of functions offered on a device has increased (e.g. component miniaturization has contributed to a reduction in power consumption and more enhanced portability of mobile devices). Electronic control of devices has not only allowed for more freedom in positioning them but also reduced the size to quite an extent. Moreover, electronic control has also allowed for a reduction in the number of buttons needed to operate a radio such as the one shown in Figure 14.3. This radio contains just three buttons:

- an on/off switch;
- a volume control button with which the user can select one of the two pre-programmed volume settings;
- a button with which the radio bandwidth can be searched for the next radio channel. Pressing the button once instructs the radio to locate

Figure 14.3 Radio with only three buttons.

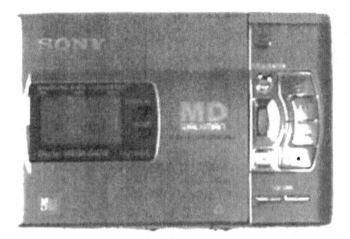

Figure 14.4 Real-size image of the Sony MR50 minidisc player.

the next radio channel. If the end of the bandwidth has been reached, the search continues at the beginning of the spectrum.

The disadvantage of smaller buttons and smaller devices is that they affect the visibility of affordances and constraints. Moreover, buttons are becoming so small that they reach the point where a user can no longer use his fingers to operate them. For example, a small pin is supplied with the Canon Ixus L1 camera with which some of the less frequently used buttons can be operated. In Figure 14.4 a real-size picture of a portable Sony minidisc player is shown. Surprisingly, frequently used buttons such as volume control only measure 8 × 2 mm.

Small controls are not expected to enhance the mobile use of these devices especially when the user is actually moving. Mobile use of portable devices is, furthermore, hindered by the fact that these devices are often kept in a pocket. Using these devices without taking them out of a pocket requires the user to memorize the button positions and their respective functions. Some buttons do contain some sort of tactile markings but no evidence exists of their effectiveness. Different button shapes as used on the minidisc player in Figure 14.4 are probably most effective. Alternatively a remote control on, for instance, the cable to the headphones could be used as an alternative to facilitate mobile usage of the device. Options for visual feedback on these remote controls are even more limited than on the device itself. The remote control of the Sony MR50 therefore incorporates auditory feedback (beeps) to indicate whether a button has been successfully pressed. This kind of feedback is limited to confirmation of successfully pressing a button rather than full-status feedback. The same beep is used for all other buttons which prevents the user from determining whether he has pressed the right button.

In general, it can be concluded that:

> *Conclusion 4   The increased miniaturization of components has reached a point where the size of buttons and displays starts to hinder the usability of mobile devices.*

It can also be questioned whether today's mobiles are truly optimized for use on the move. As was concluded earlier, digital mobile devices present their feedback in a visual manner. This does not necessarily support every context of use the user will encounter whilst being on the move. Moreover, whilst on the move, the context of use changes all the time (Vanderheiden 1997) and often at unpredictable moments. The context of use changes in terms of:

- the tasks executed by the user (the user might be walking, sitting, driving, reading, etc.);
- the physical environment (noise levels, lighting conditions, temperature, etc.).

Ideally a device should be usable under an arbitrary context of use and even between changes (e.g. a user might be listening to the news whilst driving a car). As soon as he hits a traffic queue, reading the news on a display could be more appropriate because his eyes are not occupied anymore. In an ideal world he (she) should be able to make this switch instantly without having to use separate devices that can only be used in isolation.

> *Conclusion 5   Current digital mobile devices are not optimized for use under arbitrary, ever changing (especially non-visual) contexts of use.*

14.2   Non-visual user interfaces for the visually impaired

It is obvious that there is a clear need for alternative additional input and output techniques for mobile devices. Traditional UIs have been primarily focused on visual interaction. Alternative techniques that exploit the potential of the tactile and auditory senses were mainly developed for the visually impaired. We believe that these techniques can also be applied in UIs for able-bodied users.

Able-bodied people often have the impression that visually impaired people have better ears. This is, however, not true in a physical sense. Visually impaired people are more aware of the presence of certain sources of auditory information and have more experience in exploiting these sources because these are the only sources of information they have on certain objects or events that are out of reach (Edwards 1989). Able-bodied people perceive the same auditory information but this is often not noticed because it is redundant to the visual information.

A second misconception is that all visually impaired users are able to read Braille. This is not the case. Research carried out in Germany (Nater 1990) showed that, of the population of visually impaired below the age of 66, about 52 per cent are able to read Braille. For the rest of them, UIs employing (synthetic) speech and non-speech sounds have been developed. The principles of these devices are also applicable to UIs for able-bodied users.

Over the years, several so-called access systems have been developed for visually impaired users that help them to access visual applications on computers. Systems for allowing non-visual access to Graphical User Interfaces (GUIs) are the most extreme case. These access systems present textual, spatial and graphical information as well as the UI elements in a non-visual form.

Non-visual perception[1]

To get a better understanding of the potential of our non-visual senses, let us see how visually impaired people interact with their environment.

Research has shown that the visual sense is predominant over the other

---

1 The majority of this section is taken from Poll (1996).

senses (Matlin and Foley 1992). This means that if information about a certain object or event can be perceived via multiple senses, the information acquired by the visual sense is taken to be the primary and most trustworthy source. This notion contributes to the misconception that able-bodied persons have about the superiority of a visually impaired person's non-visual senses.

Visually impaired persons who are completely deprived of vision have to exploit the potential of the four non-visual senses:

- hearing;
- touch;
- smell;
- taste.

Smell is a quite powerful sense providing information partly redundant to sight. On entering a room one can immediately notice, by the smell alone, that someone is eating tomato soup. Apart from smell-augmented cinema, there has been not much use of smell in the context of electronic products.

The non-visual perception of objects can have two goals, namely localization and identification of objects. In the following we will discuss the tactile and auditory senses with respect to localization and identification as well as some of the cognitive issues involved.

The visual sense is the sense par excellence for the perception of spatial information. Perceiving spatial information by the auditory or tactile sense is much more difficult. With the tactual senses, spatial information can be obtained about objects that are within reach of the user. Tactually acquiring spatial information is a serial process in contrast to the parallel spatial information perceived by vision. This serial process poses a large memory load on the user. Apart from this, it is also slower than the visual process which can be carried out 'in a wink'.

The tactile perception of spatial information is called kinesthesis or proprioception. Kinesthesis literally means the sense of movement, but also includes the awareness of positions and movements of the limbs and other body parts. On a much smaller scale cutaneous perception can help in perceiving spatial distributions on object surfaces (Lederman and Loomis 1981).

Spatial information can also be perceived by means of the auditory senses. However, many objects within our surroundings do not emit continuous sounds. An advantage of auditory spatial-information perception compared with tactual perception is that multiple sources can be perceived simultaneously without necessitating locomotion. A characteristic of auditory information is its temporary nature. Moreover, the spatial resolution of auditory information is poor compared with visual and tactile spatial perception. Sounds are, in general, perceived 'in time and over space' in contrast to visual information which is perceived 'over time and in space' (Gaver 1986).

Apart from acquiring spatial information, visually impaired persons have to acquire information about the identity of objects by exploiting the auditory and/or tactual senses. The manner of identifying an object's identity by touch alone is called haptics. For this identification of an object, sensations perceived by the cutaneous as well as the kinesthetic senses are used. The haptic identification of an object becomes difficult if the physical properties of an object do not permit haptic exploration of the object without locomotion, unless the object can be identified by exploring a part that is within reach. It is, for instance, impossible to distinguish between a skyscraper and a medium-sized building by haptic perception alone, nor is it possible to determine the exact size of the flame of a candle.

Until now we have only spoken of the tactual and auditory senses as separate means of acquiring information. However, when combining the information acquired with auditory and tactual senses, more information about the presence of objects could be obtained than is possible with one of the two senses alone. Previously gained knowledge and experience also play an important role in this identification process (e.g. a visually impaired person is not able to identify a bridge by touch alone, nor is he or she capable of identifying a bridge by the wind that blows through it). However, on the basis of touching the steel material of a bridge, hearing and feeling the wind, hearing the sound of other typical neighbouring objects like the river, previously gained knowledge and sensations, etc. he or she is able to identify the object as a bridge.

Non-visual user interfaces

The previous subsection mentioned that visually impaired people primarily perceive information via the auditory and tactile senses. As a result many input and output devices have been developed that address these senses. Some of them, such as Braille output elements, require quite a long learning period and are therefore not suitable for use in interfaces for able-bodied users. The main output modalities are:

- tactile displays (e.g. pin matrices, joystick with force feedback);
- speech (e.g. synthetic speech and pre-recorded speech);
- non-speech sounds (e.g. earcons [short melodies], audicons [everyday sounds], etc.).

Each of these modalities has its own characteristics and the suitability for presenting graphics, abstract information (text) and spatial information differs. In this subsection we will discuss a selection of devices that have been applied in systems for the visually disabled.

Speech has, since the advent of tape recorders, been used to provide visually impaired users with an alternative to written material such as books. Currently, many books have been read aloud by selected speakers and recorded on tape. Technical advances in the area of synthetic-speech generation allowed for the use of synthetic speech for the automatic generation of a spoken version of text. One of the first applications of synthetic speech for the visually impaired was aimed at making written material accessible without the need for intermediate able-bodied persons to do the visual to non-visual translation. Systems such as the Kurzweil Reading Machine (Kamentsky 1983) scan a document after which Optical Character Recognition (OCR) is used to digitize the text. This text is then presented to the user with synthetic speech.

Synthetic speech can also be used to access text-based computer applications. With synthetic speech approximately 200 words per minute can be read, which equals the average normal speech rate. Some visually impaired users are able to recognize speech at rates of three to four times the normal speech rate. When using speech to present written computer information a certain amount of information is lost in current computer-access systems as a result of parallel to serial conversion:

- spoken text is less accurate than written text (e.g. spelling errors are much harder to spot);
- typographic information such as font size, type face are lost;
- the exact layout of the text, which provides the context, is lost.

This does not mean that this information cannot be conveyed with speech. A number of these items listed above can be expressed by clever use of, for instance, intonation and pauses in and between sentences. To be able to do this a system will need to be able not only to identify the start and end of a sentence but also have a deeper understanding of the relationship between sentences and preferably of the content communicated in them. This is inherently difficult and computing intensive. To listen to speech at a rate higher than normal, consistency in pronunciation is considered to be more important than naturalness. Naturalness of speech is, however, an important requirement for non-frequent users for whom synthetic speech is not the only means to access electronic information.

Most of the research on presenting graphics to visually impaired persons has been focused on tactile picture presentations. A problem with the recognition of the content of pictures by visually impaired persons is that in most instances pictures are two-dimensional representations of three-dimensional visual objects (Jansson 1992). This requires a transformation from a three-dimensional picture to a two-dimensional representation which often results in a picture representation on a smaller scale. This hinders the ability to recognize pictures if the actual size of the object represented is too large to perceive as a whole in a non-visual manner, as is the case in the example of the bridge mentioned earlier (Hatwell 1993). The late-blind are in this respect able to benefit from their earlier visual experiences and newly gained haptic experiences (Heller 1989).

The applicability of tactile displays for the non-visual presentation of computer graphics not only depends on the experience, education and the functional capabilities of the tactile sense of a visually impaired person, but also on the physical properties of the tactile display (Brule 1990). At the University of Stuttgart (Schweikhardt 1985), a pin-matrix device was developed for presenting computer graphics. This device was originally intended for making video text accessible to the visually impaired. This device consists of 7,000 pins that can be lowered or raised. However, compared with the amount of pixels on computer screens that currently contain a minimum of $480 \times 640 = 307,200$ pixels, the amount of pins is relatively small. Application of these devices in mobile appliances is limited because of high power consumption and weight.

Other developments in the area of tactile displays try to compensate for the cognitive issues related to recognizing the objects represented by the pictures. A difficulty that was mentioned earlier in this chapter was that most haptic pictures are a two-dimensional representation of a three-dimensional object. Research in Japan focuses on the development (Shinohara 1992) and perceptual issues (Saida 1992) of a three-dimensional tactile display. Tests with this device show that the better the size of the tactile-object representation matches the size of the real object, the better the haptic identification becomes. No difference in performance was found between born-blind, late-blind and blindfolded subjects. This means that application of these devices is not limited to impaired users.

Research on the auditory presentation of pure graphics is scarce. In Meijer (1992) a system is described that translates arbitrary images recorded by a video camera in real time into the auditory domain. In this transformation, bit pixels from the screen that are bright are transformed into a high-pitched sound whereas dark pixels are transformed into a low-pitched sound. Brightness is converted into loudness following the rule: the brighter the light, the louder the sound. This system has, however, never been evaluated with visually impaired users so no evidence exists about the feasibility of this approach.

Another method for conveying graphics to the visually impaired is the verbalization of the graphics content. This requires the extraction of features from the graphic with the help of an optical character recognition (OCR) system. The components of the graphic are then uttered to the visually impaired user by means of synthetic speech. The applicability of an OCR system is, however, limited to graphics from which the components can be extracted automatically. This is the case in graphics representing, for instance, pie charts, bar charts, etc. (Mansur *et al.* 1985). The screen content of a GUI also contains a fixed set of components formed by the set of GUI elements and can thus be processed by an OCR system (Poll and Waterham 1995). The need for automatic graphic-feature extraction could be overcome if the designer of the graphic includes a verbal description of the graphic that is especially intended for the visually impaired user.

Because of the unavoidable parallel to serial conversion associated with the transfer of visual information into a non-visual form, the speed of information acquisition becomes very important. As Blenkhorn (1992) stated, computer-access systems should allow a visually impaired user to obtain the required information quickly, accurately and efficiently. This is especially important when considering computer use in employment where the visually impaired person has to compete with his able-bodied colleagues. The option to include a verbal description of a graphic is included in HTML, the hypertext mark-up language. This has a potential application in non-visual web browsers.

To allow fast access to computer-screen information, a computer-access system should facilitate the following:

1. Fast information transfer;
2. Effective information-selection mechanisms.

As shown in the previous subsection, the fastest information transfer rates can be achieved with synthetic speech which, when used by an expert, exceed normal speech rate by three to four times. However, the accuracy of the textual content suffers from this high transfer rate. Accuracy of information transfer is guaranteed when Braille lines are used but in this case the speed of information transfer suffers, which might be a problem when an electronic newspaper is being read and spelling errors are of minor importance.

The UI driver that controls the devices could contain mechanisms that enhance the accuracy and/or speed of the medium applied (e.g. mechanisms could be available that allow the user to control the speed of the utterance or allow the user to enter some kind of spelling mode whenever the spelling of words becomes important). Reading speed can be further improved by exploiting the fact that most people can recognize a word after hearing only the first fraction (Marslen-Wilson 1979). The UI could include a method that allows the user to skip the remaining utterance of a word after hearing the first fraction and move on to the next.

In addition to UI mechanisms that increase the speed or accuracy of the utterances, the selection mechanisms available for selecting the wanted information are also of importance. These mechanisms should allow the user to define a starting point for the serial information transfer in the original parallel visual information. In computer-access systems aimed at making character-based displays accessible, these selection mechanisms are line based. The TASO (Tactile Acoustic Slider Output) system devised by the German company Frank AudioData (Figure 14.5) is an example of a line-based selection mechanism. In this system the starting point for the speech utterance can be selected by two sliders that are placed next to and below

Figure 14.5 Reomote control for the TASO screen-access system. The starting position of the utterance of the textual screen content can be selected using and shuttles sliders.

the keyboard. When the sliders are moved, the pitches of the auditory beeps indicate whether the line currently selected is full or empty or whether the character current selected is in upper case, lower case, etc. Once found, the text will be uttered for as long as the user holds down one of the buttons that are positioned on top of the sliders.

An important aspect that has to be taken into account when designing non-visual UIs is the *memory load* put on the user. Compared with parallel information acquisition the process of serial information acquisition puts an unavoidable additional memory load on the user. Text-based computer applications also contain a specific UI of which the user has to build a mental model which poses a load on long-term memory. In the case of non-visual access to text-based applications, the user also needs to have a mental model of the access interface mechanisms which results in an even larger memory load. The strain caused by this memory load can be reduced by introducing methods for repeating or easily reselecting information that was uttered before.

So far we have only spoken of the non-visual presentation of text and the associated UI techniques to optimize it. Non-visual UIs also require non-visual presentation of:

- status;
- feedback on state changes;
- UI elements such as menus and buttons.

All these elements can be presented using speech. However, this is likely to be rather slow and in many cases better results can be achieved by incorporating non-speech sounds. Status and state changes can especially be very well represented with non-speech sounds. In these two cases, identification of the sounds by the user is of lesser importance since a state change is, in many cases, the result of a user action. The presence or absence of a sound by itself provides a confirmation that something is happening or has happened. As argued in the previous subsection, portable devices have become completely silent because of the shift from mechanical to digital technology. This means that ample opportunities exist to give auditory feedback on digital devices by using sounds that are based upon the mechanical operation of the device's predecessor. *Metaphor-based sounds* are only one form of auditory feedback and a more extensive description is given in Chapter 13.

UI elements tend to be rather abstract and are therefore very hard to present using non-speech sounds based on real-world events. More abstract sounds can be used but these need to be memorized and thus require a certain learning period. Speech in this case offers a more unambiguous solution but can also be a cause for confusion if only one voice is used (e.g. in GUI-access systems the object window could be represented by the utterance of the word 'Window'). However, the word 'window' also commonly appears as a menu option, which could be confusing. To avoid confusion, multiple voices can be used for different purposes in a non-visual UI. Care should be taken to use voices that are easily distinguishable. Moreover, the number of voices should be kept to a minimum.

14.3 Some general guidelines

At the start of this chapter it was mentioned that because of the shift from mechanically to electronically operated devices many non-visual feedback mechanisms have disappeared. In general, all current mobile devices present their information and feedback primarily in a visual way. It was argued that non-visual output will need to be (re)introduced if these devices are to be used under arbitrary and ever changing contexts of use. In the previous section, several tactile and acoustic output methods were discussed that have been applied in systems for the visually impaired. From this discussion the following guidelines can be extracted that are also applicable in mobile applications for the able-bodied:

Guideline
- *Make sure that the user is always in control of the sequential information presentation.*

This can be realized by:

1. *Allowing a user to stop utterances and/or sound at all times* (this is the golden rule of speech interfaces which is often ignored by many telephone systems):

- *including mechanisms to speed up or slow down an utterance,*
- *allowing a user to skip (part of) the utterance of a single word* (part of a word is often enough to grasp the meaning – having to wait until, for instance, a menu option finishes can be irritating especially when the option is not what the user is looking for),
- *providing mechanisms for selecting arbitrary starting points in the serial stream of information* (a good example is the TASO system where the user could select a starting point in the text by moving a vertical and horizontal slider),
- *providing different levels of accuracy* (e.g. if spelling is important, a spelling mode can be provided).
2. *Using non-speech sounds to present status information and state changes.*
3. *Using tactile modalities to convey spatial properties wherever possible.*

Many of the guidelines given can be applied in the short term in non-visual interfaces. Most of the systems developed for visually impaired users are based upon full-fledged PCs. Embedded personal devices have lacked such computing power in the past which, amongst other reasons, prevented the effective application of non-visual UIs (e.g. simulating the sound made by the motor of the original Walkman on a solid-state player requires computing power and memory). However, the technological revolution still obeys Morse's law, which means that many digital portable devices will, in the very near future, have sufficient computing power to incorporate more advanced non-visual UIs. Some devices such as mobile phones and PDAs have more computing power then the first Apple Macintosh and advanced UIs can already be employed on these systems.

The guidelines given in this chapter are general ones and application very much depends on the systems under development. In general, it can be concluded that non-visual systems developed for visually impaired users can provide a valuable source of information and inspiration and should therefore not be immediately dismissed just because they were originally developed for impaired users.

Part V
Important issues

15 Standards in user-interface design

Jennifer Weston

15.1 About standards

What is a standard?

There are many different definitions of 'a standard' but, in essence, a standard is a document describing an agreed level of performance or quality. Standards cover many aspects of our lives, from the buildings we live in, to the size of the paper we write on. Compliance to a standard can only be measured when it contains requirements (i.e. 'shall' statements). Most user-interface (UI) standards contain guidelines (i.e. 'should') statements. UI design is, in many respects, not well suited to prescriptive standards. When writing a standard for a paint coating, it is relatively easy to think of criteria and measurable values; we could talk about the reflectivity, the chemical constitution or the durability of the paint. In the discipline of UI design, life is not so simple. However, as the discipline matures, so do the standards. Many UI standards are now being written as goal-based or process-based standards (e.g. ISO 9241: part 11) rather than as a definitive description of required product attributes. This suits the discipline of UI design much better.

Standards and legislation

Most standards are not enforceable by law. However, some standards are closely linked to legislative requirements. As an example, in 1990 a European directive (number 90/270/EEC) introduced minimum safety and health requirements for work with display screen equipment. This directive was implemented in the UK by the Health and Safety (Display Screen Equipment) Regulations 1992. An example requirement from these regulations is, 'the image on the screen should be stable, with no flickering or other forms of instability'. In the guidance booklet on regulations (HSE 1992) it recommends that standard ISO 9241 part 4 is used to meet the display requirements. So, as can be seen from this example a standard is very closely linked with legislative requirements.

Standards and patents

Obviously, a standard cannot specify that something should be done a certain way if that method is covered by a patent, because this would give the company holding the patent an unfair advantage, or even make it impossible for other companies to meet the standard. This means that standards sometimes have to be limited in their specification.

15.2 How do standards get developed?

Standards come from various sources. Some, such as systems of measurement, are set down by major national or international bodies. Some (e.g. Internet protocols) are industry standards that are agreed between a number of partners. There are also internal company standards.

Standards in the electronics industry

In an ideal world, the standard would be developed before the product. However, in practice this is often impossible, especially in the electronics industry which is moving so incredibly quickly and is highly competitive. Companies do not want to risk losing market share by disclosing their secrets or delaying the product launch to wait for a standard. What has tended to happen in practice is that a new product is introduced and, if the product is successful, this becomes the basis for a standard which other products can follow. Recently, this has started to change as companies have realized the importance of standards and there are many industry groups that are working together to come up with technical standards.

Standards organizations

Tables 15.1 and 15.2 list some of the major standards organizations. Contact details and website addresses of these organizations are given at the end of this chapter.

These tables are not exclusive. There are many more standards organizations worldwide. You should be aware of those in your country and in the countries where you market your products. A good starting point to find other organizations is the ISO website.

Table 15.1 International organizations.

IEC	International Electrotechnical Commission
ISO	International Organisation for Standardization
ITU	International Telecommunications Union

Table 15.2 National organizations.

ANSI	American National Standards Institute
BSI	British Standards Institute
CEN	Comité Européen de Normalisation [European Committee for Standardisation]
DIN	Deutsches Institut für Normung
ETSI	European Telecommunications Standards Institute

15.3 Why do we need standards?

Standards are becoming increasingly important in the world of electronics. With the explosion of connectivity and networking, a seamless integration between products is essential. In order to do this, there must be standards. There are three main reasons why we need standards:

- ease and quality of use;
- reduced development time and risk;
- market advantage.

1 Ease and quality of use

Standards are based on current best practice and are written by experts. This means that by following a standard, the UI designer can benefit from the knowledge and experience of others.

The use of standards can improve usability in many ways (e.g. even with the best designed interface, it can take people a while to become familiar with a way of interacting). If there are standards covering the interaction method, then it will improve consistency between products, and hence improve the usability (e.g. pen-based computer interfaces which use a kind of shorthand so that the user can give instructions, such as 'delete' or 'undo'). Without a standard covering these gestures, each manufacturer might use different gestures which could lead to confusion for users.

2 Reduced development time and risk

One persuasive reason for using standards is to save development time. When designing a portable product with batteries, choices can be made about whether to use a standard battery size, or to develop a custom battery to suit the new product. If a standard battery size is chosen, the requirements for the size of the battery compartment and therefore some constraints on the product are easily established. In addition, since the product will be using standard battery components, then suppliers for these parts can be sourced early in the development cycle.

Related to saving time is reducing risk. The more that a product follows existing standards, the more likely it is that the end result will be of a higher quality. It also has the assurance that equipment from different manufacturers will work reliably together.

3 Market advantage

Consumer pressure is an important factor. In recent years, people have started demanding much more in terms of ease of use.

Compliance with standards can be used as a marketing tool. This is particularly true if your product is intended for a commercial (as opposed to consumer) market. There are often a number of similar products on the market. Compliance with certain standards can differentiate your product from others. In some cases, it can even command a premium price.

The World Trade Organization (WTO) has produced the Agreement on Technical Barriers to Trade, Annex 3 of which is the Code of Good Practice for the Preparation, Adoption and Application of Standards. This shows that the WTO recognizes the importance of proper development and use of standards.

15.4 How to use standards

This section describes how you should approach the use of standards during the development of a product. The process is described in a few simplified steps, as shown in Figure 15.1.

Step 1 Decide what standards might be needed

You should consider what the UI of your product consists of. This covers both hardware and software. Will your product have a display? What input device(s) will be used? What software platform will be used?

You should make sure that you consider all the potential users of your product including, for example, disabled people. Consider all phases of the product's life including unpacking and set-up as well as maintenance and disposal.

Consider other products that may be used in conjunction with your product. (e.g. you may be designing a portable electronic organizer but, if this product is to connect with a standard desktop PC, then you should also consider if there are any applicable PC standards that might impact on your product).

You should also look at your market and competitor products. You may be forced to comply with a standard if the majority of your competitors are doing so.

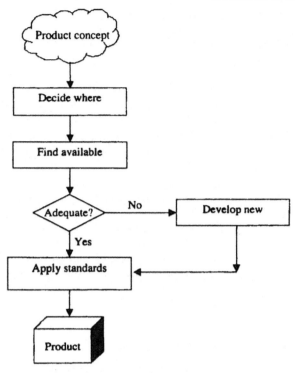

Figure 15.1 Approach to the use of standards during development of a product.

Step 2   Find out what standards are available

When you have decided where you might need standards for your product, you should find out what standards already exist. There is a list at the end of this chapter to guide you, but you should also do your own research. Remember that you can use standards for other types of products as guidance for your type of product.

It can be time consuming to find applicable standards, so it is worthwhile to have an ongoing activity in your organization to keep up-to-date information about standards.

Step 3   Is this adequate?

Once you have identified what standards exist, you should decide if this is adequate for your product. If there are areas of the UI not covered by a standard, you should consider whether it is appropriate to trigger the development of a new standard. You may wish to create an internal company standard or you may think it is important to work with other companies or a standards organization.

Once you have a list of standards, you need to decide which (if any) you will use. Section 15.5 gives a more detailed consideration of the things that you should consider when deciding whether to use a standard or not.

Step 4   Apply standards

Standards documents can be quite long and complicated and the whole of the standard may not be applicable to your product. Therefore, it is a valuable exercise to translate any standards, that you want your product to comply with, into an internal design-requirements document that can easily be followed by the development team. Always make sure that you have the latest version of the standard. Check with the issuer of the standard if they offer a subscription service so that you will be notified of revisions.

You should test at appropriate points in the product-development cycle whether you are meeting the standards. When you have a final product, you may need to have a test done by an external agency before you can claim compliance to the standard. The International Laboratory Accreditation Cooperation (ILAC) can provide contact details for accredited agencies in your country.

15.5   How to decide whether to use standards

While following standards has a number of major advantages, it is not always the best thing to do. This section discusses some of the factors you should consider when deciding whether to follow a standard or not.

Maturity and applicability of the standard

Of course, standards have to be widely accepted and used to be of value. Taking the example of the IEC standard 60417 which specifies symbols for use on equipment, some symbols, such as the one shown in Figure 15.2 for power standby, are widely used, and it would be foolish to use a different symbol, even if you think you have a better one.

Some standards, for various reasons, are not widely taken up. So, the standard loses some of its value. In a case such as this, the best solution for your product may be achieved by deviating from the standard.

Figure 15.2 Symbol for power standby.

Sometimes the only standard that exists will have been developed for a specific product type. This does not mean that you cannot use it for your product, but you should consider whether the standard is applicable or not.

Innovation

Of course, if all everybody did was follow standards, we would not have very exciting products and the huge advances in electronics products in recent years would not have happened so quickly. Standards are important, as already discussed in this chapter, but they should not be followed blindly.

Market

You should consider the demand from your customers for compliance to standards. Corporate customers are likely to be more knowledgeable and demanding about products meeting certain standards than domestic customers. Also, consider what your competitors do.

Cost of meeting the standard

The use of standards can reduce costs, but there is also the possibility that meeting the standard will incur some cost either to the product itself or in terms of development effort. Consider whether you need to have your product tested by a recognized authority before you can claim compliance with a standard and how much this will cost.

15.6   How to obtain user-interface standards

At the end of this chapter, there is contact information for the major standards bodies together with other useful websites. These sites all contain a wealth of information. Major libraries will conduct searches for you. A useful starting point for finding information is the World Standards Service Network, which has been set up with the intention of coordination of standards information. The International Classification for Standards (ICS) (International Organization for Standardization 1999) provides a standard way to identify the type of standards (e.g. 'ergonomics' is 13.180 and 'electronic display devices' is 31.120).

Copies of standards can be bought directly from the standards organizations, or in some cases from bookstores.

It is very valuable to get involved with the standards bodies that are applicable to you and the countries in which you trade, because then you will get advance knowledge of standards that are being developed.

Table 15.3 Standards.

Reference Number	Title
ISO 9241	Ergonomic requirements for work with visual-display terminals, (note: in 17 parts; to date, only some have been issued as standards)
ISO 13406	Ergonomic requirements for visual display based on flat panels
ISO 13407	Human-centred design processes for interactive systems: guidelines for the design of consumer products to increase their accessibility to people with disabilities or who are ageing
ISO/IEC 13714	Information technology – document processing and related communication – UI to telephone-based services – voice-messaging applications
ISO/IEC 14754	Information technology – pen-based interfaces – common gestures for text editing with pen-based systems
ISO 14915	Software ergonomics for multimedia UIs (note: in preparation, not yet published)
ISO/IEC 15412	Information technology – portable computer keyboard layouts
ISO 18789	Ergonomics of human–system interaction – ergonomic requirements and measurement techniques for electronic visual displays
IEC 60417	Graphical symbols for use on equipment

List of applicable standards

The availability of UI standards is limited. Table 15.3 lists some currently available standards. Even though this book is concerned with the design of UIs for electronic equipment, and not desktop computers, the list contains standards outside this scope since some of the information from standards intended to cover PCs can also be applied to smaller devices.

Architecture and software standards

There are various emerging standards dealing with technical issues applicable to electronic products. Whilst not strictly UI standards, these are of course very important in the design of a usable product. A list of some contacts and websites for further information is given at the end of this chapter.

15.7   References and sources of further information

Summary and guidelines
- *Consider standards early in the product-development process.*
- *If a standard is not available, then think whether it would be in your interests to collaborate in developing one.*
- *Adhere to available standards as much as possible, but not without questioning – do not follow them blindly.*
- *Have your own internal standards to make the application of external standards simpler and more efficient.*

Table 15.4 Standards organizations.

Organization	Web	Address
American National Standards Institute (ANSI)	www.ansi.org	11 West 42nd Street 13th floor NEW YORK, NY 10036 USA +1 212 642 49 00
British Standards Institution (BSI)	www.bsi.org.uk	389 Chiswick High Road LONDON W4 4AL United Kingdom +44 181 996 90 00
Comité Européen de Normalisation (CEN)	www.cenorm.be	Rue de Stassart 36 B-1050 BRUXELLES Belgium +32 2 519 68 11
Deutsches Institut für Normung (DIN)	www.din.de	D-10722 BERLIN Germany +49 30 26 01-0
European Telecommunications Standards Institute (ETSI)	www.etsi.org	650, Route des Lucioles F-06921 SOPHIA ANTIPOLIS CEDEX France +33 4 92 94 42 00
International Electrotechnical Commission (IEC)	www.iec.ch	3, rue de Varembé Case postale 131 CH-1211 GENÈVE 20 Switzerland +41 22 919 02 11
International Organization for Standardization (ISO)	www.iso.ch	1, rue de Varembé Case postale 56 CH 1211 GENÈVE 20 Switzerland +41 22 749 01 11

Table 15.5 Other organizations.

www.itu.int	ITU (International Telecommunications Union)
www.ilac.org	International Laboratory Accreditation Cooperation
www.wssn.net	World Standards Services Network
www.wto.org	World Trade Organization
www.w3.org	World Wide Web Consortium
www.mpeg.org	The Moving Picture Experts Group (a working group of ISO/IEC)
www.usb.org	Universal Serial Bus
www.bluetooth.com	Bluetooth is wireless technology especially designed for communication between small portable devices
www.java.sun.com	Java is a software language that was designed to be independent of the device on which it is running

16 Usability evaluation

Bruce Thomas

16.1 Purposes of evaluation

Designing a product according to the guidelines provided in this book is not necessarily a guarantee that it will be usable. Just as with hardware and software components, it is always advisable to test the user interface (UI) to check that it works. This chapter describes three of the main techniques which can be used to evaluate the usability of a UI.

An evaluation of the UI of an electronic device might, however, be carried out for a number of reasons. The purpose of an evaluation will influence the methods and techniques to be used, and what is to be measured. Van Vianen *et al.* (1996) identify some fundamental and generic questions that should be addressed regarding the purpose of an evaluation:

- *Are we trying to find out what people want with regard to user interfaces?* A starting point can be to determine the user's needs and wishes, which can be translated into a user-requirements specification
- *Are we trying to find out what people accept with regard to user interfaces?* Taking a given device, the goal of a test might be to determine the acceptance of such a device by its users.
- *Are we trying to find out what people are able to use?* Another goal might be to test to what extent users are able to use a device with which they are confronted, irrespective of the aesthetic qualities.

16.2 When to evaluate

A usability evaluation can take place at any time during the development process. It need not be limited to testing the final design; indeed, earlier evaluation is generally more productive, since by the time the product has been designed and developed, there is very little opportunity to make improvements based on the evaluation results. The earlier a problem is detected, the easier and cheaper it is to solve.

Further, evaluating usability need not be a single event in the development process. As a result of an evaluation, changes can be made and the UI evaluated again. This is an iterative process.

In order to conduct an evaluation, some kind of representation of it is required. In early phases of development, only paper-based representations of the interface may be available (e.g. storyboards). These are not interactive. Users will therefore only be able to form a sketchy impression of the interactions, and it will generally only be possible to obtain feedback on the overall design concept rather than specific problems. In such early stages of development, it can often be better to rely on the judgement of an independent expert rather than to test with users.

With a new interface concept, it is generally advisable to create a dynamic simulation as early as possible. This enables early testing with users, and thus makes it possible to detect potentially major problems at an early stage. It should be noted that the simulation does not have to be a completely faithful representation of the interface, as long as the main interactions can be visualized. It is important to maintain a degree of flexibility in simulation, so that rapid changes can be made and retested until a suitable solution is found.

It is sometimes attempted to create a simulation which can later be used as a specification for the product, and even to try to generate software code from the simulation. This is generally inadvisable. The time and effort required to create a 'perfect' simulation means that usability testing will inevitably be delayed, and the flexibility required to try out alternative solutions is lost.

As the development process nears completion, product prototypes and early production models become available. Testing these gives an opportunity to make a final usability check, but at this stage there is often little opportunity to do anything about any problems found. Testing at this stage is primarily of value to predict service requirements, and also to identify improvements for the next generation.

Van Leeuwen and Thomas (1997) illustrate the powerful effect that iterative test and development cycles can have. They describe the development of the Philips Fizz mobile phone, and show the evolution of the UI through the various design stages. It is clear that the results of feedback from users significantly affected the final design.

16.3   Tools and techniques for usability evaluation

There are a wide variety of tools and techniques available for usability evaluation, and it would be impractical to give a full description of all of them here. An overview of some methods and their strengths and weaknesses is shown in Table 16.1, taken from van Vianen *et al.* (1996).

A number of methods which have proven to be useful in industrial contexts are presented in Jordan *et al.* (1996). The methods presented in

Table 16.1 Overview of strengths and weaknesses of some testing techniques (adapted from van Vianen *et al.* 1996).

Technique	Strengths	Weaknesses
User workshop A group of users is put together to discuss more generic issues in relation to product(s), is asked to prioritize functions, redesign specific parts, etc.	User involvement in finding solutions Intense and effective initial concept generation Assess appropriateness of initial target group and define it in more detail Combine evaluation/ solution generation	Much effort required to set up More expensive initially, but provides more in-depth information at an earlier phase
Focus group A group of users is put together to discuss particular issues related to a product (stimulation)	Can discover the unexpected Revealing and prioritizing issues that are important to the user Rapid information gain from large numbers of users	Poor basis for decision making Information mainly from opinion leaders Little measurable data
Usability test Formal test in which participants perform tasks and in which effectiveness, efficiency and satisfaction are measured	Can measure task performance User involvement	More expensive, but provides more in-depth information at an earlier phase More time consuming, but can save discussion time
Informal test Short test in which internal people perform tasks	Quick	Measurements can be unreliable Users often not from target group
Inventory of usage Log the use of features and functions	Objective information Easy to process	Need for special equipment
Inventory of comments/ remarks Users keep a record of their experience with the final product	Input from users based on end product	Self-selective Dependent on motivation

that book range from very formal to very informal techniques, and vary in the degree to which they have been scientifically validated. All, however, have the merit of being used in practice.

It is not the purpose of this chapter to summarize all possible methods of usability evaluation. Instead, it will describe in some detail the principles of conventional usability testing, both formal and informal, as well as a technique for evaluating usability in situations where access to users might not be feasible (e.g. because of time and budget constraints).

Tools for evaluating some of the more intangible aspects of UIs are presented by Jordan in Chapter 17, 'Pleasure with products'.

16.4   Formal usability testing

Formal usability testing can be carried out to meet three main objectives (Thomas 1996):

- to discover particular user problems and potential flaws in design concepts;
- to determine how usable existing products are, with a view to setting baseline criteria for new development; and
- to determine whether predefined criteria have been met.

For all these objectives, measures of usability are required. Such measures are identified in ISO 9241: part 11.

ISO 9241 defines usability as 'The extent to which a product can be used by specified users to achieve specified goals with effectiveness, efficiency and satisfaction in a specified context of use.' This definition of usability has proved to be valuable as a means to specify the measures to be applied during usability testing. This is because the components of usability can be expressed quantitatively, which facilitates comparison between products and product generations (see van Vianen *et al.* 1996).

ISO 9241 defines effectiveness as 'the accuracy and completeness with which users achieve specified goals'. It can be measured in terms of whether test participants are able to complete specified tasks. A distinction can be made between completion on the first attempt, completion on the second or subsequent attempts, completion with the use of a manual, and failure to complete.

ISO 9241 defines efficiency as 'The resources expended in relation to the accuracy and completeness with which users achieve goals.' Many investigators interpret this in terms of the time taken to operate a device (e.g. Macleod and Bevan 1993), sometimes related to the number of errors made. Thomas (1996) registers all the individual actions a test participant makes in order to complete a task, including making errors and error recovery.

ISO 9241 defines satisfaction as 'the comfort and acceptability of use'.

Measures of satisfaction vary, but all rely on subjective rating by test participants in some way. There are a few 'off the shelf' questionnaires with which measures can be made, most notably the System Usability Scale (see Brooke 1996) and the Software Usability Measurement Inventory (see Kirakowski 1996).

Participants in usability tests are selected according to profiles of product use. The participants should match as closely as possible the intended users of the device. These might be members of a particular professional group, or for devices intended to be used by the population at large, then they might be matched according to such factors as income, product experience, age and gender, and level of technical interest.

A particularly important aspect of carrying out a usability test is selecting the tasks the test participants are to perform. The tasks should be representative of all system levels, but should be realistic in terms of the scenario presented during the test and of what the end-users might be expected to do.

The context in which the test is carried out might not necessarily correspond to the ultimate environment in which the device is used. Many tests are carried out in an artificial environment – a usability laboratory. A weakness of such laboratories is a lack of realism. Nevertheless, the salient aspects of the expected environment (e.g. lighting levels, furnishing, presence of related devices) can be simulated, and as a result a remarkable degree of realism and participant involvement can be achieved (see Thomas 1996). Nevertheless, a usability laboratory is simply a tool for facilitating usability testing. Perfectly adequate tests can be carried out in other locations such as offices or meeting rooms. In order to reach particular population groups (e.g. speakers of different languages), this author has also conducted tests in a variety of hotel rooms.

16.5   Informal usability testing

Formal usability testing, as described above, is often time consuming and expensive; less expensive and less time-consuming methods for evaluation do exist, one of which is to conduct an informal test.

Informal usability tests can be described as 'stripped down' versions of formal tests (Thomas 1996). They are particularly useful during concept development, where they provide the possibility to try out a large number of ideas quickly and economically. Perhaps the greatest issue with respect to informal testing is how much of the formal-test methodology can be sacrificed without losing validity or value.

An informal test can be structured to accommodate many of the aspects of usability discussed in the context of formal testing. Effectiveness can be measured on the basis of simple notes regarding whether test participants successfully complete the tasks. Satisfaction can still be measured with the same standard questionnaires. Similarly, realistic tasks can be set to be performed in an appropriate scenario; standard tasks and even scenarios can be taken or derived from tests previously conducted. The major compromises made in informal testing concern the measurement of efficiency and the recruitment of test participants.

Measures of efficiency, whether time taken, actions made or some other measure, often require extensive *post hoc* analysis. Nevertheless, a qualitative description of the problems that participants have in performing tasks may be considered sufficient for indicating whether or not they are having to do too much or are experiencing excessive frustration in manipulating a particular product.

Recruitment of participants, even for formal testing, can sometimes pose major problems, particularly when a specialized user profile is required. If a test must be carried out within a short period, recruitment of suitable participants can be a major hurdle. It then becomes necessary to abandon a strict match to the target profile and to make the best match possible with the people available. These people in many cases will be staff at the site where the test is to be carried out. In such cases, the least suitable participants will be the people directly involved in the development of the device in question, or with an extensive knowledge of the device or similar devices. An adequate informal test can be conducted with as few as five or six test participants (see e.g. Virzi 1992). It has been shown that this number is adequate to detect most of the difficulties users will have, and all the major problems with a device.

16.6   Expert appraisal

In some situations, time and budget constraints may be too great even to perform an informal usability test. In such situations it may be feasible to conduct an expert appraisal. This may even be combined with an informal test to provide a convincing and visible picture of the strengths and weaknesses of a particular concept (see Thomas 1996). A frequently applied tool to conduct such an appraisal is a checklist, such as that developed by Ravden and Johnson (1989, see also Johnson 1996).

The Ravden and Johnson checklist has nine criterion-driven sections. Each section consists of a series of detailed questions relating to each specific criterion, along with a standard, simple response scale (always, most of the time, sometimes, never). These criterion-based sections are followed by two general sections, the first of which uses a closed-question format with the standard response categories: major problems, minor problems, no problems. The checklist concludes with a brief section containing open-ended general questions. More than 150 main items are used within the checklist.

The nine sections of the checklist cover the criteria:

- *Visual clarity*   Information presented should be clear, well organized, unambiguous and easy to read.

- *Consistency* The way the system looks and works should be consistent at all times.
- *Compatibility* The way the system looks and works should be compatible with user expectations.
- *Informative feedback* Users should be given clear, informative feedback on where they are in the system, what actions they have taken, whether these actions have been successful and what actions should be taken next.
- *Explicitness* The way the system works and is structured should be clear to the user.
- *Appropriate functionality* The system should meet the needs and requirements of users when carrying out tasks.
- *Flexibility and control* The interface should be sufficiently flexible in structure, in the way information is presented and in terms of what the user can do, to suit the needs and requirements of all users, and to allow them to feel in control of the system.
- *Error prevention and correction* The system should be designed to minimize the possibility of user error, with built-in facilities for detecting and handling those which do occur; users should be able to check their inputs and to correct errors, or potential error situations before the input is processed.
- *User guidance and support* Informative, easy-to-read and relevant guidance and support should be provided to help the user understand the system.

It should be noted that the Ravden and Johnson checklist reflects the principles of design described by Mavrommati and Martel (Chapter 5). The checklist, therefore, is a simple check of whether the product really has been designed according to general design principles.

It is generally advisable that the device is evaluated by somebody other than the developer, in order to eliminate any positive bias.

The usability checklist is only one of a number of methods available for expert appraisal. Other methods are described (e.g. by Nielsen 1993 and Lindgaard 1994). For more information on these alternatives, the interested reader should consult both these sources.

Guidelines
- *Even if all the guidelines in this book have been followed, it is advisable to test the usability of the product.*
- *Usability should be evaluated throughout the development process; problems detected early can more readily be solved.*
- *Simulations need not faithfully represent every detail of the UI; a*

flexible simulation which can rapidly be changed better enables alternatives to be tried out.

- *Most major problems can be detected with as few as five test participants.*
- *Test participants should represent as closely as possible the people who will use the device.*
- *Test a cross-section of tasks ranging from simple everyday activities to those which explore the lowest levels of the interaction.*
- *In the test environment, reproduce any conditions in the real environment which are likely to influence usability (e.g. low lighting levels).*
- *In order to avoid bias, appraisals of usability should not be carried out by the product developers.*

17 Pleasure with products
The new human factors

Patrick W. Jordan

17.1 Summary

Human factors is now well integrated within the design process of many organizations. The profession has become adept at assuring product usability. This has been achieved through an understanding of the design properties associated with usability and the accumulation of a battery of methods for usability evaluation and requirements capture.

Nevertheless, a usability-based approach is limited. Products have the potential to bring a wide range of emotional benefits – to bring pleasure. Ensuring that these benefits are delivered requires attention to more than just usability.

This chapter describes a framework within which human factors can cooperate with other disciplines in order to develop products that meet users' practical, emotional and hedonic needs. A case study is presented in order to demonstrate how this framework can be used as a basis for defining the benefits that a product will deliver to its users. Suggestions are then given as to how the design, marketing and technical aspects of a product can combine to provide those benefits.

Traditionally, human factors has concentrated on usability. New human factors aims to look beyond usability to understand the issues that make a product a positive joy to use. In order to do this three major challenges must be met. First, users must be 'humanized' – be understood as rational and emotional beings rather than being regarded as just physical and cognitive components of a system. Second, links must be established between properties of products and the emotional responses they elicit. Third, methods, metrics and tools are required that enable pleasure with products to be quantified and which facilitate requirements specification with respect to pleasure.

17.2 Introduction

This chapter will consider the role of human factors in the product-creation process and its relationship to other disciplines. It will be argued that although human factors has made a significant contribution to the betterment of product design, it has tended to be too usability based in its approach. Usability is, of course, a central tenet of good design. However, human factors as a profession has been guilty of focusing narrowly on usability. This has been to the exclusion of other issues that will affect the way in which people interact with and respond to products. These issues include, for example, product aesthetics, semantics, sensorial elements of a product and the values conveyed by a product.

It is proposed here that human factors move beyond usability-based approaches to 'pleasure-based' approaches to product design. Usability-based approaches tend to view people as *users*, with products seen as *tools* with which these users complete tasks. Pleasure-based approaches, on the other hand, look at people holistically – as rational, emotional and hedonistic beings. They look at products as *living objects* with which people have *relationships*.

A framework for facilitating pleasure-based approaches is described – the four pleasures (see Section 17.6). This framework can help to structure thinking with respect to identifying product requirements and can also act as a means of facilitating communication between the various parties involved in product creation. In particular, pleasure-based approaches require and encourage effective communication between market researchers, marketing personnel, human-factors specialists and designers.

History

Human factors has come to increased prominence over recent years. This is reflected, for example, in the growing human-factors literature and the host of national and international conferences dedicated to human-factors issues. However, the most significant indication of this is the ever-increasing numbers of human-factors specialists that are now employed by industry.

This chapter is focused on the role that human factors has played in product design. This is as opposed to, for example, physical ergonomics in the workplace and health and safety issues. In relation to product design, human factors appears to have gone through three phases over the last couple of decades.

Phase 1 – Being ignored Going back 10 years and more, few manufacturing organizations employed human-factors specialists, even among the larger companies. Those companies who did were more likely to be involved in defence work. Certainly human factors was not much of a consideration for companies making consumer products.

Phase 2 – 'Bolt on' human factors This was the era of creating a new product and then asking the human-factors specialist to help put a nice interface on it. The problem, of course, was that by this stage in the product-creation process the basic interaction structure of the product had often been decided leaving only room for comparatively superficial interface improvements. Nevertheless, this marked an era when more human-factors specialists were finding employment in industry and, although sometimes misunderstood, human-factors issues started catching the attention. Apple Computer had come to prominence largely on the basis of their user-friendly interface styles and other companies sat up and took notice. Increasingly there was a feeling that '... we better add a bit of human factors in here'.

Phase 3 – Integrated human factors And so to the present day. In a number of companies human factors has become seen as something that is inseparable from the design process. At Philips Design, for example, the departmental design procedures state that a design team must include a human-factors specialist at all stages – from initiation of the work through to evaluation of the final product. This gives the human-factors specialist the chance to influence the design right from its conception.

A probable cause of this increased integration into the design process is the perceived commercial advantage good human factors – indeed good design generally – can bring to a manufacturing organization. In many product areas, technical advances and manufacturing processes have reached a level of sophistication which makes any potential competitive advantage, in terms of functionality, reliability and manufacturing costs, marginal. Many manufacturers now see design as one of the major areas in which it is still possible to gain significant advantages over the competition. Good human factors is, of course, central to achieving excellence in design.

Indeed, customers are becoming increasingly sophisticated in terms of the level of 'usability' that they expect with a product. Whilst once they may have seen usability as a bonus, it is now becoming an expectation. Users are no longer willing to accept struggling with products as a price they must pay for 'technical wizardry'. Customers now demand technical wizardry and usability and will be antagonized by products that fail to support an adequate quality of use. In the end, of course, such products will also antagonize those who manufacture them as they will find that their customers soon start to look elsewhere.

Obviously human factors has not reached a phase-3 level of integration in all organizations yet. There will still be many manufacturing companies who are operating at phase 2 and a smaller number at phase 1. However, a great many of the companies who are manufacturing products seem to have come to accept that usability is a product quality that is essential for market success. This is reflected, for example, in the use of terms such as 'ergonomically designed' and 'user friendly' in advertising material.

17.3   Usability

The International Standards Organization (ISO) define usability as:

> ... the effectiveness, efficiency and satisfaction with which specified users achieve specified goals in particular environments.
>
> (ISO DIS 9241-11)

A fuller discussion of this definition of usability is given by Thomas in Chapter 16.

In effect, a usability-based approach to user-centred design is one which sees the product as a tool with which users try to accomplish particular tasks without wanting to have to expend unnecessary effort or endure any physical or mental discomfort.

Human factors has become very adept at assuring usable design. A battery of methodologies for evaluating usability have been established. The majority of these were originally developed in psychology (e.g. interviews, experiments [the basis of laboratory-based user trials] and questionnaires). Human factors has also taken and adapted from other sources (e.g. focus groups from marketing) as well as developing methods within the profession itself (e.g. co-discovery, task-analysis techniques).

Human factors, as a discipline and as a profession, is well equipped to assure the usability of products. However, whilst usability is essential in ensuring that users reap a product's functional benefits, it does not explicitly address issues relating to the emotional and hedonic benefits that products can bring. Tackling these issues can represent the next step for human factors – going beyond usability to develop an understanding of the full potential of the richness of product–person relationships. Not just looking at what makes a product usable, but taking the wider view – looking at what makes a product pleasurable.

17.4   Hierarchy of user needs

Abraham Maslow (1970) developed a 'hierarchy of human needs'. This model views the human as a 'wanting animal' who rarely reaches a state of complete satisfaction. Indeed, if a nirvana is reached it will usually only be temporary because once one desire has been fulfilled another will soon surface to take its place. Maslow's hierarchy is illustrated in Figure 17.1. The idea is that as soon as people have fulfilled the needs lower down the hierarchy, they will then want to fulfil the needs higher up. This means that even if basic needs such as physiological needs and safety have been met, people will still meet with frustration if their higher goals are not met. (A good overview of Maslow's work can be found in Hjelle and Ziegler 1981.)

Self-actualization needs	⇑
Esteem needs	⇑
Belonging and love needs	⇑
Safety needs	⇑
Physiological needs	⇑

Figure 17.1 Maslow's hierarchy of needs.

Pleasure	⇑
Usability	⇑
Functionality	⇑

Figure 17.2 A hierarchy of user needs.

Taking the idea of a hierarchy of needs and applying it to human factors, the model in Figure 17.2 is proposed.

Level 1   Functionality

Clearly a product will be useless to the user if it does not contain appropriate functionality. A product cannot be usable if it does not contain the functions necessary to perform the tasks for which it is intended. If a product does not have the right functionality it will dissatisfy the user. In order to be able to fulfil user needs on this level, the human-factors specialist must have an understanding of what the product will be used for and the context and environment in which it will be used.

Level 2   Usability

Once users have got used to having appropriate functionality, they will then want products that are easy to use. As was mentioned earlier in this chapter, this more or less represents the situation at the moment in many product areas – people are used to well-functioning products, now they expect usability too. Having appropriate functionality is a prerequisite of usability, but it does not guarantee usability. To assure usability the human-factors specialist must have an understanding of some or all of

Table 17.1 Emotions associated with pleasurable and displeasurable products (from Jordan and Servaes 1995 and Jordan 1996).

Pleasurable products	Displeasurable products
Security/comfort	Annoyance/irritation
Confidence	Anxiety/insecurity
Pride	Contempt
Excitement	Frustration
Satisfaction	Resignation
Entertainment	Aggression
Freedom	Feeling cheated
Sentiment/nostalgia	

Table 17.2 Properties associated with pleasurable and displeasurable products (from Jordan and Servaes 1995 and Jordan 1996).

Pleasurable products	Displeasurable products
Features/functionality	Performance (poor)
Usability	Usability (lack of)
Aesthetics	Reliability (lack of)
Effective performance	Aesthetics (poor)
Reliability	Cost (too high)
High-status product	Gimmick
Convenience	Size (inappropriate)
Size	Unpredictable
Brand name	

the design principles listed in Table 17.1 and be able to use methods such as those listed in Table 17.2.

Level 3 Pleasure

Having got used to usable products, it seems inevitable that users will soon want something more. Products that offer something extra. Products that are not merely tools, but which are 'living objects' which people can relate to. Products that bring not only functional benefits but also emotional benefits. To achieve product pleasurability is the new challenge for human factors. It is a challenge that requires an understanding of people – not just as physical and cognitive processors – but as rational and emotional beings with values, tastes, hopes and fears. It is a challenge that requires an understanding of how people relate to products. What are the properties of a product that elicit particular emotional responses in a person, how does a product design convey a particular set of values?

Finally, it is a challenge that requires capturing the ephemeral – devising methods and metrics for investigating and quantifying emotional responses.

17.5 Pleasure

In 1995 and 1996 the two halves of a study making a 'first pass' at the issue of pleasure and displeasure in product use were reported to the Ergonomics Society Conference (Jordan and Servaes 1995; Jordan 1996).

The study was interview based and involved asking participants to consider products that they found particularly pleasurable and products that they found particularly displeasurable. They were then asked about the emotions and feelings that they associated with these products and the product properties that were associated with pleasure or displeasure. Jordan (1996) defined pleasure and displeasure in product use as follows:

> *Pleasure in product use: the emotional and hedonic benefits associated with product use.*
> *Displeasure in product use: the emotional and hedonic penalties associated with product use.*

Participants chose a wide range of products to talk about – from electric toothbrushes to computer software. Table 17.1 lists the emotions associated with pleasurable and displeasurable products, while Table 17.2 lists the properties of products that the participants associated with pleasure or displeasure.

It is interesting to note that while usability is one of the properties that affects pleasure it is just that: only one such property – just as appropriate functionality will not ensure usability, usability will not ensure pleasure.

Other than this modest study, the literature yields very little evidence of human factors having addressed issues associated with pleasure. This point of view was recently echoed by Dandavante *et al.* (1996), who also noted that addressing the emotional domain was something that had remained a 'gap' in human factors.

17.6 What is pleasure?

The study summarized above provided a useful first pass at pleasure and a broad definition under which to group a number of issues. However, it does, of course, still leave human factors as a profession a long way short of a rich understanding of what pleasure can mean. In the absence of human-factors sources on the issue, looking at the literature from related disciplines – such as psychology, physiology and anthropology – seems a sensible approach.

A useful starting point may be the work of Canadian anthropologist Lionel Tiger. Tiger has made a study of pleasure and has developed a framework for addressing pleasure issues (Tiger 1992). The framework models four conceptually distinct types of pleasure – physical, social, psychological and ideological. Summaries of Tiger's descriptions of each are given below. Examples are added to demonstrate how each of these components might be relevant in the context of products.

Physiological pleasure

This is to do with the body – pleasures derived from the sensory organs. They include pleasures connected with touch, taste and smell as well as feelings of sexual and sensual pleasure. In the context of products physiological pleasure would cover, for example, tactile and olfactory properties. Tactile pleasures concern holding and touching a product during interaction. This might be relevant, for example, in the context of a telephone handset or a remote control. Olfactory pleasures concern the smell of the new product (e.g. the smell inside a new car may be a factor that affects how pleasurable it is for the owner).

Social pleasure

This is the enjoyment derived from the company of others (e.g. having a conversation or being part of a crowd at a public event). Products can facilitate social interaction in a number of ways (e.g. a coffee maker provides a service which can act as a focal point for a little social gathering – a 'coffee morning'). Part of the pleasure of hosting a coffee morning may come from the efficient provision of well-brewed coffee to the guests.

Other products may facilitate social interaction by being talking points in themselves (e.g. a special piece of jewellery may attract comment, as may an interesting household product, such as an unusually styled TV set). Association with other types of products may indicate belonging in a social group – Porsches for 'Yuppies', Dr Martin's boots for skinheads. Here, the person's relationship with the product forms part of their social identity.

Psychological pleasure

Tiger defines this type of pleasure as that which is gained from accomplishing a task. It is the type of pleasure that traditional usability approaches are perhaps best suited to addressing. In the context of products, psychological pleasure relates to the extent to which a product can help in accomplishing a task and make the accomplishment of that task a satisfying and pleasurable experience (e.g. it might be expected that a word processor which facilitated quick and easy accomplishment of, say, formatting tasks would provide a higher level of psychological pleasure than one with which the user was likely to make many errors).

Ideological pleasure

Ideological pleasure refers to the pleasures derived from 'theoretical' entities such as books, music and art. In the context of products it would relate to, for example, the aesthetics of a product and the values that a product embodies (e.g. a product made from biodegradable materials might be seen as embodying the value of environmental responsibility). This, then, would be a potential source of ideological pleasure to those who are particularly concerned about environmental issues. Ideological pleasure would also cover the idea of products as art forms (e.g. the video cassette recorder that someone has in the home, is not only a functional item, but something that the owner and others will see every time that they enter the room; the level of pleasure given by the VCR may, then, be highly dependent on how it affects its environment aesthetically).

17.7 The four-pleasure framework in the product-creation process

In this section an example will be given which demonstrates how the 'four-pleasure framework' can be used as a basis for identifying the benefits that a particular product should bring to its users. An approach to the design process is proposed which encourages those involved – particularly human-factors specialists, market researchers, marketing personnel and designers – to work together to create products which deliver a range of benefits to those who interact with them. It will be argued that the human-factors specialist can play a leading role in this process, as the multi-disciplinary nature of human factors facilitates communication between the disciplines involved.

For any product, there will be a target group. The breadth and specificity of such a group can vary widely between products. Some products – particularly professional products – can be aimed at a very narrow group. (e.g. bespoke computer software specially created for a particular company or group of professionals). At the other extreme are products which are targeted at a wide spectrum of users via mass-market sales. Often household durables such as vacuum cleaners, irons and kitchen appliances will be targeted in this way – the aim being to create a product that caters for all needs and tastes. Usually, of course, the breadth of the target group will be somewhere in between these extremes.

A product's target group will usually be specified on the advice of market researchers. It is customary to specify the group in terms of a number of demographic characteristics such as age, gender, nationality and socio-economic status (Solomon 1996). Occasionally, lifestyle characteristics might also be taken into account. These characteristics might include, for example, the level of a person's interest in the type of product and their attitudes towards technology generally.

Given that a particular target group has been identified, the product-development team must then make decisions about how the characteristics of the group will influence the design requirements of the product. The four-pleasure framework can be used as a means of structuring thought here. To illustrate this point an example of the design of a 'smart' product – in this case a camera – is given below. Smart products are those which have embedded information-technology components, the aim of which is to assist users in the tasks that they are doing or to provide additional task possibilities. Other examples of smart products include microwave ovens, video recorders, facsimile equipment and automatic teller machines (den Buurman 1997).

Case study   Designing a camera

Target group

Imagine that the target user group for the camera is women aged between 25 and 35 of high socioeconomic status. Given that this is the target group, what are the implications in terms of product requirements? What follows is a four-pleasure analysis, suggesting some issues that may be of importance in this context and considering what the associated design requirements might be.

Physiological pleasure

A camera is a hand-held device. Clearly, then, the feeling of the camera in the hand is likely to be a source of physiological pleasure to the user. It is important that the camera fits the hand well – in this case the adult female hand – and that it is nicely balanced in terms of weight distribution. In order to maximize the tactile pleasure associated with the camera, it should be light enough so that it is not too heavy to hold, yet heavy enough that it gives a feeling of solidity and quality.

The material used should be pleasant to the touch: not too slippery or too sticky, not too rough or too smooth. Any moving parts should also work in a way that provides tactile satisfaction to the user (e.g. the lens cover should slide open in a satisfying way – it should not be too stiff to slide, nor too loose). Similarly, the level of resistance offered by the shutter button when taking a shot should be appropriate and should give a satisfying 'click'. Again, these issues are likely to have a strong influence on the perceived quality of the camera.

Of course, the user also has to carry the camera around with her. Maybe she carries a handbag. The camera should fit securely in a handbag. Maybe she doesn't carry a handbag. The camera should also fit well into coat pockets, trouser pockets, jacket pockets, etc. The shape of the camera should be designed such that it causes no physical discomfort when carried in a pocket (e.g. it shouldn't have any edges that poke into the body).

When taking a photograph, the user will hold the camera to her face and look through the viewfinder. Here the camera may come into contact with parts of the face, in particular the side of the nose and the eyebrow. Again it is important that the camera is shaped so as not to poke into these parts of the face when being used.

Finally, for this example, the designers should be aware that many young women may have long fingernails. Clearly, they will not want these broken when using the camera. Small buttons and catches, which users may use their fingers to operate, should be designed such that they do not offer a level of resistance that is likely to cause fingernails to get broken. Alternatively, the design should avoid buttons and catches that will require fingernail operation.

Social pleasure

A pocket camera is something that will often be used in a social context. The user may want to take photos of her friends and loved ones. She is also likely want to take photos in public places, where other people will see her using the camera.

A major social issue here is the impression of the user that the camera gives to others. Our target users are of a high socioeconomic status. Maybe they wish this to be reflected in the design of the camera. If they have paid a lot of money for the camera, then the camera's design should reflect this. In this way the camera can act as a 'badge' saying 'I'm a successful person'. Indeed, when considering how to spend resources on development of the camera, it may be wise to emphasize design over some of the less visible technical aspects of the camera (e.g. this user group may not be particularly interested in marginal improvements in picture quality [this is in contrast to the professional photographer to whom this might be very important]). On the other hand, they may be very interested in having a nice 'design object' which radiates good taste and sophistication and which makes it clear that the camera was a costly purchase.

The social context in which the camera is used will also have implications for the functionality included in the camera. Presumably, these users will mainly be taking pictures of people they know rather than of professional models. Models may be prepared to spend a considerable amount of time getting a pose just right and waiting while the photographer makes adjustments to the camera settings in order to get the shot just right. The chances are, however, that our user's friends and families will not be prepared to wait while she goes through such a rigmarole. This indicates that the camera should have a 'point and shoot' operation – one which the user simply takes out of her pocket, points at the object or people she wishes to photograph and presses the button. This suggests that features such as auto-focus, auto-iris, auto-flash, auto-wind and auto-just-about-everything-else will be appreciated.

Another social issue connected with cameras is the amount of noise that they make when the user takes a picture. Many of these auto-functions rely on small motors to drive them. Sometimes the noise generated by these can be embarrassing. A friend of this author inadvertently terrified fellow passengers on a flight from Amsterdam to New York. He was taking pictures through the window of the plane when he came to the end of his film and the camera started to rewind the film automatically. Unfortunately, the sound of the film rewinding was so loud and grating that the passengers in his and the neighbouring rows of seats thought that something was horribly wrong with one of the aircraft engines. It was only by pointing to the camera and apologizing that he was able to calm the panicked faces around him.

Psychological pleasure

This type of pleasure is concerned with enjoyment in taking the photos and enjoyment of the outcome. Probably, this will largely come down to traditional usability issues – effectiveness and efficiency. In this case, effectiveness means good-quality photos, while efficiency is about how easy it is to take the photos.

Again, it is important to consider the quality of photographs that these users will be after. As mentioned before, they will probably be after good-quality snaps rather than professional-quality photographs. While a professional photographer may be prepared to spend hours getting a shot just right, these users will probably not be prepared to spend a long time over a shot, no matter how good the outcome. The emphasis, then, must be on getting the shot right with minimal effort. Again, the issues that arose under 'social pleasure' of needing a 'point and shoot' operation with lots of auto-functions are reinforced.

Our users are young, dynamic, successful women. They probably don't have the time or inclination to sit down and read a hefty instruction manual before using the camera. It is important, then, that the camera is 'guessable'. Guessability (Jordan *et al.* 1991) refers to first use of a product. Guessable products are those that the user can use at the first attempt without difficulty. With guessable products users do not have to go through a long learning curve in order to use the product successfully. Compared with many of the electronic and information-technology products that people now use in their daily lives, a camera seems an inherently simple product. It seems unlikely – particularly for this user group – that people would be prepared to tolerate a camera that they cannot pick up and use at the first attempt.

Of course, all the usual principles of usable design (e.g. Ravden and Johnson 1989) should be taken into account here. For example, the camera should give *feedback* about in what mode it is and how many photos are left to take. The camera's mode of operation should be *compatible* with that of other cameras and *consistent* within itself. Characters used for labelling should be legible and any symbols understandable (*visual clarity*). Procedures for operation of the camera should be explicit but should be *flexible* enough so as not to constrain the user's actions.

Ideological pleasure

Choosing to use or buy a particular product over another may often represent an 'ideological' decision. Here, those involved in the product-creation process should be aware of the potential ideological perspectives of the successful young women at whom the camera is targeted.

Consider, for example, aesthetical aspects of the form of the design. What would this group's idea of a beautiful design form be? A judgement based on the age and gender of the participants might suggest that they would prefer organic, postmodern forms to more rational modern forms. Both design precedence (Tambini 1996) and empirical evidence (Hofmeester *et al.* 1996) tend to suggest a preference for such forms among female users.

Another issue affected by living in the 'postmodern' era might be the extent to which the user would wish the product to be a reflection of her femininity. Some 20 or 30 years ago overtly feminine designs may have been seen as 'patronizing'. These attitudes are reflected, for example, in the feminist literature of the 1970s, much of which viewed popular culture – including design – as a framework within which women were controlled and patronized by men (see Strinati 1995 for an overview). In an analysis of design from 1895 to 1980, Forty (1986) concluded that products tended to reflect stereotypes of men as plain, strong and assertive and women as 'decorative', weak, delicate and sensitive. In such a climate, making an overtly 'feminine' product may have been seen as crass and demeaning, especially by a user group such as this one, which, presumably, would be well educated and progressive.

Since then, paradigms have shifted, both in the design world and in society as a whole. As women increasingly win equality and make socio-economic progress, many may also feel it less contentious to express their femininity through the products that they own and use.

Indeed, the percentage of women working in design is ever increasing. These factors have led to a climate where 'feminine' design may be seen as positive and expressive, rather than patronizing. Such attitudes are reflected in postmodern feminism which tends to support the cultural representation of gender differences to a far greater extent that did the feminist movement of the 1970s (e.g. see Paglia 1995).

Feminine design is not merely about form language. It may also have implications, for example, for the colours used, the fonts used for labelling and even the name of the camera.

Aside from 'personal' ideologies such as self-perceptions and aesthetic preferences, user pleasure with products may be affected by their social, political and religious ideologies. An issue that is likely to concern a young, well-educated group such as this is the environment. Those involved in the creation of the product might be well advised to carry out an environmental audit of the product to make sure that it represents responsible environmental standards. This might mean, for example, using recycled plastics, minimizing the use of non-biodegradable materials or designing the camera such that the user can get it repaired should a defect arise, rather than having to throw it away.

In summary, then, this pleasure analysis has yielded the suggestions outlined in Table 17.3 for user requirements.

Table 17.3 Four-pleasure analysis of product requirements for a camera aimed at young women of high socioeconomic status.

Pleasure	User requirements for camera
Physiological	Fits hand well Material pleasant to touch Moving parts give optimal level of resistance Fits nicely into handbag or pocket Doesn't poke into body when carried Doesn't poke into face when taking photos 'Fingernail friendly' catches
Social	Reflects user's high socioeconomic status Demonstrates user's good taste and success 'Point and shoot' operation Many automated functions Quiet motor drives
Psychological	Good-quality (but not professional-quality) photos 'Point and shoot' operation Easy to use Guessable Ergonomically designed
Ideological	Postmodern organic design language Feminine styling Non-patronizing design Environmentally responsible

A *multidisciplinary approach*

An analysis such as this will normally be carried out by a group involving the designer, human-factors specialist, product manager, engineer, market researcher and a representative of marketing and sales. This approach promotes a holistic view of product creation. The product's technology, design and marketing message can then be coordinated right from the start. This means that the criteria used in the design brief will be the same as the criteria used in creating the marketing story. Such integrated approaches give a far better chance that the product will be a success, both commercially and from the point of view of the user (Urban and Hauser 1993) (e.g. if it has been decided that the product must provide a particular benefit, then the design can emphasize this benefit and the marketing literature can explicitly promote it). This contrasts with approaches that tend to be 'driven' from one direction or the other. Often products may be 'technology driven', 'design driven' or 'market driven'. Such approaches run the risk of being unbalanced; overemphasis of one aspect tending to lead to problems with the others.

The human factors literature, for example, includes many examples of products that have failed in terms of their design because manufacturers have placed an inordinate emphasis on technology. Norman (1988) gives many piquant and amusing examples of such technology-driven failures including telephones, televisions and, of course, VCRs. In order that users can derive maximum benefit from technological innovation, the design must support these benefits. Adherence to the principles of ergonomic design is a vital factor here. However, other factors, such as product semantics, can also play a vital role – the form of the product expressing the benefits that the product can bring (Macdonald 1997).

Considering again the camera, this might mean a high-tech design to reflect the large number of information-technology-driven automatic functions. High-tech aesthetics reflecting high-tech functionality – giving the user a clue that there are some automatic functions in there somewhere. This is also an issue in the design of computer application programs. For example, consider a menu-based word processing package. These packages tend to offer vast arrays of functions. It is important that these are grouped in a sensible and structured way so that users can understand the benefits that are being delivered. Equally, it is important that the design 'announces' the opportunities that the technology provides (e.g. through clear and meaningful menu headings).

Outside of the realms of information technology, a product that has caught the popular imagination is the Dyson vacuum cleaner. This offers one particular advantage over the competition – it doesn't require a dust bag. The design of the vacuum cleaner signals this advantage through having a clear plastic wall in the area where the dust is collected. The technological benefit is promoted via the design. Dyson have also put this feature right at the centre of their marketing strategy – an excellent example of how an integrated approach has led to the creation of a successful product.

Just as design can be a means of expressing technological benefits, technology can be a means of delivering design benefits. Microchip technology and miniaturization has delivered benefits here. By reducing the size of the technological part of products, designers have been given a bigger 'design space' to work in. Where previously the technological aspects of a product might have strongly dictated the form, it is no longer likely to be possible to fit them into a desired form. Apple Computer provide good examples of this, both in terms of software and physical product form.

Apple first came to prominence as the user-friendly computer company by producing interfaces that relied on direct manipulation, rather than command lines for executing particular tasks. The interface was, indeed still is, based on a desktop metaphor, with icons representing, for example, folders, files and wastepaper bins. Iconic interfaces of this type are far less memory efficient than are command-line interfaces; however with the development of low-cost memory chips, efficient use of memory became less important, and Apple exploited this opportunity.

The availability of electronic memory in compact form has also enabled Apple to develop interesting computer forms – veering away from the rather dull, boxy, shapes that had previously dominated the personal-computer market. Probably the most radical departure has been the development of the PowerBook. Here, the combination of technology and form have created a unique set of benefits to the user, supporting the use of the computer in circumstances where it would not previously have been possible (e.g. in aeroplanes, trains, hotel rooms, meeting rooms) and being an aesthetically pleasing product in itself. In terms of the four-pleasure framework, Table 17.4 gives some examples of the benefits that might be associated with the PowerBook.

Table 17.4 Four-pleasure analysis of benefits associated with the Apple Macintosh PowerBook.

Pleasure type	Associated benefits
Physiological	Light weight Easy to carry
Social	High-status product Public display of computer literacy
Psychological	Computing tasks can be carried out in different environments Easy to use
Ideological	'Softer' design language Aspirational product (enhancing self-image)

Just as it is important to take an integrated approach to the design, engineering and marketing of new products, it is also important to take an integrated approach to the evaluation of products and product concepts.

Most manufacturing organizations will have procedures in place for evaluating concepts before they are put into production. These will typically include tests of use – using an interactive prototype – and tests of consumers' first impressions using models of the product. When designing these tests, it is important, once again, to refer back to the product benefits that the manufacturers are aiming to deliver.

Considering the camera example again, this would mean – given the requirements identified from the four-pleasure analysis – that user tests should be investigating, among other things, issues such as guessability, the fiddliness or otherwise of the buttons and the tactile aspects of the camera. Tests of first impressions, meanwhile, should include aspects such as design language, the status associated with the product and the impression of quality that the product radiates.

What is important, then, is that the product is evaluated in accordance with criteria that reflect the benefits that were originally aimed for. This can be in contrast to much of current practice, where 'standardized' usability and face-value tests are used, which may be investigating issues that are irrelevant or are marginal to the product's potential success. In the case of human-factors evaluations there are a number of 'off-the-shelf' questionnaires for measuring usability (e.g. the System Usability Scale [SUS] [Brooke 1996] and the Software Usability Measurement Inventory [SUMI] [Kirakowski 1996]). While these types of tools surely have a place in the evaluator's 'toolbox' (e.g. in placing perceptions about a product's usability in the context of a wider standard), addressing the issues identified by the four-pleasures analysis will require the development of bespoke evaluations for individual products. Human-factors specialists have a wide variety of methods available to them (e.g. interviews, questionnaires, user trials) that they can tailor to address the specific issues associated with a particular product (Jordan 1993; Popovic 1997).

This harmonization of the evaluation criteria with the benefits identified initially also extends to the evaluation of any marketing approaches. For example, if it has been decided that the product should radiate, say, femininity, then marketing concepts (storyboards for commercials or copy for written advertisements), should be tested with potential consumers to see if they do, indeed, promote a feminine image.

17.8 Empirical challenges for human factors

In the previous section, it was demonstrated that the four-pleasure framework could be used as a basis for setting product requirements which would be used as the basis for an integrated multidisciplinary approach that would be adopted throughout the product-creation process. An example was worked through – the development of a new camera. In the example the criteria identified were included on the basis of expert appraisal.

Having a framework for the identification and categorization of different types of pleasure can be very useful for structuring thought and for promoting cooperation and integration of the various disciplines involved in the product-creation process. However, in order to maximize the benefits of a pleasure-based approach to product creation, each of the disciplines involved may need to develop theories and tools to help implement the benefits identified. As disciplines, marketing, market research and design have a history of addressing a wide range of issues with respect to the relationship between people and products. Human factors, on the other hand, doesn't as it has tended to concentrate on usability to the exclusion of many other issues. What follows is a discussion of three challenges that human factors must face if it is to assure pleasure with products. The challenges are: understanding users and their requirements, linking product properties to emotional responses in order to fulfil these requirements, and developing methods for the investigation and quantification of pleasure. The discussion that follows suggests empirical approaches to these issues which can complement expert analyses.

Challenge 1   *Understanding users and their requirements*

Usability-based approaches tend to lead human-factors specialists to consider people as processors. Physical processors with attributes such as strength, height and weight, and cognitive processors with attributes such as memory, attention and expectations. Here, then, the user is often looked at as being simply a cognitive and/or physical component of a three-component system – the other two components being the product and the environment. It could be argued that this 'usability engineering' approach is, in effect if not in intention, dehumanizing the user, ignoring, as it seems to do, the very things that make us human – our emotions, our values, our hopes and our fears.

In order to find a way into these issues, we need to have an understanding not only of how people use products, but also of the role that those products play in people's lives. This gives a chance to understand how the product relates to the person in a wider sense than just usability and can help the human-factors specialist in gaining a wider view of the user requirements – the requirements for pleasure.

To illustrate the benefits that can be associated with pleasurable products, three very brief case studies are given below. Those mentioned were participants in the Jordan and Servaes (1995) study on pleasure in product use. Like many of the examples given in this paper, the first of the case studies reported refers to a product that did not contain any information technology. However, as with the other examples, it illustrates how the requirements for pleasure go beyond usability and gives examples of user responses that could be equally desirable with information-technology-based products.

Case study 1   Hairdryer user

This 17-year-old woman chose a hairdryer as her pleasurable product. A product which she described as being 'perfect ... the best (hairdryer) I've ever had'. The reason she was so positive about the hairdryer was because it helped her to style her hair in just the way she wanted. This made her feel attractive and gave her a feeling of self-confidence when she went out. She also mentioned that the hairdryer had an unusual design and was thus something that caught the attention of her friends, '... it is "showy", I like it when people come into my room and see it.'

Both of the pleasures she mentions fall in the social-pleasure part of Tiger's framework. They are both concerned with enhancing the image of herself that she projects to others (somebody good-looking, somebody with interesting tastes) and how she feels in the company of others (self-confident).

Case study 2   Guitar player

The guitar player was a 26-year-old man who played the electric guitar. Again, the guitar facilitated social pleasure. He regarded it as a '... status symbol particularly amongst people who know about these things.' It had also provided a talking point as it had belonged to Lloyd Cole – lead singer of Lloyd Cole and the Commotions, a Glaswegian band who had a number of hits in the 1980s.

He also found playing the guitar an exciting activity in itself. This would probably be classified as a psychological pleasure or possibly an ideological pleasure (all four categories can potentially overlap). Even just having the guitar near him gave him a feeling of reassurance – again either an ideological or psychological pleasure.

Case study 3   Video cassette recorder (VCR) user

She was a 26-year-old living alone who rented a VCR which she described as being a 'standard video'. She described the emotional benefits that she gained from the video as being a feeling of anticipation – looking forward to watching what she had recorded (an ideological pleasure) and freedom – not having to stay home in order to catch her favourite programmes (a social pleasure as she could go out with her friends).

Note that, even in this little selection, the benefits that people mentioned went well beyond 'comfort and acceptability' – the satisfaction component of usability. Only ensuring that these products were effective, efficient and satisfying to use would stop well short of providing the benefits mentioned by these three people. It is clear, for example, that in case studies 1 and 2, the person–product relationship went well beyond that of the user–tool. The hairdryer was an objet d'art, the guitar was a status symbol, a talking point, even an old friend.

Challenge 1 then is to investigate and inventorize the types of pleasure that products can potentially bring to their users based on a holistic understanding of the role which the product plays in a person's life.

It should be fairly straightforward to move ahead here. In the first instance, it may simply be a matter of asking the appropriate questions during evaluation- and requirements-capture sessions. Not simply asking about functional benefits but also about the types of pleasure that a product can bring. Certainly, pleasure is not a straightforward issue. Certainly, people may not always be able to or willing to articulate descriptions of the types of pleasure that they gain from a particular product. Certainly what people do say may often be difficult to interpret. Nevertheless, simply having an awareness of pleasure issues and asking some sensible questions will very quickly move human factors in the direction of meeting this challenge.

Challenge 2 *Linking product properties to pleasure responses*

Having established the different types of pleasure that people can get from products, the next stage is to link them to particular product properties. For example, it might be that feelings of security with a product are linked to high levels of usability and/or high levels of product reliability. A feeling of pride may be linked to, say, good aesthetics. Similarly, particular types of displeasure may also be linked to inadequacies with respect to certain product properties (annoyance might be linked to poor technical performance whilst anxiety may be related to a lack of usability).

All the above suggestions are, of course, speculations. This is an issue that it would be beneficial for human factors to address empirically and systematically if it is to make a significant contribution to the development of pleasurable products. One way in which to approach this would be to correlate people's pleasure responses to a product to the 'goodness' of the product with respect to various properties.

An empirical approach aimed at directly linking design decisions to emotional responses is called Kansei engineering (Nagamachi 1995). This technique involves manipulating individual aspects of a product's design in order to test the effect of the alteration on the user's overall emotional response to the product. This technique has been used to assist in the design of a diverse range of products. Nagamachi (1997) describes examples that range from automobiles through camcorders to brassieres. To demonstrate how the technique works, an example – reported by

Ishihara *et al.* (1997) – is summarized below demonstrating how Kansei engineering was applied to the design of cans for coffee powder.

Example of the application of Kansei engineering

In this study seventy-two alternative designs of coffee can were presented to a panel of ten subjects. The seventy-two designs represented permutations of various design variables, including, for example, colour, font styles, graphics and form. Each member of the panel was asked to rate each of the designs according to how they fitted with a series of descriptor adjectives. These adjectives are known as 'elements'. There were eighty-six of these elements, generated from brainstorms with consumers and marketing experts. Panellists rated each of the seventy-two designs according to the eighty-six elements by marking five-point Likert scales to indicate the degree to which they felt each of the designs exhibited each of the elements. This meant that each participant made a total of 6,192 responses on Likert scales! Examples of elements used in this case were: showy, calm, masculine, feminine, soft, individual, high-grade, sweet, milky, etc.

A cluster analysis was then carried out in order to determine how panellists tended to group the designs according to their elements. A number of clusters emerged which Ishihara *et al.* (1997) were able to identify as being linked to particular design features. For example, one cluster of cans emerged that were regarded by the panellists as being 'milky', 'soft' and 'sweet' – this cluster of cans was characterized by the use of beige colouring for most of their surface. Another cluster was seen as being 'masculine', 'adult' and 'strong' – these elements were all associated with having a large logo on the cans. A third example of a cluster was one that was seen as being 'unique', 'sporty' and 'individual' – this was related to the use of blue and white colouring in the designs.

Kansei engineering represents a thorough empirical approach to the linkage of emotional response and product design. It seems that it could be used to address many of the issues that fall within the social, psychological and ideological sections of the four-pleasure framework. Bandini-Buti *et al.* (1997) developed a similar approach for analysing the link between physiological pleasure and the tactile aspects of a product. Called the Sensorial Quality Assessment Tool – SEQUAM, the tool has been applied in the automotive sector by Fiat.

Similar approaches have been advocated in context of marketing campaigns and promotional material – monitoring how variations in consumer response correspond to variations in particular elements of such material. This might mean, for example, variations in, say, colour, text or graphics on visual material, or stylistic or production elements of audio or audio-visual material. It could equally refer to elements of the manner in which the product is promoted (e.g. on the basis of value for money, luxury, youth, sophistication, etc.). An overview can be found in Urban and Hauser (1993).

A problem with these formal approaches is that they can be time consuming, particularly Kansei engineering – having to design seventy-two concepts just to decide on what a coffee can should look like. Formalizing design elements in this way seems to imply acceptance of the philosophy that a design is the sum of its parts. Some would dispute this, suggesting that a 'gestalt' model is more appropriate (i.e. that the overall effect generated by a design could be less or more than the sum of individual design decisions and their associated user responses).

A less formal approach is the case study: discussing users' experiences of a product, what they perceive as that product's properties and the types of pleasurable emotions that they feel with the product. Below are some examples, again drawn from the interviews reported in Jordan and Servaes (1995).

Case study 4   CD-player *owner*

This 19-year-old man had received a stereo as a Christmas present. When asked about what aspects of the product he found particularly appealing he mentioned that the CD player was 'less hassle' than a tape recorder – he had previously owned a tape player – as the CD would play for longer than a tape without him having to turn it over. It was also possible to put the CD in repeat play mode, which again enabled him to play music for a long period without having to interact with the player at all. When asked about the emotional benefits, he said that the product gave him a sense of freedom as he was able to set it playing and then listen to the music whilst doing something else.

Here, then, there was an association between one specific feature – repeat play – and one particular emotional benefit – a sense of freedom.

Case study 5   Another hairdryer *owner*

This 19-year-old woman had been given the hairdryer as a present 11 years ago. She found the product particularly pleasurable for two reasons. Like the first hairdryer user, she said that she could use the dryer to style her hair in the way that she wanted and thus it gave her a feeling of confidence in her appearance. Unfortunately, it wasn't clear which property of the dryer had particularly contributed to making it so suitable for styling – presumably it was some combination of technical performance, functionality and usability – but this wasn't expressed clearly.

In addition, though, she also noted that the product's reliability gave her a feeling of confidence in the product itself. Here, then, there is a link between the property of product reliability and the emotional response of confidence in the product.

Case study 6   TV watcher

This 19-year-old woman chose her parents' television as her pleasurable product. She described this as being big and straightforward with an easy-to-use remote control. She said that she particularly liked the TV because of its simplicity, because of the large screen and the good picture quality, and because of the wide variety of programmes that she could choose from (her parents had a cable TV subscription). Again, though, she didn't link these qualities to any particular emotional response, other than indicating that, taken together, they led to her feeling 'satisfied' with the TV.

In addition, however, she mentioned that the TV was very reliable and that this enabled her to take the TV for granted. Here, then, there was a link between the property of reliability and a feeling of 'security' that nothing would go wrong.

Challenge 3   Developing methods and metrics for the investigation and quantification of pleasure

A major aid to the incorporation of usability issues in the design process has been the ability to quantify these issues. In the case of effectiveness and efficiency this has usually meant taking performance measures of a user with a product. These include, for example, task success and quality of output (effectiveness) and time on task and error rate (efficiency). Quantitative-attitude scales, such as the System Usability Scale (Brooke 1996) and the Index of Interactive Difficulty (Jordan and O'Donnell 1992), have also been developed to measure the satisfaction component of usability.

Being able to quantify usability has enabled human-factors specialists to set usability specifications which have then been included as part of the overall product specification. As well as giving clear usability targets to aim at, this approach has proved effective as it has given clear signals to others involved in the product-creation process as to what is meant by usability in a particular context. Talking in terms of 'usability' or 'quality of use' can seem vague and 'woolly', particularly to our technical and market-research colleagues, who are used to numerical specification. Quantifying the issues gives an unequivocal signal as to what is required and enables an equally firm judgement as to whether or not the criteria have been met. It follows, then, that if human factors as a profession is to take a leading role in ensuring product pleasurability, then the development of measures and tools for quantifying pleasure will be beneficial if not essential.

Again, one approach to this would be to develop attitude scales for measuring pleasure. This would require a knowledge not only of the potentially pleasurable responses that could be associated with product use, but also an idea of the comparative importance of each and the relationship between the responses: which responses are conceptually separate in the context of associations with products (i.e. which responses are potentially, if not always, independent of each other).

A significant difficulty with analysis of emotions and feelings lies in the vocabulary. It is not always clear whether the investigator and the respondent share common interpretations of what a particular descriptor – such as 'delight', 'comfort' or 'enthusiasm' – means. Presumably, interpretations of these descriptors may be culturally loaded, notwithstanding the individual differences in interpretation that may occur. These problems are accentuated when, for example, translating the items into other languages (there may be English words that are difficult to translate precisely into, say, French or German).

Another way to investigate pleasure with products may be to look at potential behavioural correlates to pleasure and displeasure. For example, the frequency with which a person smiles when using a product may be seen as a simple measure of pleasure. Similarly, frowning may be seen as a simple measure of displeasure. Whilst measures such as these may be simple to take and appear to have a degree of objectivity that is lacking in questionnaire or interview-based approaches, they still appear, a priori, to have a number of drawbacks associated with them (e.g. there are some positive emotions with which smiling may not be associated – many of the ideological pleasures). Further, there are many reasons for which a person may smile that have nothing to do with what they are experiencing with the product. Perhaps their mind wanders and they think of something else. Perhaps they are amused about how awful the product is!

Despite these drawbacks, facial expressions do seem a promising way forward. There has been a history of research into these issues, both in the psychology profession in general and in the human–computer interaction (HCI) community in particular. For an example of the former see Ekman and Friesen's (1978) proposed coding system, while Oatley and Ramsay (1992) provide a good example in the context of HCI.

Other simple metrics of pleasure may be the frequency with which people use a particular product – where use of the product is voluntary – and hopefully, from the point of view of those who manufacture pleasurable products, purchase choice.

Of course, creating pleasurable products requires more that just effective evaluation methods. It is also very important to have methods available for capturing user requirements with respect to pleasure as a starting point in the product-creation process. Similarly, methods for early concept evaluation are also needed.

Human factors already has many methods that are likely to be effective here (e.g. the interview, focus group and questionnaires would surely be staples for anyone addressing these issues). Nevertheless, it would be prudent to look at other potential methods for addressing these issues. One promising new method is the Private Camera Conversation, developed by de Vries *et al.* (1996). This involves participants giving monologues to a camera in response to written questions. De Vries *et al.* claim that this method can provide rich information about users' relationships with products – participants cover matters that they may seem reticent to talk about face to face with an investigator for fear that it may seem 'silly' to talk about relating to a product emotionally.

Another potentially promising method is the repertory grid (Kelly 1955). Baber (1996), for example, has adapted this to human factors for determining people's ideas about desirable design features. He also suggests that the approach would be useful as a means of defining users' conceptions of usability. Presumably, the same could hold true for investigating pleasure with products.

17.9   Conclusions

Human factors has come a long way over the last couple of decades: from being something that was often totally ignored, to being fully integrated in the design process within many organizations. More often than not human factors' contribution to the design has been to incorporate usability issues. The profession has become sophisticated in linking design decisions with usability and, thanks to the development and acquisition of a variety of methods, in the investigation of usability issues.

However, a usability-based approach is limited. Products have the potential to bring a wide range of emotional benefits – pleasure. Ensuring that these are delivered requires attention to more than just usability. A four-pleasure framework could be a useful basis for an integrated approach to creating pleasurable products, by providing a structure within which potential product benefits can be identified.

Other disciplines involved in the product-creation process traditionally take a fairly wide perspective on the issue of users' relationships with products. This is not so with human factors, which has tended to focus narrowly on usability. In order to take a proactive role in the development of pleasurable products, human factors should meet three main challenges:

- understand users and their requirements – understand the emotional and hedonic benefits that a product can bring to users;
- link product properties to emotional responses – which properties of a product elicit which responses;
- develop methods and metrics for the investigation and quantification of pleasure – the ability to quantify usability has been a central reason for the ever-growing influence of human factors. A similar approach is required if our colleagues in other disciplines are to be persuaded to take pleasure seriously as an issue.

The last couple of decades have been good ones for human factors. The discipline has extended its influence and, through a usability-based approach, has contributed enormously to the creation of products that are not only functional, but which provide a high quality of use. Now the profession has the opportunity to move a stage further, to move beyond usability, to ensure pleasurable products that will delight.

18 National cultures and design

Patrick W. Jordan

In this chapter, five dimensions are described, along which national cultures can be differentiated. Cultural clusters are identified. These clusters are groups of countries who are culturally similar. It is suggested that culture may be linked to aesthetic preference and that it is important to understand these links when designing products for global distribution. Cultural-clustering approaches differ from traditional geographical-based approaches to global-market segmentation. Research into cultural–aesthetic links is reported, along with a summary of links established to date. It is concluded that there is much research still to be done. This represents a major challenge which human factors as a profession must embrace if it is to move beyond usability and contribute to the creation of products which are a positive pleasure to own and use.

18.1 Introduction

As a discipline, human factors has long been aware of human diversity, Indeed, it has been the need to understand human diversity which has been central to the role of human factors in the product-creation process. Just because a designer is able to use the product that he or she has designed doesn't necessarily mean that others can. A role of the human-factors specialist is to understand the characteristics of those who will use the product and to specify what properties the design must have in order to fit these user characteristics. Traditionally, human factors has tended to concentrate on fitting products to the physical and cognitive characteristics of those who will use them. Basically, the idea has been that if a person can use a product without experiencing an uncomfortably high physical or cognitive load, then the product is 'usable' and thus 'fits' the person for whom it was designed.

A limitation of usability-based approaches to human factors is that they tend to encourage the view that users are merely cognitive and physical components of a system comprising of the user, the product and the environment of use. The premise on which these approaches are based appears to be that the product must be designed such that the cognitive and physical demands placed on the users are minimized – that the demands do not exceed the person's processing capacity. The problem, then, with usability-based approaches is that they encourage a limited view of the person using the product. This is – by implication if not by intention – dehumanizing.

New human-factors approaches have tended to be 'pleasure based'. Such approaches look at people not merely as users, but holistically. In addition to understanding the cognitive and physical characteristics of people, such approaches take into account characteristics such as lifestyle, attitudes, values, hopes, aspirations, identity – the very things which make us human.

It has been argued that in order to understand people, it is necessary to understand three levels at which people are influenced – human nature, culture and personality (Hofstede 1994). Human nature refers to characteristics that are inherited as part of our 'instincts'. This may have implications for design in that our instincts may encourage particular responses to different shapes or colours (e.g. many people have fear responses to snakes and spiders). These are, of course, potentially dangerous animals to humans and we have, it seems, been preprogrammed to have a fear of them. This fear reaction may be triggered by such cues as the shape and colour of the snake or spider. Indeed, many people will have fear reactions to photographs of snakes or spiders. This is in contrast to reactions to photographs of guns or cars, which rarely incite fear. Of course, vastly greater numbers of humans are now killed by guns and cars than by snakes or spiders. This seems to suggest that such reactions are indeed instinctive and are indeed triggered by visual cues. It may then be possible to create designs which trigger particular instinctive responses. For example, Macdonald (1999) reports on the use of form and layout in car design in order to trigger such responses.

The second level is that of culture. People are bound to be influenced by the cultures by which they are surrounded: the culture in which they live and, perhaps more significantly, the culture in which they grew up. Different cultures may have different norms, values and customs. They may also respond to particular aesthetics in a way that is related to these norms, values and customs.

The top level is personality. People have personality traits which may incline them towards certain lifestyle choices and value judgements, as well as towards particular aesthetic preferences. Research has shown that people often think of products as having personalities of their own. Perceptions about product personality tend to be linked to the aesthetic properties of a product. Interestingly, the research showed that people tended to prefer products which they felt had similar personalities to themselves.

The focus in this chapter is on the effect of culture on aesthetic preference.

In the late 1960s and early 1970s Dutch Anthropologist Geert Hofstede conducted a study in over seventy countries in order to establish a set of dimensions by which national culture could be defined. He found five separate dimensions along which national cultures may differ from each other (Hofstede 1994). Following a questionnaire-based survey, Hofstede scored the culture of each country with respect to each dimension. The dimensions are described in the following sections.

18.2   Power distance

Power distance is the extent to which people accept that power is – and should be – distributed unequally.

The USA and most Northern European countries tend to have comparatively low power-distance cultures, whereas South American and Asian countries tend to have comparatively high power-distance cultures.

In high power-distance cultures parents tend to teach their children obedience and children, in turn, tend to show a great deal of respect for their parents. In low power-distance cultures parents tend to treat children as equals, something which is reciprocated by the children. These attitudes resurface later in the workplace. In high power-distance cultures subordinates expect to be told what to do. The ideal boss in such cultures is a benevolent autocrat. In low power-distance cultures, by contrast, subordinates tend to prefer to be consulted and the ideal boss is a resourceful democrat.

In high power-distance cultures it is accepted, indeed expected, that wealth and authority should go together. Privileges and status symbols are both expected and popular. This contrasts with low power-distance cultures where privileges and status symbols are frowned upon. Whilst flaunting wealth might be considered appropriate behaviour in high power-distance countries, people in low power-distance countries might see such behaviour as either distasteful or laughable.

18.3   Individualism

Individualism is the extent to which people see themselves as separate from others in society. In countries with individualistic cultures people tend to be brought up to look after themselves and their immediate family only. Identity is based on the individual. This contrasts with collectivist cultures, in which people tend to be born into extended families and identity is based on the social network to which a person belongs.

In collectivist countries, there tends to be pressure to maintain harmony and avoid direct confrontation, avoiding loss of face is seen as important. In individualist countries, by contrast, individual assertiveness may be admired and respected.

In the workplace hiring and promotion decisions are supposed, in individualist countries, to be based mainly on ability. This contrasts with collectivist countries, where it is expected that people will be treated differently dependent upon whether or not they are in the right in-group. It may also be the case that in individualist countries the focus in the workplace is on the tasks that need to be done, whereas in collectivist cultures the focus tends to be more on the relationship between employees.

North American and Northern European countries tend to have individualistic cultures, whereas South American and Asian countries tend to be collectivist.

18.4   Toughness

Toughness is the extent to which achievement and success are valued. Hofstede refers to this dimension as 'masculinity'; however, 'toughness' seems a more appropriate descriptor given the definition of the dimension (see Hofstede 1994 for a detailed description of the dimension).

The USA and the UK are examples of Western countries with tough cultures. In such countries success and achievement are highly valued and the social ideal may be seen as a high-performance society. The dominant values in society are material success and progress.

This contrasts with countries with tender cultures, such as the Netherlands and Sweden, where the dominant values in society are related to caring and where a welfare society is seen as ideal.

In tough countries qualities such as ambition and assertiveness are highly valued, whilst in tender countries people may put a higher value on modesty.

In education, failure can be seen as a disaster in tough countries – an attitude reflected in the high number of academic suicides in the UK and USA. In tender cultures academic failure would be seen only as a minor setback.

Tough cultures promote competition and performance. Conflicts tend to be solved by fighting them out. Tender cultures stress equality and quality of life. Conflicts tend to be solved by compromise and negotiation.

The emergence of women's rights tends to become manifest in tough cultures by the admission of women to positions which would previously have been occupied mainly by men. In tender cultures it has become manifest by men and women taking equal shares both at home and at work.

18.5   Uncertainty avoidance

Uncertainty avoidance is the extent to which people feel threatened by ambiguity. The USA, UK and Scandinavian countries are examples of nations with low uncertainty-avoidance cultures. In these countries people tend to accept uncertainty as a normal feature of life and each day is accepted as it comes. This contrasts with high uncertainty-avoidance countries, such as Japan and most South American countries, where the uncertainty inherent in life is seen as a continuous threat which must be fought. In low uncertainty-avoidance countries what is different may be seen as being interesting, whereas in high uncertainty-avoidance cultures what is different may be seen as a threat.

High uncertainty-avoidance countries tend to be conservative and may worry about changes in society. This can lead to a negative view of young people. Low uncertainty-avoidance countries tend to be tolerant and embrace change. They tend to put a high value on youth.

18.6 Long-term orientation

Long-term orientation is the extent to which people are future oriented. The North American countries and the UK are examples of short-term oriented countries. In such countries people tend to want instant or quick rewards for their efforts (e.g. people may want to spend any money that they make quickly, buying the latest things, even if they cannot really afford it). People save very little of their earnings and credit facilities are often used.

People in long-term oriented countries – which include the vast majority of Asia Pacific countries – tend to save a considerably higher proportion of their incomes and may be more critical in terms of purchase choices.

In terms of attitudes towards work, people in short-term oriented countries tend to want very quick results from their actions, whilst those in long-term oriented cultures may be more willing to persevere towards slow results.

18.7 Cultural groupings

By correlating the cultural-dimension scores of each of the surveyed countries it is possible to cluster countries into cultural groupings. From a correlation analysis, eight separate cultural groupings emerged (Jordan 2000). These are described below. The countries in each cluster are listed and the mean correlation between the cultural dimensions of the countries in each cluster are listed.

Democrats

Australia, Canada, Ireland (Republic), New Zealand, South Africa, the UK, the USA
(Mean inter-country correlation = 0.91; $n = 7$)

Perhaps the most striking feature of this cultural cluster is its emphasis on the rights and welfare of the individual. Indeed, the constitution of the USA could be seen as archetypical of the values of this group, protecting, as it does, the freedom of the individual against the power of the state. The countries in this cluster also tend to be low power-distance cultures, putting a high value on democracy and being disapproving of ostentatious displays of status. The cultures in these countries are also likely to be ones in which little respect is given to status. This is also the most short-termist of the clusters (e.g. people in these cultures tend to put a comparatively low proportion of their incomes into savings and investments).

These cultures are comparatively comfortable with uncertainty and may exhibit a dislike for rules and regulations and a positive attitude towards flexibility and initiative. They tend to tread the middle ground in terms of toughness versus tenderness – while social justice is important, it is felt that those who are successful are entitled to significantly higher rewards than those who are less so.

Meritocrats

Austria, Germany, Israel, Italy, Switzerland
(Mean inter-country correlation = 0.90; $n = 5$)

Like the democrat cultural cluster, this group tends to emphasize social justice and rewards for success in equal measure (moderate toughness). In practice this means, for example, a taxation system which raises sufficient money for welfare programmes whilst at the same time avoiding punitive levies on the incomes of the better off. The cultures in this cluster tend to be individualistic, but far less so than the democratic cultures. The freedom of the individual will be balanced against the demands of others. This may lead to laws which give the state a significantly greater say in how people live their lives than in the democratic countries. For example, in Germany it is forbidden for people to wash their cars on a Sunday in case this annoys their neighbours. These countries also tend to have far more rules and regulations governing work and business practices than do countries in the democrat cluster. Indeed, this is often cited as a cause of the failure of many European countries to compete with the USA in terms of development of IT.

These cultures tend to be uncomfortable with uncertainty and people in these cultures will often feel more secure when behaviour is governed by procedures and rules. However, like the democrat cultures, they tend to prefer low power distances – indeed, this cluster has the lowest mean power-distance coefficient – and to be very short-termist in orientation.

Egalitarians

Denmark, Finland, The Netherlands, Norway, Sweden
(Mean inter-country correlation = 0.89; $n = 5$)

Like the democrat cluster the egalitarian cluster consists of cultures that tend to put a strong emphasis on both equality (low power distance) and the rights of the individual (high individualism). However, the most striking feature of this cluster is the emphasis on tender values over tough ones – this cluster is the most tender by a significant margin. Again, the taxation and welfare systems provide a strong illustration of this. These countries tend to have excellent social-welfare systems, but at the expense of taxation rates that Anglo-Saxon countries might see as punitive.

These cultures tend to be comparatively short-termist, but significantly less so than the democrat countries, with a significantly higher percentage of income being saved or invested. Like the democrat cultures, these cultures tend to be comfortable with uncertainty and will be uninclined to enjoy a high degree of reliance on rules and regulations.

Supportives

Belgium, France, Portugal, Spain
(Mean inter-country correlation = 0.86; $n = 4$)

Like the meritocrat cultures, these cultures tend to be very uncomfortable with uncertainty and tend to feel most comfortable when rules and regulations make the demands on behaviour very clear. Indeed, this cluster is the most uncertainty avoiding of all. This cluster also tends to be comparatively tender – significantly more so than the democrat and meritocrat cultures, but far less so than the egalitarian cultures. This cluster also has by far the highest power-distance coefficient of the predominantly Western cultural groups. This may mean that people who belong to these clusters will be drawn to strong charismatic leaders who have clear personal visions – this may apply both in the political arena and in the workplace. They may tend to prefer authoritarian leaders and may settle differences through struggle and conflict rather than consensus and negotiation. This might mean, for example, that strikes and other industrial action may play a significant role in workplace culture.

Libertarians

Jamaica
(Mean inter-country correlation = 1.00; $n = 1$)

This cluster contained only one of the countries in Hofstede's study – in this case Jamaica. However, because Jamaica is a comparatively small country and is surrounded by other Caribbean islands, it has been assumed that its culture may be fairly typical of its immediate geographical region.

The most striking feature of this single-country cluster is its extremely low uncertainty-avoidance coefficient. This suggests that people in this culture tend to have a relaxed attitude to life and are able to take uncertainty and unfamiliarity easily in their stride. It would also seem to imply a strong dislike for unnecessary rules.

The toughness coefficient for this cluster is also comparatively high, implying that assertiveness and self-expression will be valued over modesty and care for the less well off. However, the low individualism coefficient indicates that people will be aware of their responsibilities to their immediate social groups, such as the family or colleagues.

This cultural cluster has a moderate power-distance coefficient, implying that a balance is sought between consultative and authoritarian leadership. No data was available on long-term versus short-term orientation.

Planners

Japan
(Mean inter-country correlation = 1.00; $n = 1$)

Japan emerged from the analysis in a cluster of its own. It appears to have a mix of cultural elements that is not substantially shared by any of the other countries in Hofstede's survey. This culture appears to be extremely tough and uncertainty avoiding, yet it has only a moderate power-distance coefficient and tends to balance the rights and welfare of the individual evenly against those of the group.

This combination of factors appears to be reflected in, or possibly derived from, the strict behavioural codes that are prevalent in Japanese society, rather than in or from legislation or formal rules or regulations. In Japanese culture, people may tend to see themselves as part of a group; however, they may also have a very well-defined role within the groups to which they belong. The groups may also have very clear and well-established rituals (hence, or because of, low uncertainty avoidance) and letting the group down may be seen as very shameful (toughness).

This one country cluster is the most long-term oriented of all clusters. People tend to be prepared to work very hard today for the promise of a brighter future.

Collectivists

Arab countries, Argentina, Brazil, Chile, Colombia, Costa Rica, East Africa, Ecuador, Greece, Guatemala, Iran, Mexico, Pakistan, Panama, Peru, El Salvador, South Korea, Taiwan, Thailand, Turkey, Uruguay, Venezuela, Yugoslavia
(Mean inter-country correlation = 0.85; $n = 23$)

Although the majority of the cultures which fell into this cluster were South American, the cluster also includes cultures from many other parts of the world, including Arab countries, East African countries, Greece and Turkey and Pakistan. Perhaps the factor most commonly present in the non-South-American countries in the cluster is that they tend to be predominantly Muslim cultures.

This cluster is the most collectivist of all. Again, there will tend to be a large emphasis on the extended family and the rights of the individual will be subsumed to those of the group. These cultures tend also to be characterized by high power distances and high uncertainty avoidance. This suggests that people like to know what the future will bring and want strong leaders with clear and methodical plans for the future.

These cultures tend to value toughness and tenderness in equal measure and are neither particularly long-term nor particularly short-term oriented.

Authoritarians

Hong Kong, India, Indonesia, Malaysia, Philippines, Singapore, West Africa
(Mean inter-country correlation = 0.87; $n = 7$)

This cluster is primarily characterized by its large power-distance coefficient (the highest of all the clusters), its collectivism (this cluster has the second lowest individualism coefficient) and its comfort with uncertain situations (second lowest uncertainty-avoidance coefficient). The high power distance and low uncertainty avoidance might mean that people in this culture are inclined to prefer strong charismatic leaders who they will trust to deal with matters when uncertainty arises. The leader, be he or she the president of the country or the boss in the workplace will not necessarily have to have a clear plan – as long as he or she seems strong and charismatic then people are likely to put their trust in him or her.

These cultures tend to be highly collectivist with emphasis on the extended family and the welfare of the group as opposed to individual. The cultures in this cluster tend to tread the middle ground between toughness and tenderness and between long- and short-term orientation.

A full listing of the countries in each cultural group are given in Table 18.1, along with the mean inter-country correlation on Hofstede's dimensions for the countries within each group.

18.8 Implications for design and human factors

A challenge for human factors is to investigate whether or not there are systematic links between a country's position on a cultural dimension and aesthetic preferences within that country. This is a particularly salient issue in the context of multinational companies who are designing products for distribution in different markets.

Traditionally, manufacturers have tended to cluster their markets geographically (e.g. a range of products might be marketed for Europe, another for Asia Pacific and another for North America).

An implication of the outcomes of the cultural groupings is that, if aesthetic preferences are indeed linked to culture, geographic regional

Table 18.1 Detailed data on the cultural dimensions in national culture clusters as found by Hofstede (1994).

Cluster name ians	Democrats	Meritocrats	Egalitarians	Supportives	Libertarians	Planners	Collectivists	Authoritarians
Variable								
Power distance	36	29	30	63	45	54	67	82
Individualism	81	64	71	56	39	46	24	26
Toughness	61	66	14	43	68	95	46	52
Uncertainty avoidance	45	70	43	93	13	92	81	37
Long-term orientation	28	31	39	–	–	80	57	56
Mean inter-country correlation	0.91	0.90	0.89	0.86	1.00	1.00	0.85	0.87
Cluster size	7	5	5	4	1	1	23	7
Cluster members	Australia Canada Ireland New Zealand South Africa UK USA	Austria Germany Israel Italy Switzerland	Denmark Finland The Netherlands Norway Sweden	Belgium France Portugal Spain	Jamaica	Japan	Arab countries* Central America* East Africa* Ex-Yugoslavia* Greece Iran Pakistan South America* South Korea Taiwan Thailand Turkey	Hong Kong India Indonesia Malaysia Philippines Singapore West Africa*

* Cluster members consisting of more than one country.
Note that Hofstede did not measure Long Term Orientation for all countries in his survey. Therefore scores on this dimension are missing from some clusters.

approaches to design and marketing may be flawed. For example, the UK falls into the same cultural group as the USA and Canada, but does not cluster with any of its European neighbours. This suggests that it may be more appropriate – culturally – to treat the UK as a North American country than as a European nation. This might mean that aesthetics used on North American products might also be appropriate for UK markets, whereas aesthetics which might prove popular in other European countries would not win favour in the UK.

18.9   Research so far

Marieke de Mooij correlated scores on Hofstede's cultural dimensions against data gathered from surveys of consumer behaviour and attitudes conducted in sixteen separate countries (reported by Jordan 2000). On the basis of the outcomes of these correlations, links were made between the five cultural dimensions and people's preferences and tastes with respect to what a product design should communicate through its aesthetics. Some of these links are summarized in Table 18.2.

De Mooij illustrates these links with a series of examples in the context of car design. In tough cultures, for example, people are likely to be more concerned about the performance aspects of their cars than in tender cultures. When people in a number of different countries were asked how big the engine of their cars was, a negative and highly significant correlation was found between toughness and the percentage of people who did not know the size of their car's engine. In other words, people in tough cultures are far more likely to know their car's engine size than are people in tender cultures. This, then, seems to support the assertion that product performance is of a greater interest or importance to people in tough cultures than to people in tender ones. This assertion is further supported by the outcomes of a survey conducted by Eurodata (reported by Jordan 2000) which suggest that people in tender cultures were more likely than those in tough cultures to choose a car on the basis of whether they felt it looked nice, whereas, again, those in tough cultures were more likely to buy on the basis of performance.

Table 18.2 Aesthetic preferences associated with high and low positions on cultural dimensions.

Cultural dimension	High	Low
Power distance	High status	Youthfulness
Individuality	Expressiveness	Familiarity
Toughness	Performance	Artistry
Uncertainty avoidance	Reliability	Novelty
Long-term orientation	Timelessness	Fashionableness

This would suggest that, when designing for tough cultures, manufacturers should employ aesthetics which emphasize the power of the vehicle, whilst when designing for tender cultures, it may be more important to ensure that the aesthetics fit the tastes and lifestyles of the people for whom the car is designed.

18.10 Cultural clusters vs. geographic clusters

The cultural clusters that emerged indicate that geographical neighbours may not be cultural neighbours. This raises questions about the wisdom of marketing and design strategies which are based on geographical regions.

Typically, multinational firms will tend to think of their markets in geographical terms. For example separate strategies and product/service ranges may be implemented for North America, Europe and Asia Pacific. Considering these regions in terms of their cultures, the North American countries do appear to share a similar culture – the USA and Canada are both democrat – however, in Europe and Asia Pacific the story is more complex (the UK and Ireland cluster with the USA and Canada in the democrat cluster while the other European nations are spread across the meritocrat, egalitarian and supportives clusters). Indeed even within Europe, geographical neighbours are not always cultural neighbours (e.g. although The Netherlands has geographical borders with Germany and Belgium, it doesn't cluster culturally with either of these nations, but rather with the Scandinavian nations in the egalitarian cluster). The situation is equally complex for the Asia Pacific nations. These are spread across the planner, collectivist and authoritarian clusters.

A danger, then, of geographical design/marketing strategies is that they may fail to take into account the cultural diversity which can exist within a particular region. Indeed, Soloman (1996) notes that big cultural differences can occur even within countries (e.g. many European countries have significant immigrant populations who bring their own customs and values with them, whilst the USA has always been a rich cultural melting pot).

18.11 A note on the dimensions and clusters

The clusters were formed using a correlation matrix and countries were grouped with others if their cultural dimensions correlated with a coefficient of 0.8 or more. A striking feature of the clusters is the size of the collectivist cluster as compared with the others. It is also striking that none of the countries in this cluster is Western. A possible explanation for this is that the dimensions identified are more appropriate for highlighting differences between Western cultures than between non-Western ones. Indeed, as Hofstede himself concedes, the cultural background of the anthropologist is always likely to have an effect on the way that he or she views the cultures which they study (Hofstede is Dutch).

18.12 Conclusions

Understanding the culture of others may be an essential element in creating products which are appreciated by people in other parts of the world. It is essential to 'new' human-factors approaches, which attempt to gain a holistic understanding of people to look not simply at people's cognitive and physical attributes, but also at their values, aspirations and lifestyles. An understanding of and respect for cultural differences is central to this approach. A challenge for human-factors is to link a nation's cultural characteristics with the aesthetic preferences within that nation. New human-factors approaches strive to understand people holistically – this includes understanding their culture. The aim is then to link people characteristics to the aesthetic and functional product properties which they prefer. Such holistic approaches go beyond usability, contributing to the creation of products that are not only user friendly, but also a positive pleasure to own and use.

Part V
Appendices

Appendix 1
Summary of guidelines

Konrad Baumann

The following is a selection of the guidelines presented in the chapters of this book. The rationale for each guideline can be found in the chapter from which it is taken. The chapter numbers are indicated in the headings.

Making a picture of the users (Chapter 2)

- Define and limit the circle of potential users as much as possible before beginning the development of a device.

Good communication (Chapter 2)

- Be sure that the users' subject-specific point of view is taken and their terminology used. When designing a user interface (UI), always think of the potential users, their problems and the way in which the device can help them solve these problems.
- All terms used in a UI and user manual (instructions for use) should be taken from common language. Otherwise they will have to be explained in a terminology understood by the potential users.

Adding new functions (Chapter 2)

- The UI of a processor-controlled device must fulfil the same ergonomic demands (e.g. creation of an overall picture of the system, self-descriptiveness) as a UI having purely analogue input and output elements.

Controls (Chapters 2 and 7)

- For each UI use some analogue or pseudo-analogue control or display elements for important functions which give an indication of the purpose of the appliance and its use, and will be remembered more easily.
- Use a wide variety of existing control forms to make the UI unmistakable.

Complexity (Chapter 4)

- Reduce the complexity of the UI by:
  1. Limiting the functions to the necessary;
  2. Grouping of controls and display elements; and
  3. Automation of functions.

Starting the design process (Chapter 4)

- First of all define the necessary functions and elements also in the abstract (i.e. in the form of variables with certain characteristics).
- At the beginning of development, alternate the usage of creative and analytic methods.
- For every UI design project set up a requirements document.
- Defined procedures are a good way of achieving quality. Optimize the development procedures.
- Pay attention to the details.

Dialogue design (Chapter 5)

- The human–machine dialogue should be kept simple. No irrelevant information should be exchanged.
- The strain on the user's memory should be kept as low as possible. The human–machine interface should be consistent. The same terms, representational forms and actions must always have the same meaning and same effect.
- It must be possible to leave the current system status or interactive step at any time. All actions must be reversible.
- Error messages should be polite, precise and constructive. Error prevention is better than error handling.

Task oriented as opposed to function oriented (Chapter 5)

- Base the dialogue design on the user tasks fulfilled with the appliance and not on the technical functions of the appliance.

Self-explanatory user interfaces (Chapter 5)

- Structure the UI in such a way that the overall picture of the system is easy to understand (internal compatibility).
- Look for possible analogies between the functions of the input and output devices, and make these discernible (external compatibility).

Feedback (Chapter 5)

- For every single operation the appliance should give obvious and immediate feedback to the user.
- For every feedback, it is advisable to use the same information channel (tactile, acoustic or visual) as the relevant user input.
- All controls, particularly those for steering and control procedures, should have tactile feedback (e.g. a pressure point or perceptive steering sensitivity).
- It should be possible to switch off acoustic signals used as feedback or output. Warning tones should not be affected by this.

Response time (Chapter 5)

- For a key-press, the recommended feedback time is less than 0.1 seconds.
- For the results of an action (e.g. involving computing), feedback should be given within 3 seconds.
- For starting or rebooting a computer system, a delay of up to 1 minute is acceptable.
- The response time should always stay the same for a specific task. A constant time between every action and the corresponding feedback is even more important than how long this time actually is (Spinas 1983).
- There is also a minimum recommended speed for scrolling. It must be possible to perform scrolling slowly enough to keep control over it and stop it at any given event.
- If a delay is longer than the recommended time, the system must show that it is still working by a constantly updated feedback message. Preferably the remaining waiting time is displayed.

Windows (Chapters 5 and 12)

- Windows should be used if the available display space is not sufficient for a clear presentation of the necessary information.
- Windows should also be used in particular for creation of analogies wherever there is insight into other levels, files, procedures, etc.
- The window frame should contain a title (upper edge), 'close' and 'minimize' icons (upper left or right corner), one or two scroll bars (right and lower edge) and a virtual handle for adjusting the size (lower right corner).

Analogies (Chapter 5)

- Whenever something is described as being 'like' something then an analogy is being made.
- Metaphors should be applied carefully, otherwise you do not get a good and consistent mental model.

Menus and menu trees (Chapter 6)

- An interactive on-screen structure should be no more than three levels deep for novice users and not more than five even for experienced users.

Screen design (Chapter 6)

- The recommended font size is $h = 0.0052d$ for a given reading distance d.
- Plan and conduct regular screen tests.
- Prioritize buttons in an on-screen interface using screen-design elements which attract the eye appropriately (e.g. colour, size, or animation).
- Use locations close to edges of the usable screen for standard navigation buttons throughout the application.
- Provide enough empty space in the screen design.
- Start your design by using a grid.
- Provide safety areas and empty spaces.
- Take the symbolism of different colours into account.
- Do not leave it till the last minute to test the colour-design in the actual screen.
- Design for all viewers, including the mature and the colour-blind.
- Test and fine-tune contrast by converting your design into greyscale.
- Make sure you know the range of light intensity and design accordingly (day screens, night screens, backlight, etc.).
- Plan and take time for iteration on the design, in order to achieve an optimal balance, accommodating user needs, aesthetic concerns and the application requirements.
- Text should be large enough for the expected target audience and appropriate to the device screen(s) that may be used.
- For a UI with a high number of functions use text-based navigation, soft keys, animated navigation techniques and more graphics (e.g. scrolling panes, three-dimensional rotating objects).

Analogue or digital (Chapter 7 and 12)

- Use analogue displays for supervision and controlling of one or more values. Use digital displays for exact reading of a value.
- Analogue controls (joystick, toggle switch, rotary control) instead of keys justify the higher expense through their added ergonomic value.

Virtual controls (Chapter 7)

- As virtual controls permit a great deal of freedom with the design, it is very important to adhere to certain standards and conventions in order to give the users the feeling of familiarity. In the case of virtual controls, allow several methods of operation as an alternative (e.g. operation of the menu using cursor and mouse/using a keyboard shortcut/using arrow keys; digital/analogue control of a virtual shuttle).
- Use symbols and direct manipulation wherever possible.

Keyboards (Chapter 8)

- For the input of digits, both the telephone-type and the calculator-type keyboard are suited equally well.
- For the input of larger quantities of text, use the conventional QWERTY typewriter keyboard. The keyboard should not be arranged linear-alphabetically.
- For the input of smaller quantities of text or in the case of lack of space, use a frequency-centred keyboard.
- Text input on a numeric (telephone) keyboard can be enhanced by a word-recognition method.

Input of figures using arrow keys (Chapter 8)

- Setting a given value by means of arrow keys or an analogue control should in any case not take longer than 10 seconds.
- In the display, digits with a rate of change of more than 5 Hz should not be continually updated nor set at value 0. Update frequencies of more than 5 Hz are no longer visible.

Key-press repeat function (Chapter 8)

- The initial repeat frequency immediately after a key-press should be around 2 Hz.
- In the case of multilevel repeat functions, the next higher frequency should occur after approximately 4 s. The second frequency should be

between 4 Hz and 10 Hz, with 5 Hz as the optimum. A multilevel repeat function should be similar to a quadratic graph.

Speech-control applications (Chapter 10)

- A key feature of the successful, operational speech-control applications is that they are not safety critical.
- When applications that are safety critical use speech control, it is important that some means of backup (i.e. reversion) are designed into the system.

Speech-control technology (Chapter 10)

- Match the type of work the speech recognizer is intended to perform with the characteristics of the available technology.
- Get to know the hardware and software environment in which the speech-control application is intended to function – the recognition system needs to work in conjunction with other parts of the application.
- According to the application decide whether the technology needs to be speaker dependent or speaker independent, and whether it needs to handle isolated or connected (continuous) speech.
- Because of the problems of end-point recognition, isolated item recognition should be used in safety-critical applications.
- Recognition is more successful with smaller vocabularies (e.g. less than one hundred items). However, the introduction of syntax can increase the vocabulary set with no loss of recognition performance.
- Involve users in evaluating the usability of the system.

Speech-control users (Chapter 10)

- Speaker-dependent recognizers have a greater likelihood of achieving successful recognition since the users are cooperative in that they have to train to use the technology. However, the choice of speaker-dependent versus speaker-independent systems will depend on the application.
- Ensure cooperation of the users (i.e. dedicated users who are motivated to use the technology).
- Take into account the characteristics of the user group: individual differences in terms of physical, cognitive and linguistic capabilities, response to stressors and level of experience.
- Develop training regimes that provide a robust set of information for the recognition system, if speaker-dependent technology is to be used.
- Ensure that enrolment takes place in the same environment as the working environment of the recognizer.
- Consider ways to accommodate voice fatigue.

Speech-control tasks (Chapter 10)

- Select tasks carefully taking into account the criticality of achieving successful recognition.
- Decisions relating to the acceptable level of recognition performance will have to be made on an individual basis with respect to the nature of the task (e.g. regarding issues of safety).
- Carefully select vocabulary size and type.
- Carefully design dialogue, feedback, error-recovery procedures and reversion techniques. Take into account the nature of the task and the criticality of achieving correct input (e.g. in some applications it might be possible to ignore occasional errors as the efforts involved in correction may not be justifiable).
- Take into account task-related characteristics such as wearing protective clothing that may muffle speech, the embarrassment of users being seen talking to a machine in a public place, etc.

Speech-control environment (Chapter 10)

- Take into account the amount and type of noise in the environment.
- Take into account other stressors in the environment.
- Take into account the amount of noise produced by the machine itself.
- Make a cost–benefit analysis regarding the use of speech control for the specific application.

Analogue displays (Chapter 12)

- Analogue displays are suitable for quick control and monitoring of several variables.
- Between two numbered main marks on a scale there is either a median or a median and eight small marks. Other scales are not recommended.
- On a scale, the height of the main marks H is calculated from the reading distance D and the proportionality factor P, where $90 < P < 150(H = D/P)$.
- The pointer must not cover scale marks and numbers.
- For circular displays, the scale should begin between the 6 and the 9 o'clock position.
- Displays should be legible even if the user does not look orthogonally at their surface.
- Colour can be used for redundant coding of display sectors.

Digital displays (Chapter 12)

- Digital displays are suitable for exact reading of values.
- Binary display elements which show different variables should be

clearly distinguishable in at least one feature. The features are size, shape, colour, brightness and frequency of flashing.

- On a numeric display there should be a larger gap between groups of three figures.
- An alphanumeric display for one character should consist of a matrix with a minimum of 5×7 dots.
- When using a 5×8 or 5×9 dot matrix for display of a character, the representation of descenders becomes possible and legibility improves.
- An $N \times M$ matrix ($N > 25$, $7 > M > 9$) is suitable for the representation of proportional typeface and ticker-tape displays. Proportional typeface is easier to read, looks better and needs less space.
- Ticker-tape displays should be shifted dot by dot to the left or upwards. The scroll should not be faster than half the maximum reading rate, nor be stopped more than once per line.

Grouping (Chapter 12)

- UI elements (controls, signals) appear to be connected and are perceived as one unit if at least one of the following relations exists between them: proximity, good continuation, similarity, unity.
- On grouping of UI elements, the following criteria should be considered: simplicity, regularity, symmetry, equilibrium.

Symbols and icons (Chapter 12)

- Symbols and icons can be understood much quicker than text. They should replace terms which occur frequently. For novice users the equivalent full text should be added or appear if the user focuses the attention (cursor) on the symbol.
- All the symbols and icons of a UI should be the same size. The majority of them should have the same, virtually square contour and be in grey tones with a standard layout.
- Only a few, very significant symbols and icons of a user interface should differ considerably in colour or shape from the others.
- Symbols and icons should be used for all prominent, important and frequently used items, but not overused for every single unimportant command. This is to avoid screen clutter and memory overload.

Auditory displays (Chapter 13)

- Auditory displays are particularly suitable if the visual-information channel is already overloaded or inactive (e.g. acoustic warning signals, speech output in a car and acoustic coding of flight direction in an aeroplane).

Tactile displays and speech output (Chapter 14)

- Make sure that the user is always in control of the sequential information presentation. This can be realized by:
  1. Allowing a user to stop utterances and/or sound at all times; this is the golden rule of speech interfaces which is often ignored by many telephone systems
     - including mechanisms to speed up or slow down an utterance,
     - allowing a user to skip (part of) the utterance of a single word (part of a word is often enough to grasp the meanings; having to wait until, for instance, a menu option finishes can be irritating especially when the option is not what the user is looking for),
     - providing mechanisms for selecting arbitrary starting points in the serial stream of information,
     - providing different levels of accuracy (e.g. if spelling is important, a spelling mode can be provided).
  2. Using non-speech sounds to present status information and state changes.
  3. Using tactile modalities to convey spatial properties wherever possible.

Standards (Chapter 15)

- Consider standards early in the product-development process.
- If a standard is not available, then think whether it would be in your interests to collaborate in developing one.
- Adhere to available standards as much as possible, but not without questioning – do not follow them blindly.
- Have your own internal standards to make the application of external standards simpler and more efficient.

Evaluation (Chapter 16)

- Even if all the guidelines in this book have been followed, it is advisable to test the usability of the product.
- Usability should be evaluated throughout the development process. Problems detected early can more readily be solved.
- Simulations need not faithfully represent every detail of the UI. A flexible simulation which can rapidly be changed better enables alternatives to be tried out.
- Most major problems can be detected with as few as five test participants.
- Test participants should represent as closely as possible the people who will use the device.

- Test a cross-section of tasks ranging from simple everyday activities to those which explore the lowest levels of the interaction.
- In the test environment, reproduce any conditions in the real environment which are likely to influence usability (e.g. low lighting levels).
- In order to avoid bias, appraisals of usability should not be carried out by the product developers.

Appendix 2
Guide to further reading

Susan Coles

A2.1   What this is

This appendix gives suggestions from this book's contributors about books and other sources on user interface (UI) design; specifically, what are considered classics, especially handy to have or outstandingly inspiring.

The emphasis is on the needs of busy professionals in industry not necessarily familiar with UI design. So the list is highly selective. This appendix has the following structure:

- *Section 3*, an overview of the material in the guide and suggestions for those new to UI design on how to use it;
- *Section 4*, recommended books;
- *Section 5*, recommended other sources (websites, conferences).

A2.2   What this is not

This appendix is not a comprehensive listing of all available publications and other sources in UI design, nor an academic critique of the literature, nor a selection based on an intensive and detailed survey of the literature.

A2.3   General overview and suggestions on how to use the material

A2.3.1   *Choice guide: one top source*

One book probably could not cover everything about UI design, let alone one as an introductory guide. Potentially it could, but it would be a very big (and very expensive) book that was never quite finished.

Forced to make a choice of one source, I would recommend physically attending to the yearly conference on Computer–Human Interaction (see Section A2.5: 'Other sources'). This gives an overview of what is going on in the computer-mediated world of 'design' in its broadest sense. If not possible, the conference proceedings and videos are easily available.

A2.3.2 *Choice guide: design principles last longer than technology*

Some published material is long-lived, and is recommended as better value for money. Such material includes the fundamentals of methods (the basic philosophy), principles of design, how people think, and what motivates, entertains or interests people. Why? Technology changes rapidly. Principles don't. If you know the principles, they can be applied across whatever current technology exists or happens to be capable of.

Most of the references in the 'books' section (Section A2.4) contain good material in this category, especially the handbooks and general introductory texts. Concise ones with headlines and bullet points include books by Donald Norman, Jakob Nielsen and Ben Shneiderman.

Principles and lessons can be derived not only from the summaries and introductions. Case-study examples showing how others have solved method and design issues are useful too. This book includes case studies mostly for small electronic devices. Examples from other application areas are also valuable both to the design of such devices and to the design of larger scale systems. The case studies in Kirwan and Ainsworth (1992) on task analysis might appear at first sight to be not very relevant, covering as they do issues such as nuclear chemical plants. However, ideas can be drawn from them about the design of other systems and their associated equipment. Similarly, it is worth checking out developments in the military and space-systems domain. The chapter by Meister in the *Handbook of Human–Computer Interaction* (Helander [ed.] 1988) gives some insights into the defence-systems field, which is advanced in terms of some of its methods, knowledge and certainly clear adoption of human-factors methodology.

A2.3.3 *Choice guide: information specific to a form of technology will (probably) date quickly*

Relatively short-lived publications include guidelines on UI design for specific forms of technology. Long-lived knowledge is not how to make a webpage for a PC or the small interface on a mobile phone or an interface using windows standards; but rather to find out what it is that needs to be made and given the principles of design therefore how it should be made and on or in what.

One year the window on the world is some box sitting on a desk with multiple windows, a keyboard and mouse; the next year you ask your car as you are driving to work did you really unplug the iron before you left the house.

Many books with UI design in the title are actually about what standard widgets or other screen elements for a particular platform are and when to apply them (style guides). This information on its own does not make a useful or usable device or interface; except to the extent that the items are applied consistently, which is only one aspect of usability. The resulting interface could be consistently inappropriate for what users want to do and consistently difficult to use.

A more useful buy in this category, in terms of learning design principles, could be something like *Designing Web Usability* (Nielsen 2000). Within the context of an existing form of technology (PCs, phones, fridge doors, etc.) and largely the existing concept of navigating through pages, examples of how to improve usability are given using principles such as depth versus breadth, clear indications of where you have been, where you can go and what you will get there, etc. The technical capability of the web will change, the way that the technology is presented and on what will (hopefully) change, the technology may even be replaced. The fundamentals of good design will live on to be adapted and reapplied. The limitations of existing technology place limits on the form that design solutions can take. But knowing what is needed and wanted by users can and should drive developments in technology as well as the form of interface solutions.

A2.3.4 Choice guide: What have fairy tales or plays got to do with user interface design?

People enjoy them, they endure through history (well, some anyway) and they are designed experiences, happening over time. The point is that across quite different domains involving a designed 'something', inspiring parallels can be drawn. *Computers as Theater* (Laurel 1993) explicitly does this, as does Kim Binsted's closing plenary presentation to the CHI 2000 conference 'Sufficiently advanced technology: using magic to control the world'.

Other examples of cross-fertilization across domains in the selection are *Experiential Marketing* (Schmitt 1999) and Patrick Jordan's chapter in the present book on 'National cultures and design'.

A2.3.5 Choice guide: published techniques and methods are examples

Specific techniques for UI design (e.g. hierarchical task analysis, scenario-based design, private-camera conversation, visual mapping) are interesting, but every project, every design problem is different. Techniques may need to be adapted or new ones created. The driver for choice of method is the design problem, not trying to match the problem to a menu of techniques, or worse using a technique then discovering what the design problem is later. Spending 3 months carefully decomposing tasks into neat hierarchies doesn't help unless it actually addresses the design problem that needs to be addressed and efficiently progresses the solution.

If you are new to UI design, the apparently large number of techniques might feel a bit overwhelming. It's important to be aware of major trends in how others are solving problems, but basically the issue is simple: what needs to be known and what's a good way of finding out.

In the books listed below the handbooks and introductory texts (e.g. Nielsen 1993, 2000) are a useful start on what types of methods have been developed and when they help. *Usability Evaluation in Industry* (Jordan *et al.* 1996) contains a large collection of examples. In Chapter 4 in the present book Irene Mavrommati gives an introduction to creativity techniques.

In understanding methods and techniques for UI design, hearing what others have used and when helps, but otherwise it's a case of learning by experience.

A2.3.6   *Choice guide: justifying investment in user-interface design*

A good book on this is *Cost-Justifying Usability* (Bias and Mayhew 1994). It gives clear examples of savings on development projects and check lists for how to work out costs and benefits.

If you have to cost-justify the need for a multidisciplinary approach to design/development, Georg Rakers' chapter on the interaction design process gives some good advice (Chapter 3 in the present book).

Methods for UI design range from 'quick and dirty' (but valid) to extensive and detailed. What is used needs to be carefully balanced with the needs of a project/product within the context of the business. *Usability Engineering* (Nielsen 1993) gives examples of 'quick and dirty' usability methods and their payback value in terms of certainty of results.

A2.4   Recommended books

These are recommendations for people new to the field of UI design, made by the contributors to this book (contributor's name in brackets) on what are considered classics, especially handy to have or outstandingly inspiring. For each book is given:

- the most recent publication information at the time of writing (July 2000), so some books were originally published earlier than shown here;
- a brief description of most of the books given by the person or persons recommending it.

Contents of most of the books, together with reviews and other information on them, can be found at www.amazon.com

Baber, C. (1997) *Beyond the Desktop. ISBN* out of print; *ASIN* 0120695502.
Balentine, B., Morgan, D. P. and Meisel, W. S. (1999) *How to Build a Speech Recognition Application – A Style Guide for Telephony*

Dialogues, Enterprise Integration Group. ISBN 0967 127815; *pages* 319.

Bergman, E. (2000) *Information Appliances and Beyond*, Morgan Kaufman Publishers. ISBN 1-55860-600-9; *pages* 400. [Konrad Baumann] Edited book featuring an interview with Don Norman and contributions from the creators of several good and well-known UIs (e.g. R. Hactani [PalmPilot], A. Marcus [design of a vehicle navigation system], E. Strommen [MS Barney]).

Beyer, H. and Holtzblatt, K. (1997) *Contextual Design: A Customer-Centered Approach to Systems Designs*, Morgan Kaufman Publishers. ISBN 155 860 411 1; *pages* 480.

Bias, R. G. and Mayhew, D. J. (eds) (1994) *Cost-Justifying Usability* Academic Press, ISBN 012 095 810 4; *pages* 334. [Susan Coles] This book describes how to make a business case for usability engineering. Included are case-study examples with calculations of return on investment, and checklists for how to work out cost–benefit balance sheets. Apart from the purely financial side, advice is given on how to market usability.

Birren, F. (1978) *Colour Psychology and Colour Therapy*, Lyle Stuart. ISBN out of print; *pages* 293. [Irene Mavrommati] This book from the pre-digital age, 1950, contains a lot of useful facts about colour and colour psychology. Because most parts refer to non-digital use of colour one has to be critical when using principles or guidelines mentioned there; however, it is useful when it comes to the digital medium.

Dertouzos, M. and Gates, B. (1998) *What Will Be: How the New World of Information Will Change Our Lives*, Harperbusiness. ISBN 006 251 540 3; *pages* 288. [Irene Mavrommati] General-interest book. How the new world of information will change our lives. Predictions about interfaces and subsequent lifestyle changes of the future.

Dul, J. and Weerdmeester, B. A. (1993) *Ergonomics for Beginners: A Quick Reference Guide*, Taylor & Francis. ISBN 074 840 079 6. [Available in Dutch: *Vademecum Ergonomie: Een praktische inleiding in de ergonomie* ISBN 90-267-1358-4]. [Bruce Thomas] This book gives a short and sweet general introduction to ergonomics for non-professionals. It has a wide scope, covering the design of tasks and the environment, as well as the principles of information display and control.

Gershenfeld, N. A. (1999) *When Things Start to Think*, Henry Holt & Company, Inc. ISBN 080 505 874 5; *pages* 244.

Helander, M. G., Landauer, T. K. and Prabhu, P. (eds.) (1997) *Handbook of Human–Computer Interaction*, Elsevier Science Ltd. ISBN 044 481 862 6. [Leo Poll] Excellent reference book which has been through a number of revisions over the years. In general, it deals with a lot of different UI issues varying from UI processes to design guidelines.

Hofstede, G. (1997) *Cultures and Organizations: Software of the Mind, Intercultural Cooperation and Its Importance for Survival*, McGraw-Hill. ISBN 007 029 307 4 (paperback); *pages* 279. [Patrick Jordan] A look at cultural differences through the eyes of one of the world's leading anthropologists. Although the book doesn't explicitly address design or human-factors issues, the analysis of culture provides food for thought for anyone involved in design for global markets. The theories outlined here underpin the approach to global design outlined in my chapter 'National cultures and design'.

Jordan, P. W. (1998) *An Introduction to Usability*, Taylor and Francis. ISBN 074 840 794 4; *pages* 111. [Irene Mavrommati] Textbook aimed as an introduction for students and practitioners approaching usability for the first time.

Jordan, P. W., Thomas, B., Weerdmeester, B. A. and McClelland, I. L. (1996) *Usability Evaluation in Industry*, London: Taylor and Francis. ISBN 074 840 314 0 (cased); *pages* 245. [Bruce Thomas] An overview of some of the methods currently applied to evaluate usability in industrial settings. It identifies issues of concern and ways of dealing with them. Methods range from formal to informal. Although not all of the methods have been validated, all have the merit of actually being applied in industrial contexts.

Kirwan, B. and Ainsworth, L. K. (eds) (1992) *A Guide to Task Analysis*, Taylor and Francis. ISBN 074 840 057 5; *pages* 417. [Susan Coles] This book gives practical information on specific techniques in task analysis (e.g. activity sampling, questionnaires, hierarchical task analysis, timeline analysis), where and how task analysis fits into the development lifecycle, and a section with case-study examples.

Laurel, B. (1993) *Computers as Theater*, Addison-Wesley Publishing Company. ISBN 020 155 060 1; *pages* 227. [Irene Mavrommati] A book dealing with the Aristotle theory of drama, and how it would work when placed in the context of human–computer interaction.

Lichty, T. (1994) *Design Principles for Desktop Publishing*, Wadsworth. [Konrad Baumann] Introduction to layout for books and posters, principles applicable to screen design (How-To Book of the Year 1994).

McKim, R. H. (1980) *Experiences in Visual Thinking*, Brooks/Cole Publishing Company. ISBN 081 850 411 0; *pages* 183. [Irene Mavrommati] Describes relationships between ideas, sketching and imagining, visual perception in general.

Naisbitt, J., Naisbitt, N. and Philips, D. (1999) *High Tech, High Touch: Technology and Our Search for Meaning*, Broadway Books. ISBN 076 790 383 8 [German translation (1999) *High Tech, High Touch*, Signum. ISBN 385 436 265 X; *pages* 274].

Nielsen J. (1993) *Usability Engineering*, New York: Academic Press. ISBN 012 518 406 9; *pages* 362. [Bruce Thomas] Perhaps the standard work on usability methods. [Susan Coles] Written in an easy-to-read style. It explains what usability is, how to incorporate it in the software development cycle, the cost–benefit of doing so and how to do it. There is a section on discount usability engineering, and for beginners exercises complete with hints printed upside down. A large bibliography contains details of many other sources of information with comments from the author.

Nielsen, J. (2000) *Designing Web Usability*, New Riders Publishing. ISBN 1-56205-810-X; *pages* 419. [Adrian Martel] Jakob Nielsen is perhaps the best known personality dealing with the usability of websites. This book is in many ways a handy compilation of the information in his Alertbox column on his website (www.useit.com/alertbox) which is also well worth reading. The main message is 'the practice of simplicity' and the main issues are discussed in an easy-to-read way with many pictures and examples. This is virtually required reading if you are designing a website which you want to make sure people can really use.

Norman, D. A. (1994) *Things That Make Us Smart: Defending Human Attributes in the Age of the Machine*, Perseus Press. ISBN 020 162 695 0. [Irene Mavrommati] The theme of the book is that technology can indeed enhance human intelligence, but only if it is properly built to fit human abilities and needs. Alas! All too often it is not. All too often people must conform to the technology. The proper way is, of course, for technology to conform to people.

Norman, D. A. (1988) *The Psychology of Everyday Things*, Harpercollins. ISBN 046 506 709 3; *pages* 272. [Patrick Jordan] A highly readable text which gives a great introduction to the discipline of human factors. Full of lots of examples of products whose crazy designs make them difficult to use. A fun read, without trivializing the issues, this book is the ideal place to start for those who want to know what human factors is all about. [Leo Poll] This is a very interesting book which is a pleasure to read. It is full of nice examples that clearly highlight some obvious shortcomings of everyday designs. The book highlights in general that we live and always have lived in a physical world. This notion could become more and more important with the advent of the virtual digital world.

Norman, D. A. (1988) *The Design of Everyday Things*, Doubleday Books. ISBN 038 526 774 6; *pages* 257. The same text as above with a different title. [German translation (1989) *Dinge des Alltags*, Campus. ISBN 3-593-34134-4.]

Norman, D. A. (1999) *The Invisible Computer: Why Good Products Can Fail, the Personal Computer Is So Complex, and Information Appliances Are the Solution*, MIT Press. ISBN 026 264 041 4; *pages* 320. [Irene Mavrommati] A general-interest book by the guru D. Norman, dealing with information appliances that fit people's needs and lives, and addressing issues such as the way companies must change the way they develop products, starting by understanding people.

Pheasant, S. (1996) *Bodyspace*, Taylor and Francis. ISBN 074 840 067 2 (cased), 074 840 326 4 (paperback); *pages 275.* [Bruce Thomas] Not directly related to our topic, but this is 'the bible' on anthropometrics. Extremely useful for all those issues to do with body sizes and dimensions or workplaces. [Susan Coles] Contents include a large number of anthropometric tables and clear descriptions of the mathematics of anthropometry (e.g. for estimating figures where none are directly available).

Picard, R. W. (1997) *Affective Computing*, MIT Press. ISBN 026 216 170 2; *pages 304.* [Konrad Baumann] By a guru of the emerging discipline of intelligent agents.

Rubin, J. (1994) *Handbook of Usability Testing: How to Plan, Design, and Conduct Effective Tests*, John Wiley & Sons. ISBN 047 159 403 2; *pages 352.* [Adrian Martel] This book is excellent at doing just what it tells you on the cover: 'How to plan, design and conduct effective tests', and it does this in a refreshingly clear, easy-to-read and non-technical way. It takes you through the whole process including deciding where and how to test, preparing the materials and equipment, testing and analysing the results afterwards. This can be applied whether you are testing a website, a VCR or an exercise bicycle and there are some examples of familiar situations scattered around. It also addresses dealing with ambiguity – an intrinsic part of the job and something I have not seen tackled in this way elsewhere.

Salvendy, G. (ed.) (1997) *Handbook of Human Factors and Ergonomics*, John Wiley & Sons. ISBN 047 111 690 4; *pages 1600.* [Bruce Thomas] An extensive book summarizing a wide range of human-factors material. This is a valuable reference book; if you are looking for specific information in any area of ergonomics, this is a good place to start.

Schmandt, C. (1993) *Voice Communication With Computers: Conversational Systems*, Van Nostrand Reinhold Computer. ISBN out of print.

Schmitt, B. H. (1999) *Experiential Marketing: How to Get Customers to Sense, Feel, Think, Act, and Relate to Your Company and Brands*, Free Press. ISBN 068 485 423; *pages 256.* [Patrick Jordan] This book describes the shift in contemporary marketing away from a cost–benefit model of marketing to an experiential model. Experiential marketing is about creating products and services which not only provide a practical benefit to the customer, but which also carry values and lifestyle associations with which customers can identify. As human factors moves towards pleasure-based approaches, as described in my chapter 'Pleasure with products', there is much that our discipline can learn from experiential marketing approaches.

Shneiderman, B. (1997) *Designing the User Interface: Strategies for Effective Human–Computer Interaction*, Addison-Wesley Publishing Company. ISBN 020 169 497 2; *pages 638*. [Irene Mavrommati] A classic book with guidelines on UI design. [Konrad Baumann] By one of the gurus of this discipline. Despite the size of the book it is easy to navigate within the well-structured text.

Tognazzini, B. (1989) *Apple Human Interface Guidelines: The Apple Desktop Interface*, Addison Wesley. [Konrad Baumann] These guidelines by the UI guru Bruce Tognazzini set the basis for the consistent style of software applications for the Apple Macintosh. This is more than a company style guide and is cited in many other books (e.g. Lauter 1987, Schmidtke 1993, Norman 1989, Helander [ed.] 1988).

Tufte, E. R. (1990) *Envisioning Information*, Graphics Press. *ISBN 096 139 211 8; pages 126*. [Irene Mavrommati] All books by this author are about visualization of information in traditional printed form. Keywords: visual thinking, data graphics, statistic graphics, and information design. [Susan Coles] As with all of Edward Tufte's books this is both informative and beautifully presented. Illustrations of depicting meaning are shown, using examples from throughout history.

Tufte, E. R. (1983) *The Visual Display of Quantitative Information*, Graphics Press. ISBN 096 139 210 X (reprint edition 1992); *pages 197*.

Tufte, E. R. (1997) *Visual Explanations: Images, Quantities, Evidence and Narrative*, Graphics Press. ISBN 096 139 212 6; *pages 157*.

Weinschenk, S., Jamar, P. and Yeo, S. C. (1997) *GUI Design Essentials, ISBN 047 117 549 8; pages 344*. [Adrian Martel] A practical guide to the design of graphical UIs and though most of the examples are drawn from the world of computer applications it also covers the creation of websites. There are a huge number of useful and relevant guidelines, excellent at helping you choose how to lay out your information and controls.

Woodson, W. E. (1987) *Human Factors Reference Guide for Electronics and Computer Professionals*, McGraw-Hill Engineering Reference Guide Series. ISBN out of print; ASIN 0070717664. [Konrad Baumann] A lot of useful examples on how to design controls and displays.

Woodson, W. E., Tillman, B. and Tillman, P. (1991) *Human Factors Design Handbook Information and Guidelines for the Design of Systems, Facilities, Equipment and Products for Human Use*, 2nd edn, McGraw-Hill Inc. ISBN 007 071 768 0; *pages 846*. [Susan Coles] This is a reference guide for a wide variety of human-factors topics including component and product design, controls and displays. The book's coverage is very broad. The content is checklist and bullet-point oriented, with multiple diagrams and tables which make it easy to pick out the information you need quickly (e.g. a 15-point checklist for designing instructions). The information is useful as reminders and as a starting point. There is a large section on human response (e.g. to temperature and vibration) and measurement data.

A2.5   Other recommended sources

Some other most information-rich and quick sources are given below.

Conferences and organizations

Currently CHI and SIGGRAPH are the biggest conferences in the UI design world.

CHI (in full ACM SIGCHI, Association for Computing Machinery Special Interest Group on Human Computer Interaction) The CHI website (www.acm.org/sigchi) contains information on CHI and other conferences, publications for sale, including videos and details of local SIGCHI groups, which exist in many countries. There is a section on recommended reading in the area, and multiple links to other sources.

SIGGRAPH (Association for Computing Machinery Special Interest Group on Computer Graphics and Interactive Techniques) The website is www.siggraph.org and contains extensive information on conferences and other resources.

Other large conferences include:

- *HCI International:* International Conference on Human–Computer Interaction of the British Computer Society. The website is www.bcs.org.uk/hci
- *HFES/IEA:* The International Ergonomics Association (IEA) and Human Factors and Ergonomics Society (HFES) meeting. The websites for these organizations are www.ergonomics-iea.org and www.hfes.org

In addition to the above, the website of the British Ergonomics Society (www.ergonomics.org.uk), is an equally good place to start searching for information.

Websites on user-interface design

The sites and comments are from Irene Mavrommati.

www.asktog.com The website of Bruce Tognazzini, excellent reference! Guidelines, articles, principles, in a very good and regularly updated website. Most recommended parts:

www.asktog.com/basics/firstPrinciples.html

www.asktog.com/columns/022DesignedToGiveFitts.html

www.415.com

www.useit.com   The website of Jakob Nielsen. A very good reference website regarding usability and evaluation. Recommended part: www.useit.com/papers/heuristic/heuristic_list.html

www.cs.ucl.ac.uk/staff/a.sasse/b123/b123book/chp11/chp11.html
Textbook bridging the gap between analysis and design presenting a framework where several analysis and design techniques can be applied. Recommended reading, a bit more academic.

www.symbian.com/epoc/papers/styleguide/styleguide.html   Psion application style guide, includes generic guidelines and principles, converted in a manner to fit the requirements of Psion.

www.tiac.net/users/jasiglar/MMASFAQ.html   Reference to a list of multimedia authoring systems.

www.infoworld.com/cgi-bin/deleteframe.pl?story = /. . ./000529hnhuman. xm   About the contribution of sociologists, anthropologists and other social scientists in the development of new media.

www.acm.org/cacm/AUG96/antimac.htm   Article by Don Gentner and Jakob Nielsen.

References

Ainsworth, W. (1988) *Speech Recognition by Machine*, London: Peter Peregrinus.

Akematsu, M. and MacKenzie, I. S. (1996) 'Movement characteristics using a mouse with tactile and force feedback', *International Journal of Human Computer Studies* 45: 483–493.

Alfano, P. L. and Michel, G. F. (1990) 'Restricting the field of view: perceptual and performance effects', *Perceptual and Motor Skills* 70: 35–45.

Andre, E. and Paiva, A. (2000) *User Modeling and User-Adapted Interaction; Special Issue on User Modeling and Intelligent Agents*, Kluwer.

Andrews, K. (1993) 'Constructing cyberspace: virtual reality and hypermedia', paper given at VR Vienna '93.

Apple Computer, Inc. (1989a) *Human Interface Guidelines: The Apple Desktop Interface*, Addison Wesley.

Apple Computer, Inc. (1989b) *Hyper Card Stack Design Guidelines*, Addison Wesley.

Baber, C. (1991) *Speech Technology in Control Room Systems: A Human Factors Perspective*, Chichester: Ellis Horwood.

Baber, C. (1996) 'Repertory grid theory and its application to product evaluation', in P. W. Jordan *et al.* (eds) *Usability Evaluation in Industry*, London: Taylor and Francis.

Baber, C. (1997a), *Beyond the Desktop: Designing and Using Interaction Devices*, San Diego, CA: Academic Press.

Baber, C. (1997b) *Beyond the Desktop: Designing and Using Interaction Devices*, New York: Academic Press.

Baber, C. and Hone, K. S. (1993) 'Modelling error recovery and repair in automatic speech recognition', *International Journal of Man–Machine Studies* 39(3): 495–515.

Baber, C. and Noyes, J. M. (1996) 'The use of automatic speech recognition in adverse environments', *Human Factors* 38(1): 142–155.

Baber, C., Usher, D. M., Stammers, R. B. amd Taylor, R. G. (1992) 'Feedback requirements for ASR in the process control room', *International Journal of Man–Machine Studies* 37(6), 703–719.

Baber, C., Mellor, B., Graham, R., Noyes, J. M. and Tunley, C. (1996) 'Workload and the use of automatic speech recognition: the effects of time and resource demands', *Speech Communication* 20(1–2): 37–55C.

Baber, C., Johnson, G. I. and Cleaver, D. (1997) 'Factors affecting users' choice of words in speech-based interaction with public technology', *International Journal of Speech Technology* 2(1): 45–60.

Baber, C., Haniff, D. J., Knight, J., Cooper, L. and Mellor, B. A. (1998) 'Preliminary investigations into the use of wearable computers', in R. Winder (ed.) *People and Computers XIII*, London: Springer-Verlag, 313–326.

Baber, C., Haniff, D. J. and Woolley, S. I. (1999a) 'Contrasting paradigms for the development of wearable computers', *IBM Systems Journal* 38(4): 551–565.

Baber, C., Knight, J., Haniff, D. and Cooper, L. (1999b) 'Ergonomics of wearable computers', *Mobile Networks and Applications* 4: 15–21.

Baber, C., Arvanitis, T. N., Haniff, D. J. and Buckley, R (1999c) 'A wearable computer for paramedics: studies in model-based, user-centred and industrial design', in M. A. Sasse and C. Johnson (eds) *Interact '99*, Amsterdam: IOS Press, 126–132.

Baber, C., Harris, T. and Harrison, B. (1999d) 'Demonstrating the concept of physical hyperspace in an art gallery', in S. Brewster, A. Cawsey and G. Cockton (eds) *Human Computer Interaction '99*, volume II, Swindon: British Computer Society.

Baecker, R. M., Buxton, W. A. S. (eds) (1987) *Readings in Human-Computer-Interaction: A Multidisciplinary Approach*, Morgan-Kaufmann.

Bandini-Buti, L., Bonapace, L. and Tarzia, A. (1997) 'Sensorial quality assessment: a method to incorporate perceived user sensations in product design. Applications in the field of automobiles', paper given at IEA '97 Proceedings, Helsinki: Finnish Institute of Occupational Health, 186–189.

Bass, L. (1996) 'Is there a wearable computer in your future?', in L. J. Bass and C. Unger (eds) *Human Computer Interaction*, London: Chapman and Hall, 1–16.

Bass, L., Siewiorek, D., Smailagic, A. and Stivoric, J. (1995) 'On site wearable computer system', paper given at CHI '95, New York: ACM, 83–84.

Bass, L., Kasabach, C., Martin, R., Siewiorek, D., Smailagic, A. and Stivoric, J. (1997) 'The design of a wearable computer', CHI paper given at '97, New York: ACM, 139–146.

Baudrillard, J. (1991) *Das System der Dinge: Über unser Verhältnis zu den alltäglichen Gegenständen*, [The System of Things: Our Relationship with Everyday Objects], Campus.

Baumann, K. (1996) 'Berührungsloses Antennensystem für eine Kraftfahrzeug-Wegfahrsperre', [Contactless Antenna System for a Vehicle Immobilizer] European patent.

Baumann, K. (1999) 'Matrix evaluation method for planned usability improvements based on customer feedback', paper given at HCII '99 proceedings in H.-J. Bullinger, J. Ziegler (eds) *Human Computer Interaction*, Lawrence Erlbaum.

Baumann, K. and Lanz, H. (1998) *Mensch-Maschine-Schnittstellen elektronischer Geräte* [Human–machine interfaces of electronic devices], Berlin: Springer.

Bergman, E. (ed.) (2000) *Information Appliances and Beyond – Interaction Design for Consumer Products*, Morgan Kaufmann.

Binsted, K. (2000) *CHI 2000*, plenary speech, held in Amsterdam, April 2000.

Birren, F. (1978) *Colour Psychology and Colour Therapy*, Lyle Stuart.

Blattner, M. M., Sumikawa, D. A. and Greenberg, R. M. (1989) 'Earcons and icons: their structure and common design principles', *Human Computer Interaction* 4(1): 11–44.

Blenkhorn, P. L. (1992) 'Requirements for screen access software using synthetic speech', paper given at Computers for Handicapped Persons: Proceedings of the 3rd International Conference, Munich: R. Oldenbourg Verlag and Vienna, Österreichische Computer Gesellschaft, 31–37.

Bocher, R. (1989) *Kommunikationsergonomie: Benutzerfreundlichkeit in der Bürokommunikation*, [Communication Ergonomics: Usability in Office Communication] TÜV.

Böcker, H. D. (1993) *Mensch-Maschine-Kommunikation. Benutzergerechte Systeme auf dem Weg in die Praxis*, [Human–Machine Communication. Usable Systems on the Way to Practice] Berlin: Springer.

Boff, K. R. and Lincoln, J. E. (eds.) (1988) *Engineering Data Compendium: Human Perception and Performance*, Harry G. Armstrong Aerospace Medical Research Laboratory.

Bolt, R. A. (1980) 'Put that there: voice and gesture at the graphics interface', *Computer Graphics* 14: 262–270.

Boud, A. C., Baber, C. and Steiner, S. (in press) *Virtual Reality: A Tool for Assembly*, Presence.

Brewster, S. A. (1997) 'Using non-speech sound to overcome information overload', *Displays* 17: 179–189.

Brewster, S. A. (1998) 'The design of sonically-enhanced widgets', *Interacting with Computers* 11(2): 211–235.

Brewster, S. A. and Crease, M. G. (1999) 'Correcting menu usability problems with sound', *Behaviour and Information Technology* 18(3): 165–177.

Brewster, S. A. and Leplatre, G. (2000) 'Designing non-speech sounds to support navigation in mobile phone menus', paper given at International Conference on Auditory Display 2000 (publication of the International Community for Auditory Display), 190–199.

Brooke, J. (1996) 'SUS, a "quick and dirty" usability scale' in P. W. Jordan, B. Thomas, B.A. Weerdmeester and I. L. McClelland (eds) *Usability Evaluation in Industry*, London: Taylor & Francis.

Brule, J. D. A. (1990) 'Workstation for tactile graphics', paper given at Computers for Handicapped Persons: Proceedings of the 2nd International Conference, Munich: R. Oldenbourg Verlag and Vienna: Schriftenreihe der Oesterreichischen Computer Gesellschaft, 19–29.

Budde, R., Kautz, K., Kulenkamp, K. and Zällighoven, H. (1992) *Prototyping. An Approach to Evolutionary System Development*, Springer.

Bullinger, H-J. (ed.) (1995) *Design interacktiver Produkte. Dialog zwischen Mensch und Produkt* [Designing interactive products. Dialogue between man and product], Stuttgart: IRB Verlag.

Burandt, U. (1986) *Ergonomie für Design und Entwicklung*, Schmidt.

Bürdek, B. E. (1991) 'Design', *Geschichte, Theorie und Praxis der Produktgestaltung*, DuMont.

Buurman, R. den (1997) 'Designing smart products: a user-centred approach', paper given at IEA '97 Proceedings, Helsinki: Finnish Institute of Occupational Health, 3–5.

Card, S. K., Moran, T. P. and Newell, A. (1983) *The Psychology of Human-Computer Interaction*, Lawrence Erlbaum.

Carroll, J. M. (ed.) (1991) 'Designing interaction', *Psychology at the Human-Computer Interface*, Cambridge: Cambridge University Press.

Charwat, H-J. (1992) *Lexikon der Mensch-Maschine-Kommunikation*, [Handbook of Human–Machine Communication] Oldenbourg.

Cohen, H. S. and Ferrell, W. R. (1969) 'Human operator decision making in manual control', *IEEE Transactions on Man-Machine Systems* 10(2): 41–47.

Commodore, Inc. (1991) *Amiga User Interface Style Guide*, Addison Wesley.

Conant, R. C. and Ashby, W. R. (1970) 'Every good regulator of a system must be a model of that system', *International Journal of System Science* 1: 89–97.

Cooper, L., Johnson, G. I. and Baber, C. (1999) 'A run on sterling – personal finance on the move', paper given at 3rd International Symposium on Wearable Computers, Los Alamitos, CA: IEEE Computer Society, 87–92.

Cresswell Starr, A. F. (1993) 'Is control by voice the right answer for the avionics environment?', in C. Baber and J. M. Noyes (eds) *Interactive Speech Technology: Human Factors Issues in the Application of Speech Input/Output to Computers*, London: Taylor & Francis, 85–97.

Dandavante, U., Sanders, EB-N. and Stuart, S. (1996) 'Emotions matter: user empathy in the product development process', paper given at Human Factors and Ergonomics Society 40th Annual Meeting, Santa Monica: Human Factors and Ergonomics Society, 415–418.

Davis, K., Biddulph, R. and Balashek, S. (1952) 'Automatic recognition of spoken digits', *Journal of the Acoustic Society of America* 24(6): 637–642.

DIN 19 227: Graphische Symbole und Kennbuchstaben für die Prozeßleittechnik.

DIN 19 235: Meldung von Betriebszuständen.

DIN 2137 Teil 10: Büro und Datentechnik – Tastaturen.

DIN 2137 Teil 6: Büro- und Datentechnik – Tastaturen.

DIN 30 600 Teil 2: Bildzeichen; Übersicht.

DIN 33 414 Teil 4: Ergonomische Gestaltung von Warten; Gliederungsschema, Anordnungsprinzipien.

DIN 40 100: Bildzeichen in der Elektrotechnik.

DIN 66 234 Teil 1: Bildschirmarbeitsplätze; Geometrische Gestaltung der Schriftzeichen.

DIN 66 234 Teil 8: Bildschirmarbeitsplätze; Grundsätze ergonomischer Dialoggestaltung.

DIN ISO 4 136: Anwendung von Pfeilen.

DIN VDE 0199: Codieren von Anzeigen und Bedienteilen durch Farben und ergänzende Mittel; identisch mit IEC 16(Sec)302 und EN 60 073.

Dix, A., Finlay, J., Abowd, G. and Beale, R. (1993) *Human–Computer Interaction*, Hemel Hempstead, UK: Prentice Hall Europe.

Doddington, G. R. and Schalk, T. B. (1981) 'Speech recognition: turning theory to practice', *IEEE Spectrum* (September): 26–32.

Dumas, J. S. and Redish, J. C. (1993) *A Practical Guide to Usability Testing*, Ablex.

Eberts, R. E. (1994) *User Interface Design*, Englewood Cliffs, NJ: Prentice-Hall International, Inc.

Edwards, A. D. N. (1989) 'Soundtrack: an auditory interface for blind users', *Human–Computer Interaction* 4: 45–56.

Ekman, P. and Friesen, W. V. (1978) *Facial Action Coding System*, Palo Alto, CA: Consulting Psychologists Press Inc.

EN 61 310 Teil 1: Sicherheit von Maschinen; Anzeigen Kennzeichen und Bedienen, Anforderungen an sichtbare, hörbare und tastbare Signale.

Engelbart, D. C. (1963) 'A conceptual framework for the augmentation of man's intellect', in P. W. Howerton and D. C. Weeks (eds) *Vistas in Information Handling*, Washington, DC: Spartan Books, 1–29.

England, R. (1992) 'Sensory-motor systems in virtual manipulation', in K. Carr and R. England (eds) *Simulated and Virtual Realities: Elements of Perception*, London: Taylor and Francis, 131–178.

England, R. (1995) 'Sensory-motor systems in virtual manipulation', in K. Carr and R. England (eds) *Simulated and Virtual Realities*, London: Taylor and Francis, 131–178.

Erman, L., Hayes-Roth, F., Lesser, V. and Reddy, D. J. (1980) 'The Hearsay-II speech-understanding system: integrating knowledge to resolve uncertainty', *Computing Surveys* 12(2), 213–251.

Farringdon, J., Moore, A. J., Tilbury, N., Church, J. and Biemond, P. D. (1999) 'Wearable sensor badge and sensor jacket for contextual awareness', paper given at 3rd International Symposium on Wearable Computers, Los Alamitos, CA: IEEE Computer Society, 107–113.

Feiner, S., MacIntyre, B., Hollerer, T. and Webster, A. (1997) 'A touring machine: prototyping 3D mobile augmented reality systems for exploring the urban environment', paper given at 1st International Symposium on Wearable Computers, Los Alamitos, CA: IEEE Computer Society, 74–81.

Fischer, G. (1986) *Methoden und Werkzeuge zur Gestaltung benutzergerechter Computersysteme*, [Methods and Tools for the Design of Usable Computer Systems] de Gruyter.

Fitzmaurice, G. W. and Buxton, W. (1997) 'An empirical evaluation of graspable user interfaces: towards specialized, space-multiplexed input', paper given at CHI '97, New York: ACM, 43–50.

Fitzmaurice, G., Ishii, H. and Buxton, W. (1995) 'Bricks: laying the foundation for graspable user interfaces', paper given at CHI '95, New York: ACM, 442–449.

Fjeld, M., Lauche, K., Dierssen, S., Bichsel, M. and Rauterberg, M. (1998) 'BUILD-IT: a brick-based integral solution supporting multidisciplinary design teams', in A. Sutcliffe, J. Ziegler and P. Johnson (eds) *Designing Effective and Usable Multimedia Systems*, Boston: Kluwer, 131–142.

Flusser, V. (1990) *Ins Universum der technischen Bilder*, [Into the Universe of Technical Pictures] *European Photography*.

Forty, A. (1986) 'Objects of desire', *Design and Society 1750–1980*, London: Thames and Hudson.

Frankish, C. R. and Noyes, J. M. (1990) 'Sources of human error in data entry tasks using speech input', *Human Factors* 32: 697–716.

Frankish, C. R. and Noyes, J. M. (1993) 'Feedback in automatic speech recognition: who is saying what and to whom?', in C. Baker and J. M. Noyes (eds) *Interactive Speech Technology: Human Factors Issues in the Application of Speech Input/Output to Computers*, London: Taylor & Francis, 121–130.

Frankish, C. R., Jones, D. M. and Hapeshi, K. (1992) 'Decline of speech recogniser performance with time: fatigue or voice drift?', *International Journal of Man–Machine Studies* 36: 797–816.

Frei, P., Su, V., Mikhak, B., Ishii, H. (2000) 'Curlybot: designing a new class of computational toys', paper given at CHI 2000, New York: ACM.

Galitz, W. O. (1993) *User-Interface Screen Design*, New York: John Wiley & Sons, Inc.

Galitz, W. O. (1997) *The Essential Guide to User Interface Design. An Introduction to GUI Design Principles and Techniques*, New York: John Wiley & Sons, Inc.

Gaver, W. W. (1986) Auditory icons: using sound in computer interfaces', *Human–Computer Interaction* 2: 167–177.

Gaver, W. (1997) 'Auditory interfaces', in M. Helander, T. K. Landauer, P. V. Prabhu (eds) *Handbook of Human–Machine Interaction*, 2nd edn, Elsevier Science, 1003–1041.

Geiser, G. (1990) *Mensch-Maschine-Kommunikation* [Human–Machine communication], Oldenbourg.

Gemperle, F., Kasabach, C., Stivoric, J., Bauer, M. and Martin, R. (1998) 'Design for wearability', *Digest of Papers of the 2nd International Symposium on Wearable Computers*, Los Alamitos, CA: IEEE Computer Society, 116–123.

Gilmore, W. E. (1989) *User–Computer Interface in Process Control*, Academic Press.

GO Corporation (1992) *PenPoint User Interface Design Reference*, Addison Wesley.

Gobel, M., Springer, J. and Luzak, H. (1994) 'Effects of tactile feedback in process control, exemplary in mouse-driven interfaces', in P. T. Kidd and W. Karwowski (eds) *Advances in Agile Manufacturing*, Amsterdam.

Gould, J. D. (1988) 'How to design usable systems', in M. Helander (ed.) *Handbook of Human–Computer Interaction*, Amsterdam: Elsevier Science Publishers BV.

Gould, J. D. and Lewis, C. H. (1983) 'Designing for usability key principles and what designers think', paper given at 1983 Computer–Human Interaction Conference, 50–53.

Grandjean, É. (1993) 'Physiologische Arbeitsgestaltung', *Leitfaden der Ergonomie*, 4th edn, Ott.

Grant, A. S. (1992) 'A context model needed for complex tasks', in P. A. Booth, and A. Sasse (eds) *Mental Models and Everyday Activities*, 2nd Interdisciplinary Workshop on Mental Model, 23–25 March, Cambridge, UK: Robinson College, 94–102.

Gunzenhäuser, R. (1988) *Prototypen benutzergerechter Computersysteme*, [Prototypes of Usable Computer Systems] de Gruyter.

Haniff, D. J. and Baber, C. (1998) 'Wearable computers for the fire service and police force: technological and human factors', *Digest of Papers of the 3rd International Symposium on Wearable Computers*, Los Alamitos, CA: IEEE Computer Society, 185–186.

Hansen, W. J. (1971) 'User engineering principles for interactive systems', paper given at Fall Joint Computer Conference, 39, Montvale, NJ: AFIPS Press, 523–532.

Hatwell, Y. (1993) 'Images and non-visual spatial representations in the blind', in D. Burger and J. Sperandio (eds) *Non-Visual Human–Computer Interactions. Prospects for the Visually Handicapped*, Colloque INSERM/John Libbey Eurotext Ltd, 228, 13–35.

Healey, J. and Picard, R. W. (1998) 'StartleCam: a cybernetic wearable camera', paper given at International Symposium on Wearable Computing, Los Alamitos, CA: IEEE Computer Society, 42–49.

Health and Safety Executive (HSE) (1992) *Health and Safety (Display Screen Equipment) Regulations*, London: Health & Safety Executive.

Heinbokel, T. (1996) 'Hausgeräte für den Menschen' [Household Appliances for People], in D. Zühlke (ed.) *Menschengerechte Bedienung technischer Geräte* [User-centred Control of Technical Devices], VDI, Bericht Nr. 1303.

Helander, M. (ed.) (1988) *Handbook of Human–Computer Interaction*, Amsterdam: Elsevier Science Publishers BV.

Hellenic Open University (1998) *Open and Distance Learning Handbook*, vol. B, Hellenic Open University.

Heller, M. A. (1989) 'Picture and pattern perception in the able-bodied and the blind: the advantage of the late blind', *Perception*, 18: 379–389.

Helm, L. (1994) *Elektronische Hilfsmittel für Unterricht, Therapie und Umfeldkontrolle*, Herzogenburg, Austria: Eigenverlag.

Henderson, A. (1991) 'A development perspective on interface, design, and theory', in: J. M. Carroll (ed.) *Designing Interaction. Psychology at the Human–Computer Interface*, Cambridge: Cambridge University Press, 254–268.

Herber, H-J. (1985) *Motivationsanalyse, Theorie und Praxis* [Motivation Analysis, Theory and Practice], Expert.

Herrmann, T. (1986) *Zur Gestaltung der Mensch-Computer-Interaktion. Systemerklärung als kommunikatives Problem* [The Design of Human–Computer Interaction. System Understanding as a Communication Problem], Niemeyer.

Herrmann, N. (1988) *The Creative Brain*, Brainbooks.

Heufler, G. (1987) *Produkt-Design – von der Idee zur Serienreife* [Product Design: From Idea to Production], Veritas.

Heuser, K. C. (1976) *Freihändig zeichnen und skizzieren*, Lehr- und Arbeitsbuch [Free-hand Drawing and Sketching, Coursebook]. Bauverlag.

Hewlett-Packard Company (1990) *Corporate Human Factors Engineering*.

Hewlett-Packard, IBM, Sunsoft Inc., USL (1993) *Common Desktop Environment: Functional Specification*, X/Open Company Ltd.

Hill, D. R. (1980) 'Spoken language generation and understanding by machine: a problems and applications oriented overview', in J. C. Simon (ed.) *Spoken Language Generation and Understanding*, Dordrecht: D. Reidel.

Hix, D. and Hartson, H. R. (1993) *Developing User Interfaces: Ensuring Usability Through Product and Process*, John Wiley.

Hjelle, L A. and Ziegler, D. J. (1981) *Personality Theories*, London: McGraw-Hill.

Hofmeester, G., Kemp, J. and Blankendaal, A. (1996) 'Sensuality in product design: a structured approach', paper given in CHI '96, New York: ACM Press.

Hofstede, G. (1994) *Cultures and Organizations: Software of the Mind*, London: Harper-Collins.

Hone, K. S. and Baber, C. (1999) 'Modelling the effect of constraint on speech-based human computer interaction', *International Journal of Human–Computer Studies* 50(1): 85–113.

Honold, P. (1999) 'Focus groups: a qualitative method to elicit culture-specific user requirements', paper given in HCII '99, H.-J. Bullinger and H. Ziegler (eds) *Human–Computer Interaction – Communication, Cooperation, and Application Design*, volume 2, Lawrence Erlbaum (Proceedings of HCII 99).

HSE (1992) *Health and Safety (Display Screen Equipment) Regulations*, London: Health & Safety Executive.

IBM (1992) *Object-Oriented Interface Design: IBM Common User Access Guidelines*, Que.

International Organization for Standardization (1999) *International Classification for Standards*, Geneva: International Organization for Standardization.

Ishihara, S., Ishihara, K., Tsuchiya, T., Nagamachi, M. and Matsubara, Y. (1997) 'Neural networks approach to Kansei analysis on canned coffee design', paper given in IEA '97, Helsinki: Finnish Institute of Occupational Health, 211–213.

Ishii, H. and Ullmer, B. (1997) 'Tangible bits: towards seamless interfaces between people, bits and atoms', paper given in CHI '97, New York: ACM, 234–257.

ISO DIS 9241-11: Ergonomic requirements for office work with visual display terminals (VDTs): Part 11: Guidance on usability.

Jacob, R. J. K. (1990) 'What you look at is what you get: eye movement based interaction techniques', paper given at CHI '90, New York: ACM, 11–18.

Jansson, G. (1992) '3-D perception from tactile computer displays', paper given at Computers for Handicapped Persons: Proceedings of the 3rd International Conference, Munich: R. Oldenbourg Verlag and Vienna: Oesterreichische Computer Gesellschaft, 233–237.

Johannsen, G. (1993) *Mensch-Maschine-Systeme* [Human–machine systems], Springer.

Johnson, G. I. (1996) 'The usability checklist approach revisited', in P. W. Jordan, B. Thomas, B. A. Weerdmeester and I. L. McClelland (eds) *Usability Evaluation in Industry*, London: Taylor and Francis.

Johnson, G. and Rakers, G. G. H. (1990) 'Ergonomics in the development of driver support systems: putting the user back in the driving seat!', paper given at 22nd International Symposium on Automotive Technology and Automation, volume 1, Croydon, UK: Automotive Automation Limited.

Johnson, H. and Johnson, P. (1991) 'Task knowledge structures: psychological basis and integration into system design', *Acta Psychologica* 78: 3–26.

Johnson, P. (1989) 'Supporting system design by analyzing current task knowledge', in D. Diaper (ed.) *Task Analysis for Human–Computer Interaction*, Chichester: Ellis Horwood.

Johnson-Laird, P. N. (1980) 'Mental models in cognitive science', *Cognitive Science* 4: 71–115.

Johnson-Laird, P. N. (1983) *Mental Models*, Cambridge: Cambridge University Press.

Jordan, P. W. (1993) 'Methods for user interface performance measurement', in E. J. Lovesey (ed.), *Contemporary Ergonomics*, London: Taylor & Francis.

Jordan, P. W. (1996) 'Displeasure and how to avoid it', in S. Robertson (ed.), *Contemporary Ergonomics*, 56–61.

Jordan, P. W. (2000) *Designing Pleasurable Products: An Introduction to the New Human Factors*, London: Taylor and Francis.

Jordan, P. W. and O'Donnell, P. J. (1992) 'The index of interactive difficulty', in E.J. Lovesey (ed.) *Contemporary Ergonomics*, London: Taylor & Francis, 397–402.

Jordan, P. W. and Servaes, M. (1995) 'Pleasure in product use: beyond usability', in S. Robertson (ed.) *Contemporary Ergonomics*, London: Taylor & Francis, 341–346.

Jordan, P. W., Draper, S. W., MacFarlane, K. K. and McNulty, S.-A. (1991) 'Guessability, learnability and experienced user performance', in D. Diaper and N. Hammond (eds.) *People and Computers VI*, Cambridge: Cambridge University Press, 237–245.

Jordan, P. W., Thomas, B., Weerdmeester, B. A. and McClelland, I. L. (eds) (1996) *Usability Evaluation in Industry*. London: Taylor & Francis.

Kahn, P. and Lenk, K. (1991) 'Designing for the computer screen', *Information Technology Quarterly* (Winter).

Kamentsky, L. (1983) 'The Kurzweil reading machine: current developments', paper given at IEEE Computer Society Workshop on Computers in the Education and Employment of the Handicapped, 97–100.

Kancler, D. E., Quill, L. L., Revels, A. R., Webb, R. R. and Masquelier, B. L. (1998) 'Reducing cannon plug connector pin selection time and errors through enhanced data presentation methods', paper given at Human Factors and Ergonomics Society 42nd Annual Meeting, Santa Monica, CA: Human Factors and Ergonomics Society, 1283–1290.

Kantowitz, B. H. and Sorkin, R. D. (1983) *Human Factors: Understanding People-System Relationships*, John Wiley.

Kelly, G. A. (1955) *The Psychology of Personal Constructs*, New York: Norton.

Kieras, D. E. (1987) *What Mental Model Should Be Taught: Choosing Instructional Content for Complex Engineered Systems*, Technical Report No. 24 (TR-87/ONR-24), 15 January, University of Michigan.

Kirakowski, J. (1996) 'The software usability measurement inventory: background and usage', in P. W. Jordan, B. Thomas, B. A. Weerdmeester and I. L. McClelland (eds) *Usability Evaluation in Industry*, London: Taylor and Francis.

Klatt, D. (1977) 'Review of the ARPA speech understanding project', *Journal of the Acoustic Society of America* 62: 1345–1366.

Kloecker, I. (1981) *Produktgestaltung* [Product Design], Berlin: Springer.

Knight, J. and Baber, C. (submitted) 'Physical load and wearable computers', paper submitted to 4th International Symposium on Wearable Computers, New York: IEEE Computer Society.

Koch, M. (1991) *Software-Ergonomie* [Ergonomic software], Berlin: Springer.

Koenig, A. (1989) *Desktop als Mensch-Maschine-Schnittstelle* [The desktop as human–machine interface], Berlin: Springer.

Kokkos, A. (1998) 'Techniques in group meetings', *Open and Distance Learning Handbook*, vol. B, Hellenic Open University.

Koons, D. B., Sparrell, C. J. and Thorisson, K. R. (1993) 'Integrating simultaneous input from speech, gaze and hand gestures', in M. T. Maybury (ed.) *Integrated Multimodal Interfaces*, New York: Academic Press.

Körndle, H. (1993) *Mensch-Computer-Interaktion: Psychologische Aspekte des Umgangs mit komplexen technischen Systemen*, [Human–Computer Interaction: Psychological Aspects of Dealing with Complex Technical Systems], Deutscher Uni-Verlag.

Kramer, G. (1993) *Auditory Display: Sonification, Audification, and Auditory Interfaces*, New York: Addison-Wesley.

Kramer, G., Lane, D. M. and Walker, B. N. (2000) 'Psychophysical scaling of sonification mappings', in P. R. Cook (ed.) *Proceedings of the International Conference on Auditory Display 2000*, International Community for Auditory Display, 99–104.

Lamming, M. and Flynn, M. (1994) 'Forget-me-not: intimate computing in support of human memory', paper given at Friend21 '94 International Symposium on Next Generation Human Interface, Meguro Gajoen, Japan: Ministry of International Trade and Industry.

Landeweerd, J. A. (1979) 'Internal representation of a process, fault diagnosis and fault correction', *Ergonomics* 22(12): 1343–1351.

Lauter, B. (1987) *Software-Ergonomie in der Praxis* [Ergonomic software in practice], Munich:. Oldenbourg.

Laurel, B. (1990) *The Art of Human–Computer Interface Design*, Reading, MA: Addison-Wesley Publishing Company, Inc.

Lederman, S. J. and Loomis, J. M. (1981) 'Tactual perception', *Handbook of Perception and Human Performance*, vol. II, chapter 31, 1-41.

Lee, J., Su, V., Ren, S. and Ishii, H. (2000) 'HandSCAPE: a vectorizing tape measure for on-site measuring applications', paper given at CHI 2000, New York: ACM.

Legg, S. J. (1985) 'Comparison of different methods of load carriage', *Ergonomics* 28(1): 197–212.

Leitherer, E. (1991) 'Industriedesign', *Entwicklung–Produktion–Ökonomie* ['Industrial Design'], Development–Production–Economics, Poeschel.

Lewis, C. *et al.* (1990) 'Testing a walkthrough methodology for theory-based design of walk-up-and-use Interfaces', paper given at CHI '90, New York: ACM.

Lichty, T. (1994) *Design Principles for Desktop Publishers*, 2nd edn, Wadsworth.

Lind, E. J., Jayaraman, S., Park, S., Rajamanickam, R., Eisler, R., Burghart, G. and McKee, T. (1997) 'A sensate liner for personnel monitoring applications', *Digest of Papers of the 1st International Symposium on Wearable Computers*, Los Alamitos, CA: IEEE Computer Society, 98–107.

Lindgaard, G. (1994) *Usability Testing and System Evaluation: A Guide for Designing Useful Computer Systems*. London: Chapman & Hall.

Loebach, B. (1976) 'Industrial design', *Grundlagen der Industrie-produkt-gestaltung* [Foundations of Industrial Production Design], Thiemig.

Long, A. C., Narayanaswamy, S., Burstein, A., Han, R., Lutz, K., Richards, B., Sheng, S., Broderson, R. W. and Rabaey, J. (1995) 'A prototype user interface for a mobile multimedia terminal', paper given at CHI '95, New York: ACM.

McCormick, E. J. and Sanders, M. S. (1987) *Human Factors in Engineering and Design*, McGraw-Hill.

Macdonald, A. S. (1997) 'Developing a qualitative sense', in N. Stanton (ed.) *Human Factors in Consumer Product Design and Evaluation*, London: Taylor & Francis.

Macdonald, A. S. (1999) 'Aesthetic intelligence: a cultural tool', in M. A. Hanson, E. J. Lovesey and S. A. Robertson (eds) *Contemporary Ergonomics*, London: Taylor & Francis, 95–99.

Macleod, M. and Bevan, N. (1993) 'MUSiC video analysis and context tools for usability measurement', in S. Ashlund, K. Mullet, A. Henderson, E. Hollnagel and T. White (eds) *INTERCHI 93 Conference Proceedings*, New York: ACM Press, 95.

Mann, S. (1997) 'Wearable computing: a first step toward personal imaging', *Computer* 30(2): 25–32.

Mansur, D. L., Blattner, M. M. and Joy, K. I. (1985) 'Soundgraphs: a numerical data analysis method for the blind', paper given at 18th Hawaii International

Conference on System Sciences, Honolulu, HI: IEEE Computer Society Press, 198–203.

Marcus, A (2000) 'User interface design for a vehicle navigation system', in E. Bergman (ed.) *Information Appliances and Beyond – Interaction Design for Consumer Products*, Morgan Kaufmann.

Markowitz, J. A. (1996) *'Using speech recognition'*, Upper Saddle River, NJ: Prentice Hall.

Marslen-Wilson, W. D. (1979) 'Speech understanding as a psychological process', in J. C. Simon (ed.) *Spoken Language Generation and Understanding*, Dordrecht: D. Reidel, 39–67.

Martin, T. B. (1976) 'Practical applications of voice input to machines', paper given at IEEE 64, 487–501.

Maslow, A. (1970) *Motivation and Personality*, 2nd edn, New York: Harper and Row.

Matlin, M. W. and Foley, H. J. (1992) *Sensation and Perception*, 3rd edn, Allyn and Bacon.

Mayhew, D. J. (1992) *Principles and Guidelines in Software User Interface Design*, Englewood Cliffs, NJ: Prentice Hall.

Mayhew, D. J. (1999) *The Usability Engineering Lifecycle. A Practitioner's Handbook for User Interface Design*, San Francisco, CA: Morgan Kaufmann Publishers, Inc.

Mehrabian, A. (1987) *Räume des Alltags: Wie die Umwelt unser Verhalten bestimmt* [Everyday Spaces: How the Environment Determines our Behaviour, Campus.

Meijer, P. B. L. (1992) 'An experimental system for auditory image representation', *IEEE Transactions on Biomedical Engineering* 39(2): 112–121.

Meister, D. (1987) 'Systems design, development, and testing', in M. Helander (ed.) *Handbook of Human–Computer Interaction*, Amsterdam: North-Holland, 17–42.

Microsoft Corporation (1992) *The Windows Interface: An Application Design Guide*, Microsoft Press.

Molich, R. and Nielsen, J. (1990) 'Improving a human–computer dialogue', *Communications of the ACM* 33(3) March.

Moran, T. P. (1981) 'The command language grammar: a representation for the user interface of interaction computer systems', *International Journal of Man–Machine Systems* 15: 3–50.

Moray, N. (1987) 'Intelligent aids, mental models, and the theory of machines', *International Journal of Man–Machine Studies* 27: 619–629.

Moray, N. (1992) 'Mental models of complex dynamic systems', in: P. A. Booth and A. Sasse (eds) *Mental Models and Everyday Activities*, Second Interdisciplinary Workshop on Mental Models 23–25 March, Cambridge, UK: Robinson College, 103–131.

Murakami, K. and Taguchi, H. (1991) 'Gesture recognition using recurrent neural networks', paper given at CHI '91, New York: ACM, 301–305.

Murray, I. R., Baber, C. and South, A. (1996) 'Towards a definition and working model of stress and its effects of speech', *Speech Communication* 20: 3–12.

Nagamachi, M. (1995) *A Story of Kansei Engineering*, Tokyo: Kaibundo Publishing.

Nagamachi, M. (1997) 'Requirement identification of consumers' needs in product design', paper given at IEA '97, Helsinki: Finnish Institute of Occupational Health, 231–233.

Nater, P. (1990) 'New changes to establish MOON by the use of converting software and a computer-driven Braille-printer', paper given at 6th International Workshop on Computer Applications for the Visually Handicapped, *Infovisie Magazine* (Leuven, Belgium) 4(3): 1–8.

National Instruments (1997) *LabVIEW User Manual*, National Instruments.

Newman, W. M. and Lamming, M. G. (1995) *Interactive System Design*, Wokingham, UK: Addison-Wesley Publishing Company.

NeXT Computer, Inc. (1992) *NeXTSTEP User Interface Guidelines*, release 3, Addison-Wesley.

Nielsen J. (1993) *Usability Engineering*, New York: Academic Press.

Nielsen, J. (1990) 'Paper versus computer inplementations as mockup scenarios for heuristic evaluation', paper given at ICIP INTERACT '90 3rd International Conference HCI, 315–320.

Nielsen, J. www.useit.com/papers/guerilla_hci.html

Nielsen, J. and Molich, R. (1990) 'Heuristic evaluation of user interfaces', paper given at CHI '90, April, New York: ACM.

Nielsen, J. and Mack, R. L. (eds) (1994) *Usability Inspection Methods*, John Wiley.

Norman, D. A. (1983) 'Some observations on mental models', in D. Gentner and A. L. Stevens (eds) *Mental Models*, Hillsdale, NJ: Laurence Erlbaum Associates.

Norman, D. A. (1986) 'Cognitive engineering', in D. A. Norman and S. W. Draper (eds) *User Centered Systems Design: New Perspectives in Human Computer Interaction*, Lawrence Erlbaum.

Norman, D. A. (1988a) *The Design of Everyday Things*, Doubleday.

Norman, D. A. (1998) *The Invisible Computer – Why Good Products Can Fail, the Personal Computer Is So Complex, and Information Appliances Are the Solution*, The MIT Press.

Norman, D. A. (1988b) *The Psychology of Everyday Things*, Basic Books.

Norman, D. A. (1989) *Dinge des Alltags: Gutes Design und Psychologie für Gebrauchsgegenstände*, Campus.

Noyes, J. M. and Frankish, C. R. (1989) 'A review of speech recognition applications in the office', *Behaviour and Information Technology* 8(6): 475–486.

Noyes, J. M., Haigh, R. and Starr, A. F. (1989) 'Automatic speech recognition for disabled people', *Applied Ergonomics*, 20: 293–298.

Noyes, J. M., Baber, C. and Frankish, C. R. (1992) 'Industrial applications of ASR', *Journal of the American Voice I/O Society* 12: 51–68.

NRC (1997) *Tactical Display for Soldiers*, Washington, DC: National Academy.

Oakley, I., McGee, M. R., Brewster, S. and Gray, P. (2000) 'Putting the feel in "look and feel" ', paper given at CHI 2000, New York: ACM, 415–422.

Oatley, K. and Ramsay, J. (1992) 'Emotions while interacting with computers', paper given at International Society For Research On The Emotions Annual Conference, Carnegie Mellon University, USA, 18–20 August, Storrs, CT: ISRE.

Ockerman, J. J. and Pritchett, A. R. (1998) 'Preliminary investigation of wearable computers for task guidance in aircraft inspection', *Digest of Papers of the 2nd International Symposium on Wearable Computers*, Los Alamitos, CA: IEEE Computer Society, 33–40.

Open Software Foundation (1993) *OSF/Motif Style Guide*, Prentice Hall.

Oppermann, R. (1992) *Software-ergonomische Evaluation. Der Leitfaden EVADIS II* [Software-eronomic evaluation. The EVADIS II Guidelines], de Gruyter.

Paglia, C. (1995) *Sex and Violence, or Nature and Art*, Harmondsworth: Penguin Books.

Peschanel, F. (1990) *Sind Linkshänder besser?*, [Are Lefthanders Better?], Universitas, paperback: 1993, Goldmann.

Perlman, G. (1989) *User Interface Development*, lecture notes, Carnegie-Mellon-University, retrievable via FTP from: archive.cis.ohio-state.edu in /pub/hci/SEI/

Perlman, G. and Gasen, J. (1993) *HCI Education Survey*, Ohio State University, retrievable via FTP from: archive.cis.ohio-state.edu in: /pub/hci/Education/

Pfurtscheller, G. (1992) 'Bewegungssteuerung mit Hilfe des EEG: Grundlagen für ein Brain-Computer-Interface' [Movement control using the EEG: Basic Elements for a Brain–Computer Interface], *TW Neurologie und Psychiatrie* 6: 834–841, December.

Pfurtscheller, G., Flotzinger, D. and Kalcher, J. (1993) 'Brain–computer–interface – a new communication device for handicapped persons', *Journal of Microcomputer Applications* 16: 293–299.

Pheasant, S. (1996) *Bodyspace*, 2nd edn, London: Taylor & Francis.

Picard, R. (1997) *Affective Computing*, Cambridge, MA: MIT Press.

Pirsig, R. M. (1978) *Zen und die Kunst ein Motorrad zu warten* [Zen and the Art of Motorcycle Maintenance], Fischer.

Pisoni, D., Nusbaum, H. and Das, S. (1986) *Automatic measurement of speech recognition performance: A comparison of six speaker-dependent recognition devices*, final report for IBM Corporation Contracts No. 435114 and No. 562010.

Poll, L. H. D. (1996) 'Visualising graphical user interfaces for blind users', Ph.D. thesis, Eindhoven University of Technology.

Poll, L. H. D. and Waterham, R. P. (1995) 'Graphical user interfaces and visually disabled users', *IEEE Transactions on Rehabilitation Engineering* 3(1): 65–69.

Polson, P. G. and Kieras, D. E. (1984) 'A formal description of users' knowledge of how to operate a device and user complexity', *Behavior Research Methods, Instruments, and Computers* 16(2): 249–255.

Popovic, V. (1997) 'Product evaluation methods and their applications', paper given at IEA '97, Helsinki: Finnish Institute of Occupational Health, 165–167.

Post, E. R. and Orth, M. (1997) 'Smart fabric, or "wearable clothing"', *Digest of Papers of the 1st International Symposium on Wearable Computers*, Los Alamitos, CA: IEEE Computer Society, 167–168.

Pötter, G. (1994) 'Die Anleitung zur Anleitung', *Leitfaden zur Erstellung technischer Dokumentationen* [The Manual for Manuals, Guidelines for Writing Technical Documentation], Vogel.

Preece, J., Rogers, Y., Sharp, H., Benyon, D., Holland, S. and Carey, T. (1994) *Human–Computer Interaction*. Wokingham, UK: Addison-Wesley Publishing Company.

Rakers, G. G. H. (1987) *The theory of 'Frame of Mind'*, IST-MEMO-87-01, Enschede: Vakgroep Instruktietechnologie, Universiteit Twente.

Rakers, G. G. H. (1992) 'Instructional format design: a teaching strategy to augment cognitive modelling', in D. G. Bouwhuis, T. Bösser, G. d'Ydewalle and F. L. Engel (eds) *Cognitive Modelling and Interactive Environments in Language Learning*, Berlin: Springer-Verlag.

Rakers, G. (1998) Several teaching modules in *PAO-Cursus Taakanalyse Map 1* [PAO-Course Task Analysis Binder 1] and *PAO-Cursus Taakanalyse Map 2* [PAO-Course Task Analysis Binder 2], Amsterdam: P.A.O. Informatica.

Rakers, G. (1999) Several teaching modules in *PAO-Cursus Ontwerpmethoden voor User Interfaces Map 1* [PAO-Course Design Methods for User Interfaces Binder 1] and *PAO-Cursus Ontwerpmethoden voor User Interfaces Map 2* [PAO-Course Design Methods for User Interfaces Binder 2] Amsterdam: P.A.O. Informatica.

Rakers, G. (2000) Teaching modules 10, 14, 17, and 20 in *PAO-Cursus Ontwerpmethoden voor User Interfaces Map 1* [PAO-Course Design Methods for User Interfaces Binder 1] and *PAO-Cursus Ontwerpmethoden voor User Interfaces Map 2* [PAO-Course Design Methods for User Interfaces Binder 2], Amsterdam: P.A.O. Informatica.

Rakers, G. and Pieters, J. (1989) 'Het Ontwerpen van Zachte Produkten' [Designing soft products], in F. R. H. Zijlstra and A. G. Arnold (eds.) *Mens-Computer Interactie in Nederland* [Man–computer interaction in the Netherlands], Amsterdam: Stichting Informatica Congressen, 41–68.

Rakers, G. and Wittkämper, D. (1995) 'Ein Multidisziplinärer Ansatz zur Entwicklung interaktiver Produkte' [A multi-disciplinary approach to the development of interactive products], in H-J. Bullinger (ed.) *Design Interactiver Produkte. Dialog zwischen Mensch und Produkt* [Designing interactive products. Dialogue between man and product], Stuttgart: IRB Verlag, 105–109.

Rakers, G. G. H., Pieters, J. M. and de Bruijn, I. (1988) 'Het leren programmeren van computergestuurde gereedschapwerktuigen: Een afbeeldingstheorie en een informatieverwerkingsmodel [Learning to program computerized machine tools. An image theory and an information processing model], *Tijdschrift voor Onderwijsresearch* 13(4): 181–200.

Rakers, G. G. H., Pieters, J. M. and Dijkstra, S. (1990) 'The role of an advance organizer on CNC program debugging performance', in S. Dijkstra, B. H. A. M. van Hout Wolters and P. C. van der Sijde (eds) *Research on Instruction, Design and Effects*, Englewood Cliffs, NJ: Educational Technology Publications Inc.

Rash, C. E., Verona, R. W. and Crowley, J. S. (1990) 'Human factors and safety considerations of night vision systems flight using thermal imaging systems', *Proceedings of SPIE – The International Society for Optical Engineering* 1290: 142–164.

Rasmussen, J. (1986) *Information Processing and Human–Machine Interaction. An Approach to Cognitive Engineering*, Amsterdam: North-Holland.

Ravden, S. J. and Johnson, G. I. (1989) *Evaluating Usability of Human–Computer Interfaces: A Practical Method*, Chichester: Ellis Horwood.

Redmond-Pyle, D. and Moore, A. (1995) *Graphical User Interface Design and Evaluation. A Practical Process*, London: Prentice Hall.

Reichert, G. W. (1987) *Kompendium für technische Anleitungen* [Handbook for Technical Instructions], Girardet.

Rhodes, B. (1997) 'The wearable remembrance agent: a system for augmented memory', *Personal Technologies* 1: 218–224.

Risak, V. (1986) *Mensch-Maschine-Schnittstelle in Echtzeitsystemen* [Human–Machine Interface in Realtime Systems], Berlin: Springer.

Ritchie, R. J. and List, J. A. (1996) 'System design practice, emerging development acceleration strategies, and the role of user-centered design', in M. Rudisill, C.

380   References

Lewis, P. Polson and T. McKay (eds) *Human–Computer Interface Design. Success Stories, Emerging Methods, and Real-World Context*, San Francisco, CA: Morgan Kaufmann Publishers, Inc.

Rohaly, A. M. and Karsh, R. (1999) 'Helmet-mounted displays', in J. M. Noyes and M. Cook (eds) *Interface Technology: The Leading Edge*, Baldock, UK: Research Studies Press Ltd. 267–280.

Roth, E. M., Woods, D. D. and Gallagher, J. M. (1986) 'Analysis of expertise in a dynamic control task', paper given at Human Factors Society 30th Annual Meeting, 179–181.

Rouse, W. B. and Morris, N. M. (1986) 'On looking into the black box: prospects and limits in the search for mental models', *Psychological Bulletin* 100(3): 349–363.

Rudisill, M., Lewis, C., Polson, P. and McKay, T. (eds) (1996) *Human-Computer Interface Design. Success Stories, Emerging Methods, and Real-World Context*, San Francisco, CA: Morgan Kaufmann Publishers, Inc.

Ryle, G. (1949, 1980) *The Concept of Mind*, Harmondsworth, UK: Penguin Books Ltd.

Saida, S. (1992) Development of a 3-dimensional tactile display for the blind-preferable presentation', paper given at Computers for Handicapped Persons, Proceedings of the 3rd International Conference, Munich: R. Oldenbourg Verlag and Vienna: Oesterreichische Computer Gesellschaft, 431–437.

Salvendy, G. (ed.) (1987) *Handbook of Human Factors*, New York: John Wiley & Sons, Inc.

Sampson, J. B. (1993) 'Cognitive performance of individuals using head-mounted displays while walking', paper given at Human Factors and Ergonomics Society 37th Annual Meeting, Santa Monica, CA: Human Factors and Ergonomics Society, 338–342.

Sarini, M. and Strapparava, C. (1998) 'Building a user model for a museum exploration and information-providing adaptive system', paper given at Workshop on Adaptive Hypertext and Hypermedia, 20–24 June, Pittsburgh, PA.

Sawhney, N. and Schmandt, C. (1998) 'Speaking and listening on the run: design for wearable audio computing', paper given at ISCW '98: International Symposium on Wearable Computing, Los Alamitos, CA: IEEE, 108–115.

Schmidt, A., Gellerson, H-W. and Beigl, M. (1999) 'A wearable context-awareness component: finally a good reason to wear a tie', *Digest of Papers of the 3rd International Symposium on Wearable Computers*, Los Alamitos, CA: IEEE Computer Society, 176–177.

Schmidtke, H. (1993) *Ergonomie* [Ergonomics], Hanser.

Schmitt, A. (1983) *Dialogsysteme, Software-Ergonomie* [Dialogue Systems, Software Ergonomics], BI.

Schweikhardt, W. (1985) 'Interaktives Erkunden von Graphiken durch Blinde', paper given at Software-Ergonomie '85, Stuttgart.

Schweizer, P. (1989) *Systematische Produkt-Entwicklung mit Mikroelektronik – Technische und psychosoziale Erfolgsstrategien* [Systematic Product Design with Microelectronic, Technical and Psychosocial Strategies]. Ott.

Scott, A. (2000) 'A glove-based interface for mobile phones', unpublished BEng report, Birmingham: University of Birmingham.

Seagull, F. J. and Gopher, D. (1997) 'Training head movement in visual scanning: an embedded approach to the development of piloting skills with helmet-mounted displays', *Journal of Experimental Psychology: Applied* 3: 163–180.

Seeger, H. (1983) *Industrie-Design. Basiswissen über das Entwickeln und Gestalten von Industrie-Produkten* [Industrial Design. The Basics of the Design of Industrial Products], Expert.

Seeger, H. (1992) *Design technischer Produkte, Programme und Systeme* [Design of technical products, programmes and systems.] Berlin: Springer.

Shinohara, M. (1992) 'Development of a 3-D tactile display for the blind: system design', paper given at Computers for Handicapped Persons, Proceedings of the 3rd International Conference, Munich: R. Oldenbourg Verlag and Vienna: Oesterreichische Computer Gesellschaft, 422-430.

Shneiderman, B. (1987) *Designing the User Interface: Strategies for Effective Human–Computer Interaction*, Reading, MA: Addison-Wesley Publishing Company.

Siegel, J. and Bauer, M. (1997) 'On site maintenance using a wearable computer system', paper given at CHI '97, New York: ACM, 119–120.

Smallwood, R. D. (1967) 'Internal models and the human instrument monitor', *IEEE Transactions on Human Factors in Electronics* 8(3): 181–187.

Soloman, M. R. (1996) *Consumer Behavior*, 3rd edn, Englewood Cliffs, NJ: Prentice Hall.

Spinas, P. (1983) *Leitfaden zur Einführung und Gestaltung von Arbeit mit Bildschirmsystemen* [Guidelines for Implementation and Design of Work with Display–based Systems], Verlag Industrielle Organisation.

Starner, T. (1996) 'Human-powered wearable computing', *IBM Systems Journal* 35: 618–629.

Stary, C. (1996) *Interaktive Systeme – Software-Entwicklung und Software-Ergonomie* [Interactive systems – software development and ergonomics], Vieweg.

Stein, R., Ferrero, S., Hetfield, M., Quinn, A. and Krichever, M. (1998) 'Development of a commercially successful wearable data collection system', *Digest of Papers of the 2nd International Symposium on Wearable Computers*, Los Alamitos, CA: IEEE Computer Society, 18–24.

Strinati, D. (1995) *An Introduction to Theories of Popular Culture*, London: Routledge.

Strommen, E. (2000) 'Interactive toy characters as interfaces for children', in E. Bergman (ed.) *Information Appliances and Beyond – Interaction Design for Consumer Products*, Morgan Kaufmann.

Sturman, D. J. and Zeltzer, D. (1994) 'A survey of glove-based input', *IEEE Computer Graphics and Applications* 14(1): 30–39.

Sun Microsystems, Inc. (1989a) *OPEN LOOK Graphical User Interface Functional Specification*, Addison Wesley.

Sun Microsystems, Inc. (1989b) *OPEN LOOK Graphical User Interface Application Style Guidelines*, Addison Wesley.

Sutcliffe, A. G. (1995) *Human–Computer Interface Design*, Houndmills, Basingstoke, UK: Macmillan Press Ltd.

Tambini, M. (1996) *The Look of the Century*, London: Dorling Kindersley.

Tan, H. Z. and Pentland, A. (1997) 'Tactual displays for wearable computing', *Digest of Papers of the 1st International Symposium on Wearable Computers*, Los Alamitos, CA: IEEE Computer Society, 84–89.

Teichner, W. and Krebs, M. (1974) 'Laws of visual choice reaction time', *Psychological Review* 81: 75–98.

Tennyson, R. D. (1990) 'Computer-based enhancements for the improvement of learning', in S. Dijkstra, B. H. A. M. van Hout Wolters and P. C. van der Sijde (eds.) *Research on Instruction, Design and Effects*, Englewood Cliffs, NJ: Educational Technology Publications Inc.

Thimbleby, H. (1990) *User Interface Design*, Wokingham, UK: Addison-Wesley Publishing Company.

Thomas B. (1996) '"Quick and dirty" usability tests', in P. W. Jordan, B. Thomas, B. A. Weerdmeester and I. L. McClelland (eds) *Usability Evaluation in Industry*, London: Taylor & Francis.

Thomas, B., Tyerman, S. and Grimmer, K. (1997) 'Evaluation of three input mechanisms for wearable computers', paper given at 1st International Symposium on Wearable Computers, Los Alamitos, CA: IEEE Computer Society, 2–9.

Thomas, B., Grimmer, K., Makovec, D., Zucco, J. and Gunther, B. (1998) 'Determination of placement of a body-attached mouse as a pointing input device for wearable computers', *Digest of Papers of the 3rd International Symposium on Wearable Computers*, Los Alamitos, CA: IEEE Computer Society, 193–194.

Thurlow, W. R. (1986) 'Some comparisons of characteristics for alphabetic codes for the deaf-blind', *Human Factors* 28: 175–186.

Tiger, L. (1992) *The Pursuit of Pleasure*, Boston: Little, Brown and Company, 52–60.

Tognazzini, B. (1990) *Tog on Interface*, Apple.

Trimmel, M. (1992) 'Auswirkungen der Mensch-Computer-Interaktion: Psychologische Aspekte', *Informatik Forum* 6: Jg. Heft 4/92.

Trimmel, M. *et al.* (1993) 'Psychological and psychophysiological effects of working with computers: experimental evidence', in H. Luczak (ed.) *Work With Display Units 92, Selected Proceedings*, North-Holland.

Tschichold, J. (1960) *Erfreuliche Drucksachen durch gute Typographie* [Pleasing Printed Text through Good Typography], Maro.

Tucker, P. and Jones, D. M. (1993) 'Voice as a medium for document annotation', in C. Baber and J. M. Noyes (eds) *Interactive Speech Technology: Human Factors Issues in the Application of Speech Input/Output to Computers*, London: Taylor & Francis, 109–116.

Tufte, E. R. (1984) *The Visual Display of Quantitative Information*, Graphics Press.

Umbers, I. G. (1979) 'Models of the process operator', *International Journal of Man–Machine Studies* 11: 263–284.

Urban, G. L. and Hauser, J. R. (1993) *Design and Marketing of New Products*, 2nd edn, Englewood Cliffs, NJ: Prentice Hall.

Usher, D. M. (1993) 'Automatic speech recognition and mobile radio', in C. Baber and J. M. Noyes (eds) *Interactive Speech Technology: Human Factors Issues in the Application of Speech Input/Output to Computers*, London: Taylor & Francis, 73–83.

Vanderheiden, G. C. (1997) 'Anywhere, anytime (+ anyone) access to the next-generation WWW', *Computer Networkers and ISDN Systems (Netherlands)* 8–13(29): 1439–1446.

van der Veer, G. C. (1984) *Readings on Cognitive Ergonomics*, Springer.

van der Veer, G.C. (1998) Most of the teaching modules in *PAO-Cursus Taakanalyse Map 1* [PAO-Course Task Analysis Binder 1] and *PAO-Cursus Taakanalyse Map 2* [PAO-Course Task Analysis Binder 2], Amsterdam: P.A.O. Informatica.

van der Veer, G. C. (1999) Most of the teaching modules in *PAO-Cursus Ontwerpmethoden voor User Interfaces Map 1* [PAO-Course Design Methods for User Interfaces Binder 1] and *PAO-Cursus Ontwerpmethoden voor User Interfaces Map 2* [PAO-Course Design Methods for User Interfaces Binder 2], Amsterdam: P.A.O. Informatica.

van der Veer, G. C. (2000) Most of the teaching modules in *PAO-Cursus Ontwerpmethoden voor User Interfaces Map 1* [PAO-Course Design Methods for User Interfaces Binder 1] and *PAO-Cursus Ontwerpmethoden voor User Interfaces Map 2* [PAO-Course Design Methods for User Interfaces Binder 2], Amsterdam: P.A.O. Informatica.

van der Veer, G. C., van Vliet, J. C. and Lenting, B. F. (1995) 'Designing complex systems – a structured activity', paper given at DIS '95, Symposium on Designing Interactive Systems, Ann Arbor, MI, New York: ACM Press, 207–217.

van der Veer, G. C., Hoeve, M. and Lenting, B. F. (1996a) 'Modeling complex work systems – method meets reality', in T. R. G. Green, J. J. Canas and C. P. Warren (eds) *Cognition and the Worksystem 8th European Conference on Cognitive Ergonomics (EACE)*, Le Chesnay cedex: Inria, 115–120.

van der Veer, G. C., Lenting, B. F. and Bergevoet (1996b) GTA: Groupware Task Analysis – modeling complexity'. *Acta Psychologica* 91: 297–322.

van der Veer, G. C. and Mariani, M. (1997) 'Teaching design of complex interactive systems. Learning by interacting', paper given at Workshop TeaDIS – Teaching Design of Interactive Systems, Schaerding, Austria, 20–23 May.

van Leeuwen, M. and Thomas, B. (1997) 'The integration of interaction design and human factors in the product creation process: a case study', paper given at 16th International Symposium on Human Factors in Telecommunications, Oslo.

van Vianen, E. P. G., Thomas, B. and van Nieuwkasteele, M. (1996) 'A combined effort in the standardization of user interface testing', in P. W. Jordan, B. Thomas, B. A. Weerdmeester and I. L. McClelland (eds) *Usability Evaluation in Industry*, London: Taylor & Francis.

Veldhuyzen, W. and Stassen, H. G. (1977) 'The internal model concept: an application to modeling human control of large ships', *Human Factors* 19(4): 367–380.

Vester, F. (1984) 'Wissen: Wege aus dem Datenfriedhof', in *Neuland des Denkens: Vom technokratischen zum kybernetischen Zeitalter*, dtv.

Virzi, R. A. (1992) 'Refining the test phase of usability evaluation: how many subjects is enough?', *Human Factors* 34: 457–468.

Vries, G. de, Hartevelt, M. and Oosterholt, R. (1996) 'Private camera conversation method', in P. W. Jordan *et al.* (eds) *Usability Evaluation in Industry*, London: Taylor and Francis.

Wandmacher, J. (1993) *Software-Ergonomie* [Ergonomic software], de Gruyter.

Want, R., Hopper, A., Falcao, V. and Gibbons, J. (1992) 'The active badge location system', *ACM Transactions on Information Systems* 10(1): 91–102.

Weiner, D. and Ganapathy, S. K. (1989) 'A synthetic visual environment with hand gesturing and voice input', paper given at CHI '89, New York: ACM, 235–240.

Weiser, M. (1991) 'The computer for the 21st century', *Scientific American*, September, 933–940.

Wickens, C. D. and Hollands, J. G. (2000) *Engineering Psychology and Human Performance*, 3rd edn, Upper Saddle River, NJ: Prentice Hall.

Wiklund, M. E. (ed.) (1994) *Usability in Practice: How Companies Develop User-Friendly Products*, Academic Press.

Willumeit, H-P. and Kolrep, H. (1998) 'Wohin führen Unterstützungssysteme?', *Entscheidungshilfe und Assistenz in Mensch-Maschine-Systemen* [Where Do Support Systems Lead? Decision Support and Assistance in Human–Machine Systems], Pro Universitate.

Wilson, J. and Rosenberg, D. (1988) 'Rapid prototyping for user interface design', in M. Helander (ed.) *Handbook of Human–Computer Interaction*, North-Holland.

Wilson, J. R. and Rutherford, A. (1989) 'Mental models: theory and application in human factors', *Human Factors* 19: 367–380.

Wittgenstein, L. (1921, 1984) *Tractatus Logico-Philosophicus*, Frankfurt am Main: Suhrkamp.

Wohinz, J. W. and Hagen, H. (1991) *Kreativitätstechniken* [Creativity techniques], TU Graz: IBL Induscript.

Wood, L. E. (ed.) (1998) *User Interface Design. Bridging the gap from user requirements to design*, Boca Raton, FL: CRC Press.

Woodson, W. E. (1987) *Human Factors Reference Guide for Electronics and Computer Professionals*, McGraw-Hill.

Youdin, M., Sell, G. H., Reich, T., Clagnaz, M., Louie, H. and Kolwicz, R. (1980) 'A voice controlled powered wheelchair and environmental control system for the severely disabled', *Medical Progress through Technology* 7: 139–143.

Zankl, G. and Heufler, G. (1985) *Lehrerhandbuch Produktgestaltung* [Product Design Handbook for Teachers], Veritas.

Zijlstra, F. R. H. and Arnold, A. G. (eds) (1989) *Mens-Computer Interactie in Nederland* [Human–computer interaction in the Netherlands], Amsterdam: Stichting Informatica Congressen, 41–68.

Zimmerman, T. G., Lanier, J., Blanchard, C., Bryson, S. and Harvill, Y. (1987) 'A hand gesture interface device', paper given at CHI '87, New York: ACM, 189–192.

Zühlke, D. (1996a) 'Menschliches Versagen – Analyse, Gründe, Vermeidungs-ansätze [Human Error: Analysis, Causes, Prevention Strategies], in D. Zühlke (ed.) *Menschengerechte Bedienung technischer Geräte*, VDI, Bericht Nr. 1303.

Zühlke, D. (1996b) 'Mensch-Maschine-Kommunikation – heute' [Human–Machine Communication Today], in D. Zühlke (ed.) *Menschengerechte Bedienung technischer Geräte. Tagungsbericht* [Proceedings, Symposium on User-centred Control of Technical Devices] Kaiserslautern 1996, VDI-Bericht Nr. 1303.

Index of Authors

Ainsworth, W. 356, 360
Akematsu, M. 183
Andre, E. 188
Arnaut, L. Y. 158
Ashby, W. R. 42

Balentine, B. 358
Bandini-Buti, L. 323
Bass, L. 209
Bergman, Eric 181, 359
Bevan, N. 298
Beyer, H. 359
Bias, R. G. 358, 359
Binsted, Kim 178, 357
Birren, F. 359
Blenkhorn, P. L. 277
Bolt, R. A. 176
Boud, A. C. 177
Braille, L. 183, 274, 276, 279
Briggs, C. 13
Brooke, C. C. 299, 319, 325
Burandt, Ulrich 132, 134, 136–8, 141, 162, 222–5, 227, 231
Buxton, Bill 177

Cleveland 244
Cohen, H. S. 42
Conant, R. C. 42
Cresswell Starr, A. F. 194

Davis, K. 191
de Mooij, Marieke 338
den Buurman 312
Dertouzos, M. 359
Descartes, René 243, 244

Dix, A. 29
Doddington, G. R. 201
Dul, J. 359
Dvorak, A. 93, 164

Eberts, R. E. 29
Edwards, A. D. N. 274
Ekman, P. 326
Engelbart, Doug 238
England, R. 183
Erman, L. 190

Farringdon, J. 212, 213
Feiner, S. 214
Ferrell, W. R. 42
Fitts, P. 78, 80, 125, 365
Fitzmaurice, G. W. 177
Flynn, M. 209
Foley, H. J. 275
Forty, A. 315
Fourier, Jean-Baptiste 254
Frankish, C. R. 194, 202
Frei, P. 178, 180
Friesen, W. V. 326

Galitz, W. O. 29
Gates, Bill 359
Gauss, C. F. 11, 13
Gaver, W. W.. 276
Geiser, G. 170
Gentner, Don 365
Gershenfeld, Neil A. 359
Gilmore, W. E. 100, 132, 163, 237
Gobel, M. 183
Gould, J. D. 46

Grandjean, Étienne 13, 132, 134, 136–8, 223, 226
Greenstein, J. S. 158

Haitani, Rob 181
Hansen, W. J. 33
Hatwell, Y. 277
Hauser, J. R. 324
Healey, J. 212
Helander, M. G. 29, 154, 158, 163, 356, 359
Heller, M. A. 277
Henderson, A. 29
Herrmann, Ned 14
Hill, D. R. 199
Hjelle, L. A. 306
Hofmeester, G. 315
Hofstede, Geert 330–2, 335, 337–40, 360
Hollands, J. G. 194
Holtzblatt, Karen 359
Hone, K. S. 204
Honold, Pia 16
Ishihara, S. 322
Ishii, Hiroshi 177, 252

Jacob, R. J. K. 175
Jamar, P. 363
Jansson, G. 277
Johnson, G. I. 33, 300, 315
Johnson-Laird, P. N. 42
Jones, D. M. 194
Jung, C. G. 14

Kamentsky, L. 277
Kelly, G. A. 327
Kieras, D. E. 42
Kirakowski, J. 299, 319
Kirwan, B. 356, 360
Klatt, D. 190
Koons, D. B. 175
Krebs, M. 194
Kurzweil, R. 277

Lamming, M. 29, 209
Landauer, T. K. 359
Landeweerd, J. A. 42
Lanz, Herwig 135–7, 165

Laurel, Brenda xvii 29, 357, 360
Lauter, B. 237
Lederman, S. J. 275
Lee, J. 176, 178, 252
Lewis, C. 46
Lichty, Tom 360
Likert, R. A. 323
Lind, E. J. 212
Lindgaard, G. 301
List, J. A. 29
Lombard, E. 206
Loomis, J. M. 275
Luczak, H. xiii

Macdonald, A. S. 317, 330
Mackenzie, I. S. 183
Macleod, M. 298
Mann, S. 213
Mansur, D. L. 277
Marcus, Aaron xv, 359
Mariani, M. 36
Markov, A. A. 199
Markowitz, J. A. 198
Marslen-Wilson, W. D. 277
Martin, T. B. 192
Maslow, Abraham 306, 307
Matlin, M. W. 275
Mayhew, D. J. 29, 46, 358, 359
McClelland, Ian L. 360
McKim, R. H. 360
Meijer, P. B. L. 277
Meisel, W. S. 358
Meister, D. 29, 356
Moore, A. J. 29
Moran, T. P. 42
Moray, N. 42
Morgan, D. P. 358
Morris, N. M. 42
Morse, Samuel 282
Murakami, K. 175
Myers, I. B. 13

Nagamachi, M. 322
Naisbitt, John 360
Naisbitt, N. 360
Nass, C. 188
Nater, P. 274
Newman, W. M. 29

Nielsen, Jakob 301, 356–8, 360, 361, 365
Norman, Donald A. 18-20, 42, 93, 164, 211, 215, 317, 356, 361
Oakley, I. 183
Oatley, K. 326
O'Donnell, P. J. 325
Orth, M. 212

Paglia, C. 316
Paiva, A. 188
Pentland, A. 214
Peschanel, Frank 14
Pfurtscheller, G. 186
Pheasant, S. 153, 362
Philips, D. 360
Picard, Rosalind W. 209, 212, 362
Pieters, J. M. 29, 43
Pisoni, D. 201
Polson, P. G. 42
Popovic, V. 319
Post, E. R. 212
Prabhu, P. 359
Preece, Jenny 29, 46

Ramsay, J. 326
Rasmussen, J. 42
Ravden, S. J. 300, 315
Redmond-Pyle, D. 29
Reeves, R. 188
Rhodes, B. 209, 212
Risak, V. 13
Ritchie, R. J. 29
Roth, E. M. 42
Rouse, W. B. 42
Rubin, J. 362
Rutherford, A. 42
Ryle, Gilbert 42

Saida, S. 277
Salvendy, Gavriel 362
Sawhney, N. 185, 214
Schalk, T. B. 201
Schmandt, C. 185, 214, 362
Schmidt, A. 213
Schmidtke, H. 12, 185
Schmitt, B. H. 357, 362
Schweikhardt, W. 277

Scott, A. 212
Seeger, H. 10, 11, 17, 220
Servaes, M. 308, 309, 320, 324
Shinohara, M. 277
Shneiderman, Ben 29, 33, 356, 363
Sholes, C. L. 164
Smallwood, R. D. 42
Soloman, M. R. 339
Spinas, P. 101, 163, 346
Starner, T. 213
Stassen, H. G. 42
Stein, R. 211
Strinati, D. 315
Strommen, Erik 180
Sutcliffe, A. G. 29

Taguchi, H. 175
Tambini, M. 315
Tan, H. Z. 214
Teichner, W. 194
Tennyson, R. D. 43
Thimbleby, H. 29
Thurlow, W. R. 183
Tiger, Lionel 310
Tillman, B. 363
Tillman, P. 363
Tognazzini, Bruce 13, 14, 363–5
Tucker, P. 194
Tufte, E. R. 363

Ullmer, B. 177, 252
Umbers, I. G. 42
Urban, G. L. 324
Usher, D. M. 192

van der Veer, Gerrit C. 33, 36
van Leeuwen, M. 296
van Vianen, Edwin P. G. 295, 297, 298
Vanderheiden, G. C. 273
Veldhuizen, W. 42
Verhoeven, P. 215
Vester, Frederic M. 220
Virzi, Robert A. 300

Wandmacher, J. 237, 238, 242, 244, 246
Want, R. 214

Warwick, Kevin 210
Waterham, R. P. 277
Weerdmester, B. A. 359, 360
Weinschenk, S. 363
Weiser, M. 216
Wickens, C. D. 194
Wilson, J. 42
Wittgenstein, Ludwig 29
Wittkämper, D. 29

Woodson, Wesley E. 100, 132, 134,
  137, 150, 222, 223, 225, 363

Yeo, S. C. 363
Youdin, M. 194

Ziegler, D. J. 306
Zimmerman, T. G. 175
Zühlke, Detlef 147

Index of Companies and Products

All brand names and product names mentioned in this book, especially the ones included in this index, are trademarks or registered trademarks of their respective companies.

415 Productions, Inc. 51, 365
Aaron Marcus and Associates xv
ACM (Association for Computing Machinery) 364
ActiMates 181
Adobe 76
AfterEffects 76
Aibo 180, 181, 252
Amazon.com 358
Amiga 369
Amsterdam University xiii
ANSI (American National Standards Institute) 287, 293
Apple Computer, Inc. 14, 28, 33, 82, 89, 90, 92, 100, 153, 180, 238–40, 242, 243, 282, 305, 318
Art Center College of Design xviii, 386
Association for Computing Machinery 364
AT&T 191
Authorware 76

Bang & Olufsen 22, 98, 102, 152
Barney 180, 252, 359
BATE 120
Bell Laboratories 191
Birmingham University xi, 184, 211, 385
Blaupunkt 21
Bluetooth 294
Bond, James 214

Borg 210
Bosch 18
Bristol University xii, 200, 387
British Aerospace xii
British Computer Society 364
British Ergonomics Society xiii, 364
British Telecom 174
BSI (British Standards Institute) 287, 293
BUILD-IT 177

Canon 272
Carin 75
CD (Compact Disc) 257, 261, 265, 269, 324
CEN (Comité Européen de Normalisation) 287, 293
CHI Conference on Human Factors in Computing Systems 178, 356, 364
Coca-Cola 96
Commodore 369
Communicator 174
Computer Technology Institute xii, 386
Contemporary Trends Institute xi
Corel 76
CorelDraw 76
CTI (Computer Technology Institute) xii, 386
Curlybot 180, 252

DARPA (Defence American Research
  Projects Agency) 190
Dayton 75
Delphi 76
Diamond 270
DIN 225
DIN (Deutsches Institut für Normung)
  225, 287, 293, 369
Doc Martin's 310
Donau-University Krems xi
  Dreamweaver 76
DVD (Digital Versatile Disc) 209
Dyson 317

EAP xii
Ergonomics Society xii
Ericsson 67
ETSI (European Telecommunication
  Standards Institute) 287, 293
EU (European Union) 174
Eurodata 338
European Directive 286
Fachhochschule Joanneum (Joanneum
  Graduate School) xi, 385
FiZZ 127, 296
Flash 76
Frank AudioData 279, 280
Frontpage 76

GameBoy 174
General Electric 40
Genie 127
GPS (Global Positioning System) 214
Grande Valse 263–5
Graz Technical University xi, xv, 186

Handspring 180, 182
Harley Davidson 257
Harpy 190
HBDI (Herrmann Brain Dominance
  Instrument) 15
HCII Conference (Human–Computer
  Interaction International) 364
Hearsay 190
HECTOR 174
Herrmann Brain Dominance
  Instrument 15
Hewlett-Packard 145, 180

HFES (Human Factors and
  Ergonomics Society) 364
HP (Hewlett-Packard) 145
HSE (Health and Safety Regulations)
  286
HTML (Hypertext Meta Language) 74
Human Factors and Ergonomics
  Society 364

IBM xiii, 180, 211
ICS (International Classification for
  Standards) 291
IEA (International Ergonomics
  Association) 364
IEC (International Electrotechnical
  Commission) 287, 290, 292
IEEE (International Association of
  Electrical and Electronics Engineers)
  xii
IID (Index of Interactive Difficulty) 325
ILAC (International Laboratory
  Accreditation Cooperation) 290, 294
Index of Interactive Difficulty 325
Interactive Barnie 180, 181, 252
Interface Consult xv
IOS 291
ISO (International Organization for
  Standardization) 285–7, 291–2, 294,
  298, 306, 369
ITU (International Telecommunication
  Union) 287
Ixus 272

James Bond 214
Java 76, 294
Joanneum Graduate School
  (Fachhochschule Joanneum) xi, 385
JVC (Japanese Victor Company) 20

Kansei engineering (a technique)
  322–4
KEMA xii
Kidcom 81
Kodak 174, 385
Krems University xi
Kuka Robots AG 147
Kurzweil Reading Machine 277

Likert Scale 323
Linux 213
Lisa 239
Lloyd Cole and the Commotions 321
London College of Fashion xi
Lotus 249
Loughborough University xi, xiii

Macintosh 14, 92, 239, 240, 242, 282, 318
Macromedia 74, 76
Mannesmann VDO, 75, 111
Massachusetts Institute of Technology 180, 252
MD (MiniDisc) 272, 273
MicroOptical 210
Microsoft 74, 76, 92, 107, 112, 180, 213, 239, 252, 359
MIDI (Musical Instruments Digital Interface) 263
MiniDisc 272, 273
MIT (Massachusetts Institute of Technology) 180, 252
Moving Picture Experts Group 261, 294
MP3 265
MPEG (The Moving Picture Experts Group) 261, 294
Myers–Briggs Test 13

Newton 153, 180
Nikon 140
Nokia 81, 174, 263–5
Nomadic Radio 185
Notes 249
NTSC (Northern Telecommunication Standards Committee) 109

Optacon 183
Owens-Illinois Corporation 192

PAL (a television standard) 109
Palm 174, 180, 182, 359
PalmPilot 174, 180, 182, 359
PalmPix 174
PC-104 210
Pentium 152, 211

Phantom 184
Philips xi–xv, 4, 9, 29, 32, 40, 75, 80, 103, 104, 120, 127, 179, 268, 296, 305, 385, 387
Photoshop 76
Porsche 310
Post-it 56, 69
PowerBook 318
Premiere 76
Psion 365

RAL 225
Ravden Johnson Checklist 300
Reading University 210
Rio 270
Robocop 210, 215
Royal Academy Educational Technology 385
Royal Conservatory The Hague xiii
Royal Philips Industries *see* Philips

Satama xii, 386
SensAble 184
Sensorial Quality Assessment Tool 323
SEQUAM (Sensorial Quality Assessment Tool) 323
Siemens xv, 9, 18, 40, 152, 175, 176
SIGCHI (Special Interest Group for Human–Computer Interaction) xiii, 364
SIGGRAPH Conference 364
Simatic 152
SNS Bank xii
Software Usability Measurement Inventory 299, 319
Sony 9, 22, 140, 175, 178–80, 209, 252, 268, 269, 272
Soundscapes xiii
SparK 127
Star 238
Star Trek 210
Stuttgart University 277
SUMI (Software Usability Measurement Inventory) 299, 319
Sun Microsystems 294
Surrey University xi, xiii
SUS (System Usability Scale) 299, 319, 325

Symbian xi, 365, 386
System Usability Scale 299, 319, 325

T9 Inc. 167
TASO (Tactile Acoustic Slide Output) 279, 280, 282
Texas Instruments 167
The Moving Picture Experts Group 261, 294
TI (Texas Instruments) 167
Trackpoint 146
TU Graz (Graz Technical University) xi, xv, 186

US Robotics 180
USB (Universal Serial Bus) 294
Utrecht University xiii

VDE (Verein Deutscher Elektrotechniker) 225, 369
VDO 75

Visor 182
Visual Basic 74

W3 (World Wide Web Consortium) 294
Walkman 209, 211, 268–70, 272, 282
Whirlpool xii
Windows 112, 211, 213, 239
Windows ACPI 213
Word 92, 107
WSSN (World Standards Services Network) 294
WTO (World Trade Organization) 288, 294
Wuppertal University 29

Xenium 9
Xerox 90, 238
Xerox PARC (Palo Alto Research Center) 90
Xybernaut 211

Index of Subjects

action-feedback sign 248, 249
adaptability 124
aeroplane steering 146, 185, 193, 199
affordance 78, 93
agent, intelligent 188, 362
air traffic control 185, 194, 199
alarm clock 258
alphabetic keyboard 164
alphanumeric keyboard 163
American simplified keyboard 164
amplification 259
amplitude 255
analogue control 21, 170
analogue display 222
analogy 89
animation 28
answering machine 266
anthropometrics 362
anthropomorphic display 252
appropriateness 79, 83, 125, 301
arrow key 143, 144, 149, 168
ASK (American simplified keyboard) 164
ASR (automatic speech recognition) 191
ATM (automatic teller machine) 95, 153, 312
audicon 276, 258
audio 185
auditory display 253, 277
auditory perception 275
Augmented reality (AR) 252
authoritarians 337
automatic speech recognition 191
automatic teller machine 95, 153, 312

automobile user interface 12, 221
balance 124
bar chart 247
benchmarking 54, 57
blinking frequency 231
block diagram 24
Braille display 183
Braille reading 274, 276, 279
brain-computer interface 186
brainstorming 55, 62
button shape 119, 132

CAD (computer-aided design) 155
calculator keyboard 99, 145, 162
camera 173, 312, 316, 319
capturing images 173
car immobilizer 179
car-navigation system 21
Cartesian coordinates 243, 244
cassette recorder 268
cathode-ray tube 109, 153
CD player (compact disc player) 269, 324
cellular phone 9, 98, 127, 167, 174, 270
character density 237
characteristic curve 24
checklist 300–1
chord keyboard 167
clarity 250, 300, 315
clicking 174
clothing 209, 212
coffee can 323
collectivists 336

colour, use of 114, 224, 359
colour-blindness 116, 224, 242
column chart 244
compatibility 301, 315
computer-access system 279
computer-aided design 155
concept creation 52
concept modelling 41
conceptual model 86
consistency 78, 104, 301, 315
constraints 93
contactless data carrier 178
contrast 84, 223
control 301
control column 146
control error 149
control knob 136, 137, 138
control signal 141, 143
coordinates 243
creativity techniques 51
CRT (cathode ray tube) 109, 153
cultural aspects 115
cultural grouping 333
culture 16, 329, 360
cursor 121
curve chart 244
customer types 10

data 219
database 220
democrats 333
decibel 255
design requirements 39
desktop metaphor 35, 212, 238
dialogue 204
dictation 205
digital display 228
direct voice input 191
disability 190, 209, 268, 274, 277, 281
display 219
dot chart 244
DVI (direct voice input) 191
Dvorak keyboard 93, 164
dynamic time warping 198

earcon 185, 276, 262
early adopter 19
echo cancellation 207

ECU (environmental control unit) 193
EEG (electroencephalogram) 186
effectiveness 298
egalitarians 334
elderly people 84, 114, 116
electroencephalogram 186
electronic control 142, 144, 157, 268
electronic fabric 212
emotion 326
end-point detection 198
enrolment 202
environmental control unit 193
environmental noise 205
error correction 205
error prevention 301
error tolerance 78, 106
evaluation 53, 123, 295–302, 360
expert appraisal 300–1
expliciteness 85, 301
eye tracking 174

facial expression 326
facsimile equipment 103, 266, 312
familiarity 85
fare collection 178
fatigue 201
fax machine 103, 266, 312
features 10, 19
feed forward 78, 83, 93
feedback 78, 99, 183, 249, 256, 258,
    265, 270, 273, 301, 315
feedback sound 265
feminism 315, 316, 319
Fitts' law 78, 80, 125, 365
flexibility 301, 315
floppy disk 94
flowchart 68
focus group 297
foil key 135
folder 89
font 122
font size 110
formant 196
Fourier analysis 255
frequency 254
frequency response 259
frequency-centred keyboard 165

Gauss' distribution 11, 13
gestalt 103, 237, 238, 324
gesture input 175
global positioning system 214
glove interface 212
GPS (global positioning system) 214
graphical user interface 91
graphics tablet 155
grid 114
grouping 102
grouping 237
guessability 314
GUI (graphical user interface) 91
guidance 301
guidelines 345
guitar 321

hairdryer 321, 324
handwriting recognition 180
head-mounted display 215, 250
head-up display 250
healthcare 174
hearing 253
hertz 254
hidden Markov model 199
hi-fi 20, 22
high-design process 32
highlight 118, 120
HMD (head-mounted display) 215, 250
hot spot 153
HTML (Hypertext Meta Language) 74, 279
HUD (head-up display) 250
hyperlink 126
Hypertext Meta Language (HTML) 74, 279

icon 28, 240, 242
image capturing 173
immediacy 84
impairment 190, 209, 268, 274, 277, 281
implantation 210
individual differences 201
individualism 331
informal test 297, 299
information 219

information appliance 211, 215, 359, 361
inscription 231
instruction model 41
intelligent agent 188
intelligent companion 188, 362
interaction design 29, 123
inventory of comments 297
inventory of usage 297
ISO 9241 291
iteration 119, 296

jargon 31
joystick 146, 154, 276
Kansei engineering 322–4
key 119, 132
key row 141
keyboard 162, 264
key-click 132
kinesthesis 275

label 231
language 199
laptop computer 318
late adopter 19
LCD (liquid crystal display) 109, 152, 153, 228, 269
learnability 91
LED (light emitting diode) 142, 228, 232, 269
legibility 84
libertarians 335
Likert scale 323
light emitting diode 142, 228, 232, 269
lighting 117
line chart 244
liquid crystal display 109, 152, 153, 228, 269
Lombard effect 206
long-term orientation 332
look and feel 40, 95
loudness 255, 278

manufacturing 192
mapping 52, 58, 258
marketing 317, 338, 357, 362
Markov model 199

measurement device 248
medical appliance 174
memory load 280
mental model 42, 230
menu structure 126
meritocrats 334
metaphor 89
miniaturization 271, 273
mobile device 268
mobile phone 9, 98, 127, 167, 174, 270
model aeroplane 146
mood board 54, 57
Morse's law 282
mouse 150
movement capturing 177
multifunctional key 144
music theory 261
Myers–Briggs test 13

national culture 329
naturalness 201
navigation 126
navigation model 41
neural network 199
new freedom 30
noise 205, 207, 257
NTSC (a television standard) 109
numeric display 233

object display 250
OCR (optical character recognition) 277, 278
one-shot key 132, 141, 142
optical character recognition 277, 278
PAL (a television standard) 109
paramedic 174, 215
pattern matching 196
PDA (personal digital assistant) 153, 167, 182, 270, 282
peak value display 233
perception 112
persona 188, 362
personal digital assistant 153, 167, 182, 270, 282
phicon 179
photography 173
physical icon 179

piano keyboard 162
pie chart 242
pilot 185, 191
pin-matrix device 277
pitch 255
planners 336
pleasure 303
pointer 151, 153, 222
pointing and clicking 174
post hoc analysis 300
power distance 331
power standby symbol 290
practice 201
presentation model 41
prioritization 101
private camera conversation 327
product creation process 311, 317, 320
product management 317
profile chart 247
progress bar 249, 259
projector 175
prosthesis 210
prototype 53, 72, 75
pulse width modulation 264
push-button 132
PWM (pulse width modulation) 264

quality control 192
questionnaire 360
QWERTY keyboard 163

radio buttons 141, 142
radio tuner 271
readability 123
redundancy 228–30
remote control 97, 120, 146, 152
repeat function 171
response time 127
ringer sound 263
robot 147
role playing 55, 63
rotary control 138
rotary switch 137, 138

safety 157
sample playback 260
satisfaction 78, 107, 298–9
scale 222, 223

scanner 265, 266
science fiction 210
screen graphics 109
screen organization 118
screen saver 250
screen-access system 280
scrolling 127
scrolling text 236
security 157
self-descriptiveness 78, 85, 345
semantic language 96
sensor 177
sequence keyboard 168
sequencer 263
shift key 144
Sholes keyboard 164
shuttle 138
simplicity 250
simulation 296
six-dimensional (6D) mouse 147
skill 201
slider 139, 140
sliding switch 140
smell 275
snowballing 55, 63
socio-technical organization model 40
soft key 145
software tools 74
solid state player 270
sone 255
sound 191, 253
sound pressure level 255
sound synthesis 261
speaker-dependent recognition 202
specification 31, 296
spectrogram 196
spectrum 259
speech control 190
speech interface 281
speech output 268
speech recognition 185, 190, 191, 358
speech synthesis 277
speech-to-text conversion 183
spider-web chart 242
SPL (sound pressure level) 255
standard 285
standby symbol 290
state diagram 23, 142, 143
statistical pattern matching 197

storyboard 65, 76, 296, 319
stressor 191, 207
structure 29
suitability 78
support 301
supportives 335
switch key 162
symbol 28, 240, 242
syntax 197
synthesizer 263

tactile display 214, 268
tactile feedback 135, 156, 183
tactile perception 275
tag 179
tangible user interface 177
tape-measure 176
task analysis 360
task model 41
task-situation model 35
taste 275
technical model 39
technology life cycle 19
telephone 195, 270
telephone keyboard 163
telephone system 281
television 207, 325
template 202
testing 296, 319, 362
text display 122
text input 164–7
text-to-speech conversion 277
thought input 186
thumbwheel 138, 139
timbre 255
timing 259
toggle switch 135, 143
tools 173
touch screen 151–3, 175
touch-sensitive device 151
toughness 332
toy 180, 181
trackball 149
trackpad 154
trackpoint 146
translation 326
transparency 78, 105
trend board 54, 58
trigger switch 136

TV (television) 207, 325
typewriter keyboard 162, 163

ubiquitous computing 216
uncertainty avoidance 332
understandability 101
upgradeability 124
URL (uniform resource locator) 179
usability laboratory 299
usability testing 297–300
user workshop 297

vacuum cleaner 317
variability 201
variable 219
VCR (video cassette recorder) 312, 321
video 174
video recorder 312, 321
viewing distance 110
virtual control 156
virtual reality 177, 183, 250
virtual touch screen 175
visual clarity 250, 300, 315

visual design 109
visual impairment 274, 277, 281
visualization 363
vocabulary 199, 326
voice control 190
voice input 191
VR (virtual reality) 177, 183, 250

waterfall model 32
waveform 254
wearable audio 214
wearable computers 209
web usability 361
WIMP (windows, icons, menus,
    pointing) 188, 239
WML (wearables meta language) 271
word recognition 167
workload 207
workshop 54, 297
wristwatch 168, 221
writing speed 167
young people 114